CARL DJERASSI

IN RETROSPECT

FROM THE PILL TO THE PEN

CARL DJERASSI

IN RETROSPECT

FROM THE PILL TO THE PEN

ICP
Imperial College Press

Published by

Imperial College Press
57 Shelton Street
Covent Garden
London WC2H 9HE

Distributed by

World Scientific Publishing Co. Pte. Ltd.

5 Toh Tuck Link, Singapore 596224

USA office: 27 Warren Street, Suite 401-402, Hackensack, NJ 07601

UK office: 57 Shelton Street, Covent Garden, London WC2H 9HE

Front cover: Photo – Karen Ostertag, Graphic – Thomas Gambke

Library of Congress Cataloging-in-Publication Data
Djerassi, Carl, author.
In retrospect : from the pill to the pen / Carl Djerassi, Stanford University, USA.
pages cm
Includes bibliographical references and index.
ISBN 978-1-78326-531-2 (hardcover : alk. paper) -- ISBN 978-1-78326-532-9 (pbk. : alk. paper)
1. Djerassi, Carl. 2. Chemists--United States--Biography. I. Title.
QD22.D54A3 2014
540.92--dc23
[B]
2014028183

British Library Cataloguing-in-Publication Data
A catalogue record for this book is available from the British Library.

Typeset by Stallion Press
Email: enquiries@stallionpress.com

Printed in Singapore

In memoriam:
Pamela Djerassi (1950–1978)
Diane Middlebrook (1939–2007)

and as always:
Dale and Alexander Djerassi
Leah Middlebrook

"The past cannot move into the present uncorrupted"

Bruce Bawer

Contents

Caveat Lector

"Let the reader beware" is a dangerous and unwise start to any book, but especially to an autobiography. Though fully cognizant of that danger, I must start in this fashion because full disclosure requires it.

I have already written one lengthy autobiography with the sonorous title *The Pill, Pygmy Chimps, and Degas' Horse*, which I started nearly a quarter of a century ago. It was published in 1992 when I was sixty-nine, an age that can be considered "conventional" for autobiographers, i.e. oldish but not too old. But a more detailed count, incapable of disguise in this age of brutally intrusive Googling that rejects any wishy-washy denials, would point to no less than three autobiographies. If true, I can expect to be rightly accused of megalomania by releasing a fourth one on my ninety-first birth year — once more shaking hands with myself through the ever dustier lens of memory with only an occasional swipe with some cleansing tissue. Or am I just displaying the first symptoms of Alzheimer's disease in not being able to remember what I had written earlier when I turned in the late 1980s from research chemist to author? Instead, let me embark on my defensive full disclosure by first describing the nature of these "autobiographies" and then outline why I intend to heap one more spoonful on this autobiographical Mount Everest.

In late 1987, I was asked to contribute to an autobiographical series. This collection, edited by Jeff Seeman, was designed to include the autobiographies of twenty-two noteworthy organic chemists from thirteen countries. After some hesitation, I agreed and the finished volume was published in 1990 by the American Chemical Society under the title *Steroids Made it Possible.*[1] Although clearly an autobiography, I claim that it should not, in all fairness, count in the present context. The purpose of the American Chemical Society series, entitled *Profiles, Pathways,*

[1]C. Djerassi, *Steroids Made it Possible.* American Chemical Society Books, Washington, DC (1990).

and Dreams, was, in the editor's words, "to document the development of modern organic chemistry by having individual chemists discuss their roles in this development." Written in the obscurity of chemical pictography, it was entirely directed at chemists and thus as incomprehensible to the general public as hieroglyphic cave drawings depicting the biannual migration of the woolly mammoth.

But however dry and ultimately estranged from the sentimental heartbeat of the general reading public, this first effort at articulating my life as a scientist placed me squarely into an autobiographical mode. Indeed, I was at that time of my life literally wedded to the idea of an autobiography. I had recently married, and my new wife, Diane Middlebrook, was infected with the vivacious curiosity that inhabits a new marriage (and especially a third one on both our sides). A highly esteemed professional biographer and Professor of English Literature at Stanford University, she wished to know something about the first fifty years in the life of her new husband; it took little persuasion to satisfy her curiosity. In addition, her observation that I had lived in an extraordinary time of European history, eminently worthy of documentation through personal case history, was persuasive. For instance, demographically speaking, my own life encompasses a period unique in world history: since my birth in 1923, the world's population has nearly quadrupled, an event that will never be repeated on our precarious Planet Earth. My professional activities, be they my involvement at age twenty-eight in the first synthesis of an oral contraceptive (the Pill) or writing a book at the age of eighty-five with the title *Sex in an Age of Technological Reproduction*,[2] had always been profoundly touched in one way or another by that monumental and horrifying demographic fact. My continuing preoccupation with its dark consequences will undoubtedly continue until my death.

Rather than complying with my wife's wishes to write a "real" autobiography in the conventional chronological manner, I chose to compose individual chapters based on various events that might be significant (such as the synthesis of the Pill), amusing (experiences as an immigrant teenager on the Midwestern church lecture circuit or my unsuccessful

[2]C. Djerassi, *Sex in an Age of Technological Reproduction: ICSI and Taboos* (with accompanying DVD). University of Wisconsin Press, Madison (2008).

forays into movie production) or utterly tragic (the suicide of my daughter). The various chapters were essentially self-contained and hence could be read in any casual order. The resulting book, *The Pill, Pygmy Chimps, and Degas' Horse*,[3] proved to be a sufficiently readable account of a seemingly complicated life to be translated into seven languages. Of these, only the Chinese and Hungarian editions are still in print.

A decade later, I moved onto the slippery slope of memoirs, which are almost always autobiographies manqué. My reflections on the fiftieth birthday of the Pill[4] included so many personal musings and comments that they must fairly be described as at least stained with autobiographical disclosures.

Thus, irrespective of how one counts, I still have to answer the legitimate question as to why I am now embarking on one more addition. Denying any self-promotional motive would be pointless, since nobody would believe this of a multiple autobiographer in spite of any firm denial. So let me turn the argument around and say that only if one has already published some self-reflections or autobiographical accounts two decades earlier is it worthwhile to revisit them and reinterpret some of their deeper meanings as the end of the author's life is rapidly approaching and with it a recognition of how big a role discontent has played in it. This is hardly original, considering what Flaubert once said: "An autobiography?... Wait 20 years to write about a painful experience." What I shall be saying in this book about the topic of discontent — a painful synonym for the darker shadows in life — is likely to be the last word, given that I am now past the age of ninety. Furthermore, a lot has happened in my last two decades — in many respects a totally new life as an author and playwright rather than as a scientist. The reasons for such a transformation are worth describing, partly as useful examples to others in a progressively more geriatric world to demonstrate that it is never too late to transform oneself and even grow, but also as warnings, because there are many things I would not do now if I had the chance to start all over again.

---

[3] C. Djerassi, *The Pill, Pygmy Chimps, and Degas' Horse: An Autobiography.* Basic Books, New York (1992). An e-book version has recently been released: http://plunkettlakepress.com/tppc.html

[4] C. Djerassi *This Man's Pill — Reflections on the 50th Birthday of the Pill.* Oxford University Press, London, New York, 2001.

But there are other reasons for still writing about myself. One is to record without much embellishment my methods of overcoming grief and disaster from which something useful may be learned. A related one is to show how I have managed to cope with an overwhelming sense of loneliness, an incurable disease at age ninety, but which I have learned to tolerate by a special form of therapy that *bona fide* psychoanalysts are unlikely to recognize, namely autopsychoanalysis. The sort of novels and plays that I have written during the past two decades have allowed me to accomplish something that is simply impossible in standard autobiography, namely to bypass one's psychic filter and in the process become both analyst and analysand.

I say this because autobiographies by definition have gaps — be they inadvertent or deliberate — for reasons of discretion, shame, embarrassment or even faulty memory. In addition, virtually all autobiographies contain automythological or even fictitious components, since they have to pass through the author's psychological filter in which one's persona is consciously or unconsciously sanitized. But the novels and plays that I have published during the past twenty-five years have allowed me, the scientist previously lacking in self-reflection, to address in the guise of fiction the indelible imprints that the culture, mores and behavioral practices of the scientific tribe to which I have belonged for over half a century have left on me. If curiosity about the persona of Carl Djerassi leads people to Google, Facebook or even Wikipedia, the truly meaningful aspects will not be encountered there. They can only be found hidden among the many characters, male as well as female, in my short stories, my five novels and my nine plays, which cumulatively are available in over twenty languages. I shall not provide a Rosetta stone to the personal secrets I have divulged there, for the simple reason that I myself have not yet completely deciphered all of them. But I am now convinced that the key themes in my literary writings have all unconsciously arisen from my desire to examine these subjects in my own life in the guise of fiction.

At the time I wrote my first autobiography, I was still a scientist for whom anything fictitious was professionally *verboten*, but who was now entering a novelist's territory where everything fictitious was allowed, indeed required. In the process of that transformation, I came to realize that only in fiction — actually in the guise of fiction — can the real truth be told. If, by definition, autobiography must be a sort of fiction, because it is written under the constraints of a psychic filter, then the type of fiction and drama I have written is the reverse, namely factual, unvarnished biography. Some fiction writers are autobiographers wearing a mask and there is no doubt in my mind that I belong to that subset.

People who complain "So what?" or "Who really cares?" should simply not bother to read further. But whether we like it or not, autobiographies have a taint of exhibitionism and readers of autobiographies are usually voyeurs who, even if they do not admit it, are interested in some scandalous or unexpected aspects of the autobiographer's life. Having written several autobiographies, I am at least aware of the problem and honest enough to make the following point. This "last" autobiography will clearly contain some personal stories catering to the voyeuristic tastes of readers, but it will also deal with a lot of "issues" which have interested me for decades and which I am now listing: the ever-increasing gulf between the sciences, the humanities, the social sciences and mass culture; the gap between the developed and developing countries, which I now call geriatric and pediatric ones; the extremes of the human population problem in terms of an emphasis on conception in the geriatric societies and on contraception in the pediatric ones; the role of theater as "edutainment;" the meaning of suicide; and many more. All of them have a didactic component, sometimes deliberately so, sometimes inadvertently. I am sure that this is the reason why more than one of my more recent friends complain that "You are lecturing to me." All I can say is *nolo contendere* because by now I am too imprinted with such faults to change.

A final reason for re-entering autobiographical terrain is that my intended readership is changing. The present volume was first released in 2013 on my ninetieth birthday in German translation, the language of my birth and early upbringing (though not the language in which I now write or dream) until my forced emigration from Austria — a language spoken primarily in some of the most geriatric countries in the world, where nearly 20% of the population is now above the age of sixty-five. While I still consider myself mentally wide awake and younger than my actual age, I am aware that I am writing from the perspective of a person who will be dead within a decade, when nearly a quarter of the inhabitants of Germany and Austria will be over the age of sixty-five. Clearly, what I have to say in this book is likely to appeal to such an older segment of the population which is also the one that is more likely to read than their younger counterparts. This in turn has prompted me to write this last autobiography backwards, starting with the present and ending up with earlier portions of my life which I have already recounted in my earlier autobiographies but which I am now revisiting with a special focus on the shadows. In the process, I shall quote extensive passages from some of my earlier, but now mostly out-of-print autobiographical writings, without, hopefully, being accused of self-plagiarism; in addition, I shall also cite specific passages from my fiction and plays to remind myself as well as to demonstrate to the reader how much of my personal life has only been divulged there, quite often subconsciously. Yet not everything will be dark, because I am still imbued by what Cynthia Ozick expressed in one of her great phrases: "These days a hot liquid of remembrance and of imagination lives in the nerve of my joy." Given my age, there is no doubt that the present volume has turned for me into "required writing," even though I am realistic enough to recognize that this does not convert it into required reading. Speaking of reading, the various chapters of this book deal mostly with specific yet unconnected subjects that have touched me deeply during the past two decades — issues that will stay with me for the limited remainder of my life. Thus, I encourage the reader to read, skim, or even skip a given chapter according to one's taste. To put it bluntly, I have followed Goethe's recommendation in the introduction to *Faust*:

> Who offers much, brings something unto many,
> And each goes home content with the effect,
> If you've a piece, why, just in pieces give it:
> What use, a Whole compactly to present?
> Your hearers pick and pluck, as soon as they receive it!

But what about acknowledgments? Virtually every scientific article contains acknowledgments to funding agencies and to colleagues who helped with the work described. In a display of humorous one-upmanship, I have at times indulged in unusual, though factually accurate acknowledgments. Thus, in a paper on the chemistry of Brazilian sea cucumbers published in the scientific journal *Tetrahedron*, I thanked the Rockefeller Foundation for financial assistance and the Brazilian Air Force for the loan of a B-27 bomber. I am also an aficionado of literary acknowledgments and always browse through any book by first searching for the author's expression of gratitude, with the latter usually featuring real or perfunctory homage to one's literary agent (I have none), to one's editor, to one's secretary (I have a marvelous one, but did not burden her since I am quite competent with a computer keyboard) or one's spouse for putting up with the author's prolonged absences (I am a widower). But there are acknowledgments that are worth reading and pondering upon. One of my favorite examples is the single sentence "My thanks are due to Professor J. W. McBain from Stanford University for giving me his fountain pen to write this book" in a 1947 monograph on the Chemistry of Muscular Contraction by Albert Szent-Györgyi, winner of a Nobel Prize for the discovery of vitamin C. To me these pithy words were so full of pregnant interpretations that they could easily provide a marvelous first sentence to a novel. No wonder that I am embarrassed to contrast my own acknowledgment in my *Four Jews on Parnassus — A Conversation*[5] (in my shameless opinion the best book I have ever written) in which the acknowledgments ranged over six full, single-spaced pages. The explanation for this verbosity is simple: *Four Jews on Parnassus* is a biography — albeit an unusual one written in dialogic style — which required extensive archival research and interviews that fully deserved recognition. No wonder that biographies in general have the longest acknowledgment sections.

But since the present volume is not just an autobiography but also an autopsychoanalysis, I need to thank or blame no one but myself. Yet I do wish to thank the brave reader who is willing to continue reading this very last autobiography by Carl Djerassi in spite of the stark warning with which I started. I only hope that curiosity and interest are tempting the prospective reader rather than some form of literary masochism or even worse: using it as a possible treatment for insomnia.

[5]C. Djerassi (illustrations by Gabriele Seethaler), *Four Jews on Parnassus — A Conversation: Benjamin, Adorno, Scholem, Schönberg*. Columbia University Press, New York (2008).

Suicide

Associated Press news report, October 30, 2023
CARL DJERASSI — COINVENTOR OF THE PILL AND AUTHOR MISSING. SUSPECTED SUICIDE.

Carl Djerassi, a co-inventor of steroid oral contraceptives and sometimes referred to as one of the fathers or mother of the Pill, has disappeared under mysterious circumstances one day prior to his 100th birthday. Djerassi was a distinguished scientist for half a century — only one of two American chemists to have received both the National Medal of Science as well as the National Medal of Technology — and long-time Professor of Chemistry at Stanford University before reinventing himself in his sixties into a prolific author of novels, plays, and autobiographies in which he wove science, scientists and literature in seamless fashion for what he stated proudly "didactic purposes."

According to his son, Dale Djerassi, his father had left his home early on October 28 for his daily visit to a San Francisco gym where he was by far the oldest client. He never appeared at the gym, but apparently drove south along the coast to one of the beaches in San Mateo County with which Djerassi was well acquainted as a long-time resident on his nearby SMIP Ranch which was also the site of the Djerassi Resident Artists Program, one of America's best-known artists' colonies. On the morning of October 29, his 100th birthday, his empty car — a rare red 1998 Volvo cabriolet — was found abandoned on San Gregorio Beach. A morning jogger had noticed in the sand not too far from the water line a shoe with missing shoelaces and a cane, later identified as Djerassi's through its unusual ebony handgrip. A search of the waters by the coast guard failed to provide any clues.

Reuters news report, November 4, 2023
CHEMIST AND AUTHOR CARL DJERASSI ASSUMED DEAD.

Following a huge Pacific storm, which caused major damage on the beaches of San Mateo County and obliterated all further traces at the site where Djerassi's car, shoe, cane and a silver pill box with the initials "CD" were found, has led to the conclusion that the scientist-turned-author had committed suicide by drowning on the eve of his 100th birthday. (An analysis of the contents of the pill container showed them to be saccharine.) In a private ceremony, a symbolic scattering of ashes in San Gregorio Creek which bisects the family's SMIP (Sic manebimus in pace) Ranch was attended by his only survivors: his son Dale, an award-winning documentary film maker residing on the family ranch; his grandson Alexander, the distinguished Samuel Dvir Professor of International Law at Georgetown University and regular NPR commentator; and Pamela Djerassi, his only great-grandchild, named after Djerassi's daughter Pamela, who had committed suicide in 1978 and in whose memory the Djerassi Resident Artists Program was founded. Djerassi's stepdaughter, Leah Middlebrook, Dean of the College of Arts and Sciences at the University of Oregon and daughter of Djerassi's third wife, the late Diane Middlebrook, also attended. The family requested that gifts in Djerassi's memory be made to the American College of Sofia, Bulgaria, the school where Djerassi learned English after his flight in 1938 from Austria prior to his immigration to the USA.

Letter to the Editor, *The New York Times*, November 6, 2023.

To the Editor,

Long obituaries in The New York Times and major newspapers in Europe reported that the distinguished chemist and author, Carl Djerassi, committed suicide by drowning on October 28, the day before his 100th birthday in spite of the fact that his body was not recovered. Your obituary cited his best known achievements — the first synthesis of an oral contraceptive in 1951 and the first synthesis in the same year of cortisone from a plant material — achievements that garnered him numerous awards including 35 honorary doctorates. The obituary also listed his 11 plays and his tetralogy of "science-in-fiction" novels, starting with Cantor's Dilemma, currently in its 41st print run.

I find it remarkable that your otherwise so extensive obituary failed to mention Djerassi's novel, Marx, Deceased, which was published in 1995 and is long out of print. The novel dealt with a famous writer's obsession for reading his own obituary, which caused him to stage his death in a sailing accident in Long Island Sound followed by secret travel to San Francisco, where he led a new literary life under a nom de plume. Djerassi's play, Ego (subsequently renamed Three on a Couch), written nearly 10 years after the novel had premieres in London and New York, was followed by a tour covering 68 theaters in Germany. It is noteworthy that Ego dealt with the same topic of a staged suicide. As the author of a literary monograph on Djerassi (Der intellektuelle Polygamist: Carl Djerassis Grenzgänge in Autobiographie, Roman und Drama, Berlin 2008) I feel justified in raising the following question: how do you know that Djerassi is actually dead? Perhaps the centenarian is now laughing at all of us from some unknown location.

Ingrid Gehrke
Graz, Austria

Mother, Alice Friedmann, ca. 1916, and aunt, Grete Friedmann, ca. 1933

What made me start with an announcement of my prospective suicide? I am not and have never been suicidal, although suicide has touched me since my childhood. My aunt Grete — a true beauty and European woman silver medal fencing champion who lived with us in my grandmother's home in Vienna — killed herself, still in her thirties, reputedly because of the death in 1935 of her lover Alexander Moissi, arguably at that time the most famous actor on the German and Austrian stage. Until our departure from Vienna in 1938, Moissi's death mask rested on the top of our piano. After our immigration to the US, my mother threatened on numerous occasions to commit suicide — a form of emotional blackmail which eventually made it unbearable for me to respond and led to our estrangement. Yet she lived until the age of ninety-one to die of dementia. And then followed the greatest tragedy of my life: my daughter's suicide which I address in the later chapter *What If*? But while I am not suicidal, I have at times thought of suicide under one very specific scenario. Even though I

Daughter, Pamela, age twenty-five

now live alone and thus would not be a burden to others, the idea of Alzheimer's disease or similar condition of mental incapacity would undoubtedly cause me to kill myself. Indeed, when I closed my laboratory in the 1990s, I took one bottle with me which I hid carefully at home, disclosing its location only to my son. It is a bottle of potassium cyanide, large enough to kill a pride of lions. I asked him to remember that hiding place and show it to me if I ever reached such a stage of mental deterioration. The problem, of course is, that at such a time I would not only have forgotten where the bottle is but would probably also forget to ask my son.

But the deeper meaning of a "just-in-case suicide" has stuck with me, so much so that I felt it worth exploring in the only place where I am comfortable openly discussing such personal topics, namely in my fiction. Thus, the very first sentences in my second novel, *The Bourbaki Gambit*, start with

> *"What would* <u>*you*</u> *use to commit suicide?"*
>
> *That's the first sentence I remember her uttering. At least, that's what I tell her now, although we both know it isn't completely true.*

And a couple of pages later I continue as follows:

> *It occurred to me suddenly that I might be dealing with a lunatic, a potential suicide — or worse. I decided to play it cool and answer as if it were something I'm asked every day. "Cyanide," I said judiciously.*
>
> *"Hm," she nodded, "I suppose so. But where would I get cyanide?"*
>
> *"You asked me how I would commit suicide. I've got plenty of cyanide in the lab."*
>
> *"Would you give me some?" she asked. She might have been asking me to pass the salt.*

> *"Of course not," I broke out laughing. That would make me an accessory." I squinted to get her into better focus. "But you aren't serious, are you?"*
>
> *"About wanting some cyanide? Dead serious. But not about committing suicide. I only want some — just in case."*
>
> *I raised my eyebrows and waited for some elaboration on "just in case," but she got up from the sand.*

Following Anton Checkov's famous adage, "One must not put a loaded rifle on the stage if no one is thinking of firing it," I did continue with the cyanide scenario but only fired the proverbial gun in the last chapter of *The Bourbaki Gambit.* Curious readers can find there Carl Djerassi's own "just-in-case" alternative.

A very different and much uglier suicide variant of not knowing whether the announced suicide was actually consummated has sufficiently haunted me that it also found its place in another one of my works — this time in the play *Ego.*

> ***STEPHEN*** *Tomorrow is my 50th birthday. I know how to celebrate it... by pushing you... who craves certainty... into the purgatory of perpetual uncertainty. Here... look at this. (Takes out cellophane envelope.) I brought some cyanide just to prove that I'm not bluffing.*
>
> ***MIRIAM*** *How dare you threaten me like that?*
>
> ***STEPHEN*** *If you think it's fake, feed it to your pet Dalmatian. As for me, you'll never find out what happened, because I've plenty more where this sample came from.*

The reason for addressing this "what if?" question is not primarily associated with my mother's suicide threats, but the horror and utter desolation I felt (as described in the chapter *What If?*) when my son-in-law read to me my daughter's suicide note without knowing for some days what had really happened since we had not located her body.

Before leaving the morbid topic of suicide, I might as well mention the only other set of conditions that might cause me to consider killing myself so late in life, namely being unable to write and read — an intolerable loss of intellectual independence that I would consider much more serious than being bedridden or otherwise physically incapacitated. But since I still barely need glasses and still write obsessively every day I shall drop this dismal subject of an unknown future and proceed in the direction that I had announced: to write a retro-autobiographical account, where the last couple of decades cover previously undescribed years of my life and then proceeding through some of the earlier years that I had already recorded in my autobiographical writings but where I now examine the hitherto unnoticed shadows. In the process, I am likely to arrive at answers to the question Paul Klee, my favorite artist who will appear several times in later portions of this book, posed through one of the very last watercolors he completed just before his own death in 1940: "*woher? wo? wohin*?" (*Whence? Where? Whither?*).

Having started this first chapter with three prospective newspaper clippings, I shall end with a real one replete with Anglicisms from the page *Kennenlernen (Getting Acquainted)*, which I came across in an issue of the distinguished German weekly *DIE ZEIT*.

> *Complicatet [sic] Mission — High Reward. Wir helfen uns beide! Ich suche eine herrliche junge Dame, Musik, Theater, Humor (viel), good english [sic], intelligenz, unkompliziert, natürlich, slim, NR Ich, ein sehr alter Mann, jewish, german/US/UK background, leichte Gehbeschwerden, Entrepreneur, very clever, möchte viel reisen und auch im Ausland leben. Ready to travel? Bildzuschriften: ZA55024 DIE ZEIT, 20079 Hamburg.*
> [Complicated Mission — High Reward. We'll help each other! I search for a marvelous young lady, music, theater, much humor, good English, intelligent, uncomplicated, natural, slim, non-smoker. I am a very old man, Jewish, German/US/UK background, minor walking restrictions, entrepreneur, very clever, fancies much travel as well as life abroad. Ready to travel? Send pictures to ZA55024 DIE ZEIT, 20079 Hamburg.]

Such personal ads in *DIE ZEIT*, rather similar in style to those in sophisticated American publications of *The New York Review of Books* type, focus on a search for companions among well-educated professionals and are written in a style that barely touches on any sexual connotations. But in view of how I started this chapter, will anybody believe that I am not also the author of this ad?

Except for two minor revisions — I have major rather than minor walking restrictions because of a fused left knee from a skiing accident and am of Austrian rather than German origin — this notice with its list of desiderata applies in every particular also to me, except that I would also have added my fierce dislike of jeans and of mobile phones. Is there a German doppelgänger of Carl Djerassi living in Hamburg? Yet I would never have dreamed of publishing such an ad, be it in a newspaper or on the web, because the desire of a "very old man" searching for a "marvelous young lady" — ignoring for the moment that the term "young lady" could cover the range from twenty-one to fifty-seven depending upon the position of the "very old" man on the sixty-nine to ninety-five years scale — can only mean that the man is either willing to subject himself to public derision or willing to assume the role of a sugar daddy. I am loath to expose myself to either alternative for the simple reason that while I am chronologically very old, I neither feel or act my age, nor am I prepared to acquire companionship purely on the basis of affluence. The alternative is accepting proud loneliness — a penumbra that I will address in more than one of the subsequent chapters.

The Bitter-sweet Pill

As already stated, this final autobiography is written retrodirectionally, by starting at the end, specifically my putative suicide in 2023. So why do I now take up the Pill, arguably an important scientific discovery with huge societal consequences, when my own involvement with it started over sixty years ago and been documented by me on various occasions, including in no less than three chapters (*Birth of the Pill, The Pill at Twenty, The Pill at Forty: What Now?*) in my earlier autobiography? Except for some selected passages, I shall not reprint this material but simply direct any interested reader to that source. So once more, why start with the Pill?

The answer is simple, since my announced perspective in this personal account focuses on the shadows in my life and thus also in my achievements. In so far as the shadows in my personal Pill story are concerned, three recent ones have perturbed me more than I expected and thus justify that I start with the chemical birth of the Pill. The events, which I would categorize as slander, occurred since 2008, in other words in the time of Google and Wikipedia, where every news driblet, however preposterous, is soaked up in seconds and permanently preserved. If these incidents had occurred two decades earlier when Google and web browsing did not exist, when even e-mail was far less prevalent than it is now, I most likely would have brushed them aside. But now, any error — inadvertent or deliberate — any insult — however crude and manipulative — indeed any factoid — true or false — cannot be erased or corrected. It is simply fixed in cyberspace and usually picked up by slipshod journalists and the huge sloppy segment of the browsing public, who consider any cyberdetritus as truths set in stone or at least grist for journalistic mills. I start with the following two examples because the first illustrates the sloppiness of many media, whereas the second demonstrates the indelible character of deliberate misrepresentation.

Early in 2009 in San Francisco, I was suddenly bombarded by queries from American reporters and radio stations about my supposed condemnation of the Pill. Initially, I took this as some sort of prank, but rapid Googling produced an avalanche of entries (still existing to this day) on sites such as *Christian and American* with the headline "Carl Djerassi, Inventor of birth control pill, condemns it" and then stating that

> *Eighty five year old Carl Djerassi, the Austrian chemist who helped invent the contraceptive pill, now says that his co-creation has led to a "demographic catastrophe." The assault began with a personal commentary in the Austrian newspaper Der Standard by Carl Djerassi [where he] outlined the "horror scenario" that occurred because of the population imbalance, for which his invention was partly to blame.*

I quickly discovered that not only fringe outlets of the *Christian and American* variety, but also mainstream newspapers such as *The Guardian* in the UK featured similar articles on their web sites, which prompted me to make the following demand on January 18, 2009 for a retraction from the British paper:

> *I was informed that Viennese Cardinal Christoph Schönborn, after referring to me as "the inventor of the Pill," quoted me on 21 December 2008 on Austrian TV (ORF) as having stated that Austria was facing a demographic catastrophe, which could only be offset by current Austrian families raising its current average of 1.4 children to 3 or else by a more intelligent immigration policy.*
>
> *While the Cardinal's broadcast ignored totally the context and overall thrust of my article, those specific statements attributed to me were correct. But the resulting conclusions spread by The Guardian in England on its 7 January 2009 web site and by various newspapers in the US and elsewhere under headlines such as "Pill inventor slams Pill" or "Co-inventor of Birth Control Pill Now Calls It A Catastrophe" are calumnies bordering on libel. Let me explain the grounds for my outrage and the preposterous nature of these libelous conclusions by sloppy or disingenuous journalists, who never checked the original source material.*
>
> *On 21 October 2008 at the Medical University of Graz and on 11 November 2008 at the University of Vienna, I spoke at the invitation of these universities about my reflections on the 70th anniversary of the Nazi Anschluss. Toward the conclusion of my speech, I decried the resurgence in Austria during its recent election of a right wing political party*

with strong xenophobic tendencies. I later summarized these remarks in a slightly edited article that appeared in DER STANDARD.[6]

Contraception, birth control, abortion, or the Pill were nowhere mentioned in my article. Instead, I accused the startlingly large xenophobic segment of Austrian voters of assuming that their small country was not situated in the middle of Europe but rather on an island where God permits them to live independently to enjoy their Schnitzels. I warned against an impending demographic catastrophe if those xenophobic Austrians did not choose to have at least 3 children (which I considered totally unlikely) or were then to reject immigration as the other solution. I drew attention to Bulgaria — a country of roughly the same current population, age distribution, and average family size — which had no immigration alternative, since nobody wants to immigrate to Bulgaria, in contrast to Austria or other western European countries. As a result, demographic estimates predict a 34% decline by 2050 of Bulgaria's current population! I also indicated that Germany's average family size, again almost identical to Austria's, requires an annual immigration of ca. 200,000 persons just to maintain the country's current population. I pointed toward the need in Austria for continuing immigration, since absent of such immigration, a country requires ca 2.1 children/family to maintain its demographic status quo. To assume that I attributed the decline in Austria's average family size (even worse in all-Catholic Italy and Spain) to the Pill is absurd. People don't have smaller families because of the availability of birth control, but for personal, economic, cultural and other reasons, of which the changes in the status and lifestyles of women during the past 50 years is the most important. Telling support for my conclusion is provided by Japan, which has an even worse demographic problem than Western Europe; yet where the Pill was only legalized in 1999 and not used widely. Japan's rampant xenophobia makes the immigration solution practiced by the USA unlikely.

The Catholic Press *network in its various articles, which were faithfully reproduced in* The Guardian *and other newspapers and then accepted as gospel truth, claim that in my article I associate Austria's and Europe's demographic problems with the Pill and that I now*

[6]December 13, 2008, p. A-3 in its Album weekend section.

> *regret the work with which I was involved nearly 60 years ago. The words in the* Philadelphia Bulletin *(9 January 2009), "A co-inventor of the birth control pill, Austrian chemist Carl Djerassi, now says his creation has led to a 'demographic catastrophe'" are repeated almost verbatim by* The Guardian. *I reject such a conclusion unequivocally and insist upon a published retraction. One only needs to read my memoir,* This Man's Pill: Reflections on the 50th Anniversary of the Pill *(Oxford University Press, 2001), to find my personal views on contraception, the Pill, and the* de facto *separation of sex and reproduction, which sooner or later the Catholic Church must face realistically and humanely.*
>
> *I need to make one additional important point. It will not be sufficient to publish a one-line retraction or correction. The story has now spread like wild fire over the web and has in the last two weeks appeared in so many publications — most of them associated with Catholic media — that a commensurate correction* must *appear that points to the exact source of the gross misrepresentation. The latter not only applies to me personally (in claiming that I now condemn the Pill) but also to the absurd conclusion that the drop in family size in Austria and elsewhere is associated with the Pill. In my rebuttal, I have indicated why this is ridiculous. I must point out that time is of the essence since this calumny is spreading too fast and too far to then be offset by just a simple retraction or even full rebuttal.*

The Guardian, as a responsible newspaper, immediately accepted the blame and published a long correction, which included steps to remove its erroneous article from the web. I also wrote similar requests for retraction to some of the main Catholic sites, but not a single one responded nor did any remove their religiously inspired slander which can be found to this day on the web.

Such trigger responses by virulent opponents of the Pill have also occurred in private missives to me, but since these were only seen by me, I was able to ignore them or reply directly to the sender. I only cite one such instance, written in 2008 just a few months prior to the above newspaper explosion, as an example that vituperative fanaticism still exists half a century after the introduction of the Pill. The day following my appearance on a popular German TV talk show (*Menschen bei Maischberger*) dealing with "The New Sexual Revolution" and reputedly watched by 1.3 million viewers, I received the following e-mail written in German, which I hereby translate:

Good day, Mr. Djerassi,

If I were in your place, I would ask how many people were not born because of you, how many people have never been created and never seen the sun's light. Excuse me please that I should be thinking so bitterly about you and your work, but I find it strange that a person should never be born solely because another person had wished that such a life should never be completed.

These days, history judges people like Hitler and his sect, and I hope that the same fate will later also be meted out to you. Let historians judge how many lives you have on your conscience; how many hours of happiness as well as pain you have precluded. Let history judge how many persons' lives you have personally taken and with it everything they could have experienced (light, sun, warmth, cold — simply everything on this planet whether beautiful or ugly).

For me personally, you are much worse than the Nazis who wreaked havoc over our Germany — excuse me, please — but in my opinion you are the greatest Fascist under the sun, you are much worse than the Nazis; you are the Chief Nazi of the entire brown clan.

You consider yourself broad-minded, enlightened and liberal? On the contrary, what you are doing is pure Fascism; had the Nazis had someone like you, who brought such an invention to the world, then the Nazis would have been the very first to use it. That you will never understand this is clear to me. You are simply the greatest luminary among all the damned Nazis, yet you will not even recognize on your deathbed what you have wreaked. Just like the Nazis, you will think until the end that you have initiated a good thing.

Disgustedly,
Prof. Dr. A. H.

Readers may wonder why I even bothered to respond to such a diatribe, but equating me — a refugee from the Nazis to that scum — simply got my goat. Here is my reply.

Mr. H.

Your e-mail is not only incredibly impertinent, but also totally illogical. In which century are actually living?

In Japan, the Pill was only legalized in 1999, yet the low number of children (1.5 children per family) had already occurred there some decades earlier. In Catholic countries like Italy and Spain, already since the 1970s, the average family size is 1.3 children. In the 1980s, less than 5% of Italian women used the Pill, yet abortion was already legal. These days, the greatest number of illegal abortions occurs in the all-Catholic countries of Latin America,

Nobody forces a woman to use the Pill nor, in my opinion, is the Pill the ideal solution for everyone. But it is an option, which has prevented millions of abortions. As I indicated in the TV interview, the roughly 130 million coital acts that occur globally every 24 hours lead to ca. 1.2 million fertilizations, of which 50% are unexpected and half of those unwanted, i.e. ca. 300,000 every 24 hours! The result: daily, 125,000 abortions and unfortunately often illegal ones as is the case in many all-Catholic countries. In my opinion, the Pill has at least partially prevented those tragedies.

I don't know what kind of a Professor you are, but to allow yourself to call me the "Chief Nazi of the entire brown clan" is as we say in America "beyond the pale." You should be ashamed of yourself! I only hope that you have already been prohibited for some time to have any contact with students. I also hope that the owner of such a poisonous tongue and mind has no children of his own.

Carl Djerassi

In the end, I was glad that I responded to that diatribe rather than ignoring it, because my explosion elicited the following response:

Forgive me — it was really impertinent of me. I was emotionally much too aroused. I sincerely ask for your forgiveness.

Prof. Dr. A.H.

Before returning later in this chapter to an even deeper shadow, I shall first describe the early history of the Pill. Only by throwing light on this object, can shadows, real or imagined, be examined. So let us start with the four-letter word PILL and ask how that generic term has become synonymous with oral contraceptives.

"The Pill": Origin of the Word

When I first encountered the capital P in some news article, I visualized its author as a macho journalist of the late 1950s — sleeves rolled up, eyes squinting through the smoke of the cigarette dangling from the corner of his mouth, two index fingers producing a machine gun rattle on the Remington — who, while writing some pithy piece on oral contraceptives, decided to capitalize the word *pill,* and thus inadvertently converted this pedestrian generic term into a powerful four-letter epithet. Since then, the Pill has been described as everything from a woman's panacea or her poison to the cause for the social emasculation of men. In the 1970s, in preparation for my first book addressed to a general audience, *The Politics of Contraception,*[7] I re-read Aldous Huxley's 1958 *Brave New World Revisited* — his magisterial reflections on his *Brave New World* of 1932. That's where I found a previously capitalized "pill", framed within startlingly relevant words of wisdom:

> [The population problem] *is becoming graver and more formidable with every passing year. It is against this grim biological background that all the political, economic, cultural and psychological dramas of our time are being played out.... The problem of rapidly increasing numbers in relation to natural resources, to social stability and to the well-being of individuals — this is now the central problem of mankind; and it will remain the central problem certainly for another century, and perhaps for several centuries thereafter.... Obviously we must, with all possible speed, reduce the birth rate to the point where it does not exceed the death rate. At the same time, we must, with all possible speed, increase food production, we must institute and implement a worldwide policy for conserving our soils and our forests, we must develop practical substitutes, preferably less dangerous and less rapidly exhaustible than uranium, for our present fuels.... But all of this, needless to say, is almost infinitely easier said than done. The annual increase of numbers should be reduced. But how?... Most of us choose birth control — and immediately find ourselves confronted by a problem that is simultaneously a puzzle in physiology, pharmacology, sociology, psychology and even theology. "The Pill" has not yet been invented.*

As I already mentioned in an earlier autobiography of mine and now repeat verbatim here, for years, I was convinced that this was the first printed appearance of "the Pill." I was disabused of that notion

[7]C. Djerassi, *The Politics of Contraception.* W. W. Norton, New York (1980).

by the second edition of the *Oxford English Dictionary*. It quotes one C. H. Rolph writing in 1957 about "the quest now going on for what laymen like myself insist on calling 'the Pill'." But who was C. H. Rolph and in what context did he write these words? I had never heard of him, but anyone credited with the first modern usage of "the Pill" certainly merited some recognition. I tracked it down in one of his memoirs, entitled *Further Particulars*.[8] I was pleased to note that someone with his wide-ranging interests was apparently the linguistic father of "the Pill." For a start, the centenary of his birth (as Cecil Hewitt) coincides with the fiftieth anniversary of the Pill. He started his career with the London City Police, from which he retired after twenty-five years at age forty-five with the rank of Chief Inspector. He began to write while still on the police force, publishing under the pen name C. H. Rolph in periodicals ranging from the *Police Reporter* to the *New Statesman*. He pursued this second career full-time as a journalist, essayist, book author, BBC interviewer and more. His tastes — for music, literature, law and liberal social policy — were so diverse that his memoir reads like a mini "Who's Who" of twentieth century Britain. His observations on the Pill appeared in the introductory chapter of an anthology, *The Human Sum*, for which Hewitt alias Rolph had managed to assemble a stellar list of authors, including Julian Huxley and Bertrand Russell. This is what Rolph had to say:

> *He* [Dr. A. S. Parkes of the Medical Research Council] *gives a modestly exciting account of the quest now going on, in biological laboratories in various parts of the world, for what laymen like myself insist on calling "the Pill"; and by this phrase, which, like all men of science, Dr. Parkes would doubtless reject, I mean the simple and completely reliable contraceptive taken by the mouth.* [In point of fact, Parkes in his chapter makes no mention whatsoever of then already on-going research with orally effective steroids, thus showing Rolph as an even more successful prophet.] *This, it can hardly be doubted, will one day become available for the control of human fertility, universally, among the most backward as well as the most advanced communities in the human race; and its tremendous implications must, in the soberer thoughts of any person with social compassion, dwarf any other consideration that this book can provide.*

[8]C. H. Rolph, *Further Particulars: Consequences of an Edwardian Boyhood*. Oxford University Press, Oxford (1987).

I could not help but be struck that this coinage of "the Pill" was prompted by A. S. Parkes, another author in Rolph's *The Human Sum*. In 1993, I had the privilege of delivering the first annual Parkes Memorial Lecture of the Society for the Study of Fertility at Cambridge University in honor of that distinguished British reproductive biologist. History is indeed circular!

Still, even if Huxley was second in print, he was certainly the most elegant proponent of the now accepted capitalized version of the Pill. Within three years of his pronouncement, the Food and Drug Association (FDA) approved the use of oral progestational steroids for contraceptive application and not long after, the Rolph/Hewitt–Huxleian usage was consecrated in both the *American Webster's* and the *Oxford English Dictionary*: "*often cap*: an oral contraceptive — usu. used with *the*."

Having learned this much about the acceptation of "the Pill" in English, I found myself wondering: what about other languages? And (more to the point) does the informal epithet a nation chooses tell us something about a country's attitude toward birth control?

I started with the languages of which I have some reading competence. French, Spanish and Italian were unexceptional variations on the Rolph–Huxley theme: "*la Pilule*," "*la Píldora*" or "*la Pillola*." But not so in German. Under "*Pille, die*," my *Duden* contains the pithy "*Arzneimittel in Form eines Kügelchens*" [medicine in form of a small ball]. For the more pregnant personal meaning of "the Pill" or its international equivalents, I had to turn to the letter "A" in the then current edition of the *Duden* to find "*Antibabypille*" with its laborious definition "*empfängnisverhütendes Mittel in Pillenform auf hormonaler Grundlage*" [contraceptive agent of hormonal basis in pill form — no charming *Kügelchen* diminutive here!]. Why the bitter, almost brutal, "*Antibabypille*"? Did the Church preempt this linguistic terrain before the journalists could settle on the pithy "*die Pille*"?

Steroid oral contraceptives were never designed as agents against babies. Since the Pill acts on the body of a woman who is not pregnant, there is no baby involved at all: the ultimate "pro-woman" development, perhaps, but unless you imagine a woman's interests as somehow "antibaby" it is hard to understand how such a term could come into use. Personally, I have no hesitation in calling it the "pro-baby Pill," because its ultimate purpose is to assure that every child is a wanted child. With the German propensity for complicated words, "the anti-unwanted-child-Pill" would not be an unprecedented coinage, but I doubt if even the *Duden* would accept as an entry "*die Antiunerwünschtebabypille*." While most public media in Germany and Austria still seem to be chained to the "*Antibabypille*" convention, street German these days has more or less dropped the pejorative "anti-Baby" prefix. Is this just another manifestation of the linguistic Coca-Colanization of the present German generation or is it a token of a more realistic attitude in a country where the Pill — anti or not — has become the most popular method of birth control?

There is another reason for striking "*Antibabypille*" from the German vocabulary. It supports the common misconception — by no means limited to Germany or Austria — that the Pill is responsible for the rapid drop of birth rates in the industrialized nations, an argument that I already rebutted in my above-cited letter to *The Guardian*.

But if the contraceptive association with the word "Pill" was created linguistically before the scientific discovery was actually made, then it behooves us now to see how in fact a pill became THE Pill. I shall start the tale in my newly chosen retrodirectional fashion by picking an arbitrary year, 1985 — seemingly quite unrelated to the history of the Pill — because it was a traumatic year in my life when I was diagnosed with colon cancer and immediately entered a hospital for a major operation. (As I shall recount in "*Writer*", this was one of the events that caused me to subsequently turn from chemistry to literature.)

Mothers of the Pill

For a time after a cancer operation, the patient is allowed morphine almost *ad libitum*. The effect of my last injection had worn off, and I had rung for the nurse to give me another one. I focused on the chocolate-colored, Teflon-smooth skin of the nurse's upper arm, and lightly touched it. "How did you end up with such superb muscles?" I asked. "I'm a bodybuilder," she replied, and continued, "Are you really the father of the Pill?"

I am often asked the question in this phallocentric way; if I had been a woman, would she have asked, "Are you the mother of the Pill?" Usually I respond in the same genealogical vein, pointing out that our phallocentric society invariably focuses on the patrimony of a scientific discovery, of a new drug, of the Pill… then searching for the "father of…." But the birth of a drug, first and foremost, requires a mother, and most of the time also a midwife or obstetrician. Every synthetic drug, including steroid oral contraceptives, must start with a chemist. Until she or he has invented it, i.e. conceived its chemical structure and then synthesized the molecule, nothing can happen. This is the reason why I maintain that in the parentage of any synthetic drug, including the ovulation-inhibiting progestational constituent of the Pill, the chemist — irrespective of gender — symbolizes the mother, with the chemical entity representing the egg. Only then does the biologist enter the picture, performing a variety of biological experiments which I equate to sperm floating around the ovum. The key experiment, confirming the anticipated biological activity or demonstrating some unexpected new one, can then be considered the sperm associated with the actual fertilization. Thus in my picture, the biologist — again regardless of gender — plays the paternal role, while the clinicians' subsequent efforts correspond to obstetrical and pediatric functions in the development and maturation of a drug. Or to put all this into plain English: drug development is an interdisciplinary effort in which no single discipline nor single individual can truly be assigned the exclusive role.

But to be accurate, one needs to retrace the Pill's genealogy at least down to the grandparents and a few uncles. By definition, every synthetic drug originates in a chemist's laboratory; what happens to this chemical entity after it has been synthesized, however, how it ultimately becomes a drug that reaches the consumer, depends very much on circumstance. Frequently, a substance synthesized in connection with some specific chemical problem is only as an afterthought, sometimes even years later, submitted for wide pharmacological screening in the hope that some useful activity will be noted as an extra bonus. The insecticidal activity of DDT was discovered through such screening decades after the substance was first synthesized in a German university laboratory. Alternatively, a substance — Viagra™ being a classic example — may be synthesized for a specific biological purpose, found to be inactive in that regard, and then exposed to wider pharmacological scrutiny in the hope that something might be salvaged. The literature of medicinal chemistry is replete with instances in which such random screening uncovered unexpected biological activity that provided the impetus for further chemical, pharmacological and clinical work.

It is hardly surprising that the modern medicinal chemist is unhappy with this state of affairs, as predictability rather than serendipity is the essence of science, and especially of chemistry. Chemists since Paul Ehrlich, who founded modern chemotherapy in the early part of the last century, have attempted to establish relationships between chemical structure and biological activity that lead to the *a priori* prediction of a potentially useful drug. To a considerable extent, the development of steroid oral contraceptives represents a successful instance of this predictive approach, in which we deliberately set out to synthesize a substance that might mimic the biological action of the female sex hormone progesterone when administered orally, since progesterone itself is essentially inactive by this route unless given in huge doses.

Some people, especially female authors and historians, have derided or ignored my description of a chemist as the mother of any drug — in this case the Pill — because they searched hard and wide for an ovary-bearing female rather than a metaphoric (i.e. gender-neutral) mother. Their irritation focused on the fact that the scientists involved in the development of oral contraceptives were uniformly male, yet ignoring that in the 1950s, science in general and reproductive science in particular were still totally male dominated for reasons which we now all deplore but which had virtually nothing to do with the nature of the research.

This belief has grated on women for decades. As the famous anthropologist Margaret Mead put it in the late 1960s:

> [The Pill] *is entirely the invention of men. And why did they do it?… Because they are extraordinarily unwilling to experiment with their own bodies… and they're extremely willing to experiment with women's bodies… it would be much safer to monkey with men than monkey with women.*[9]

While Mead's irritation may well be understandable, it nevertheless represents a gross oversimplification that ignores that nature had provided scientists with a crucial hint on which to build — women do not get pregnant during pregnancy because of the continuous secretion of progesterone — whereas no such clue to inherent natural infertility exists in male reproductive biology.

[9] *Chemical and Engineering News*, October 25, 1971.

A pervasive sense of the ironies of this historic male bias has led to a focus on two women, Margaret Sanger and Katherine McCormick, whom journalists some thirty years ago have increasingly romanticized as key figures in the development of oral contraceptives. While I find this romanticization charming and emotionally understandable, I nevertheless must emphasize that they had nothing to do with the actual science and especially not with our work. If women scientists deserve listing, they should — as I have frequently pointed out — be Elva G. Shipley, the biologist who first demonstrated the biological activity of our 19-norprogesterone and norethindrone (which I shall describe below) or Edith Rice-Wray, who contributed heavily to the clinical work. I had not even heard of McCormick until about twenty-five or thirty years after our initial chemical synthesis and the subsequent Syntex-initiated biological work, which was quite distinct from the biological work of Gregory Pincus — one of the fathers of the Pill — to whose laboratory McCormick contributed financially. Margaret Sanger as a proponent of birth control was, of course, world famous, but anointing her as some "mother of the Pill" because she reportedly told Pincus that she would like to see a contraceptive Pill in this world is analogous to calling President Nixon one of the scientific fathers in the cancer field, because during his term of office he declared a war on cancer and a hoped-for cure in ten years. Sanger made zero contribution to the science of contraception, which should not demean in any way her role and fame in political or social aspects of birth control as well as an important historical figure in contraception.

What about Katherine McCormick? She clearly contributed financially to the early efforts in Pincus's laboratory, but however commendable such philanthropy is, anointing her as one of "the indisputable mothers of the Pill" (as was done by one journalistic author, Bernard Asbell, in *A Biography of the Drug That Changed the World* and then repeated by many other journalists and historians until being set in cyber-perpetuity through Google and Wikipedia), is as far-fetched as calling John D. Rockefeller one of "the fathers of the Pill." (The Rockefeller Foundation and its offspring, the Population Council, supported and still supports much more research in reproduction and contraception than Mrs. McCormick ever did and did so over the course of many decades.) Quite bluntly, financial support, valuable as it frequently is, can never be equated with creativity; otherwise, the Medicis would be considered the greatest artists of the Renaissance. Instead, let me repeat the name of Elva G. Shipley as literally the first biologist — male or female — who established the high progestational activity of orally-administered norethindrone. If her results had been negative, we would have dropped the project and would never have sent the material to other biologists, including Gregory Pincus, who, as I will demonstrate below, can rightfully be called a "father of the Pill." A curious fact, not commented upon by any other writers is that Pincus, though dedicating his *opus magnum, The Control of Fertility* to "Mrs. Stanley McCormick" for "her steadfast faith in scientific inquiry ..." did not mention any financial support on her part in his acknowledgment section in spite of a lengthy list of funding agencies, individuals, and companies, notably G. D. Searle and Co.

The Overlooked Role of Ludwig Haberlandt

The least-known character in the Pill's story is not a woman after all. It is Ludwig Haberlandt, professor of physiology at the University of Innsbruck. As early as 1919, he carried out a crucial experiment, in which he implanted the ovaries of a pregnant rabbit into another rabbit, which, in spite of frequent coitus, remained infertile for several months — a result that Haberlandt called "hormonal temporary sterilization." (Partisans of Mrs. McCormick might take note that this and subsequent work of Haberlandt's was supported financially by the Rockefeller Foundation.) The problem with this method, of course (other than its reliance on surgery), as well as with subsequent attempts to avoid surgery by the use of "glandular extracts," was that these extracts were not the pure hormone responsible for its contraceptive effect. A mixture of hormones and other proteins, they constituted a potential problem of toxicity for the recipient. Attempts to "purify" these extracts presented the next hurdle to overcome on the way to a practical oral contraceptive.

In numerous subsequent experiments and publications over the course of ten years, Haberlandt — invariably using the first person singular, so strikingly different from today's insistence by scientists on the royal "we" — emphasized the obvious applicability of his animal experiments to human contraception. He fully recognized that the responsible factor was a constituent of the *corpus luteum* and in 1931, in a remarkable book entitled *The Hormonal Sterilization of the Female Organism* of less than 15,000 words that hardly anyone now living seems to have

Die

hormonale Sterilisierung

des weiblichen Organismus

Von

Dr. med. Ludwig Haberlandt

a. o. Professor der Physiologie an der Universität Innsbruck

Mit 6 Abbildungen im Text

Jena
Verlag von Gustav Fischer
1931

Edda Haberlandt

Ludwig Haberlandt (1925) and the title page of his 1931 monograph

read, Haberlandt outlined in uncanny detail the contraceptive revolution of some thirty years later. He pointed out that oral administration, which he actually demonstrated in mice, would be the method of choice as well as the necessity for periodic withdrawal from the hormone to allow menses to occur. He called for the use of such contraception on clinical and eugenic grounds, arguing that it would enable parents to have the desired number of healthy children. Objections that too many women would take advantage of hormonal contraception was dismissed by Haberlandt with the argument that such preparations would require a physician's prescription and would not be made available over the counter. He ended his manifesto with a visionary claim:

> *Unquestionably, practical application of the temporary hormonal sterilization in women would markedly contribute to the ideal in human society already enunciated a generation earlier by Sigmund Freud (1898). "Theoretically, one of the greatest triumphs of mankind would be the elevation of procreation into a voluntary and deliberate act."*

Haberlandt did not limit his publications to the scientific literature. He also published in the popular press and gave interviews that led to newspaper headlines like "My aim: fewer but fully desired children!" (January 20, 1927 issue of the *Acht Uhr Abendblatt, Berlin*) complete with commentary by the now-familiar chorus of physicians, lawyers and theologians. His obsession with the therapeutic potential of *corpus luteum* extracts was so well known that his students hung a banner by his home with the couplet, "*Verdirb nicht Deines Vaters Ruhm mit Deinem Corpus Luteum*" [Don't mar your father's renown with your *corpus luteum*]. But Haberlandt was not content with the visionary's role only. He contacted several pharmaceutical companies in an attempt to obtain consistently active and nontoxic *corpus luteum* and placental extracts for human clinical experiments. In his 1931 book, he finally reported success in the following words: "I have been in contact for over three years with the therapeutic firm Gideon Richter in Budapest [to this day, a company active in the steroid field] and it is likely that in the near future a suitable "sterilizing preparation" under the name "Infecundin" will be available for systemic administration in clinical experiments as I had already announced in Vienna [September 1930]." He confirmed that experiments in mice with orally administered "Infecundin" had demonstrated temporary infertility without toxic reactions, "since only in this manner does the new method have any chance for clinical success." A year later, the forty-seven-year-old Haberlandt committed suicide as a result of the incessant criticism of his work in conservative Austria, but the name "Infecundin" survived. In 1966, it became the trade name of the first oral contraceptive produced in Hungary by the very same company Haberlandt had contacted forty years earlier.

Within two years of his death, pure progesterone was isolated in no less than four laboratories in Germany, the US and Switzerland; its chemical structure established by Karl Slotta (a Hitler refugee who eventually settled in Brazil); and its synthesis from the soyasterol stigmasterol accomplished by E. Fernholz in Göttingen and by Adolf Butenandt in Danzig. Had Haberlandt lived, there is no question that he would have pursued his

Carl Djerassi (fifth from left) and Gregory Pincus (ninth from left with cup) at CIBA Foundation Colloquium, London 1952

dream of temporary hormonal sterilization in humans without resorting to *corpus luteum* extracts. But even with pure progesterone, he could have shown only that ovulation can be inhibited by injection as the appropriately named American investigator A. W. Makepeace demonstrated in 1937 in rabbits and E. W. Dempsey in guinea pigs. For oral administration, Haberlandt would have needed another steroid — not naturally occurring, but waiting to be synthesized — and that took twenty more years. Thus, nothing further happened, and Haberlandt's work fell into such oblivion that the next biologist to take it up, Gregory Pincus (who clearly should have known better), did not even feel obligated in 1965 to cite Haberlandt among the 1,459 references in his own book, *The Control of Fertility.*[10] Nor for that matter did Pincus's clinical collaborator, John Rock, whose book *The Time Has Come*[11] in 1963 quotes Makepeace's work but none of Haberlandt's pioneering earlier research! Yet if there ever was a grandfather of the Pill, the Austrian Ludwig Haberlandt above all others deserves that honor. Aside from not mentioning Haberlandt's work, these two books by Pincus and

[10] G. Pincus, *The Control of Fertility*. Academic Press, New York (1965).
[11] J. Rock, *The Time Has Come*. Knopf, New York (1963).

Rock, who are often called the two fathers of the Pill — an interesting variant of parthenogenesis — contain another surprising lacuna: zero reference to the chemical invention of the Pill, without which, of course, no biological or clinical research on today's Pill could even have started.

Since the term "father of the Pill" carries reproductive connotations, the development of oral contraceptives might well deserve some reproductive metaphor. Thus one might ask why the metaphoric fathers, Pincus and Rock, did not refer to the usual missing partner in such a process, namely a metaphoric mother, which, as I pointed out earlier, should be assigned to chemists. An interesting answer — to my knowledge, hitherto ignored by all journalists and most historians dealing with the history of the Pill — was provided at an unusual session held on a Friday morning, May 5, 1978, in an old New England mansion on the outskirts of Boston, the headquarters of the American Academy of Arts and Sciences. The Academy was holding a closed two-day session on "Historical Perspectives on the Scientific Study of Fertility." The purpose of the meeting was to have a free-flowing dialogue among some of the key scientists who had been active in the field of fertility in the US during the previous forty years (therefore, it was not surprising that, as far as I could tell, at age fifty-five, I was the youngest of that group) in order to collect a record that historians of science might draw upon in the future.

The unedited transcript of that Friday morning session reads awfully: nouns do not match verbs, tenses get mixed, punctuation is lost, and many words are misspelled or appear to be inaudible. Nevertheless, one gets a real flavor of excited human dialogue and interruptions, of hurt egos, of hitherto undisclosed vignettes. Here are two samples.

Hechter: May I take a couple of minutes?

Djerassi: I haven't finished. I'd like to continue because I've only gotten to the first half of my story.

Reed: He can have my time. This is the first really fruitful... (inaudible)

Greep: This is history from the horse's mouth, and I think it's very good.

Djerassi: I misunderstood. Did you want me to continue?

Greep: Yes.

The scientific co-chairman of the Boston Academy's May 1978 meeting was Roy O. Greep, a distinguished endocrinologist at Harvard, who had known personally most of the actors in this play. Another key participant was Oscar Hechter, who for many years had been senior scientist of the Worcester Foundation for Experimental Biology. Though not directly involved in the development of oral contraceptives, he had been an intimate collaborator of Gregory Pincus. James Reed of Rutgers University was a historian studying the birth control movement in America.

I felt that this was the one opportunity, years after Pincus's death, where I could find out why he had been so ungraciously selective in not acknowledging work of others that was crucial to the development of the Pill. John Rock, who had not behaved very differently, was in the room, but he had reached an age where it was not any more possible for him to contribute to the dialogue. His was a silent, poignant presence. But Celso-Ramon Garcia, Rock's and Pincus's closest clinical colleague, was present, which led to the following exchange:

> *Garcia: Basically, the monograph* Control of Fertility *that Pincus wrote expresses in detail what his feelings were about who contributed to what.*
>
> *Djerassi: Why did he not mention any chemists, do you happen to know that?*
>
> *Garcia: He was a biologist, the same way as you are principally presenting your story as a chemist.*
>
> *Djerassi: That's not true. That's why I submitted a paper here with biological references, including yours.*
>
> *Garcia: Well, okay, but the fact is that principally you are a chemist and your major contribution has been that of a chemist.*
>
> *Djerassi: But this would be like my describing the history of oral contraceptives without a single reference to Pincus or Rock or yourself!*

In other words, Garcia — and by inference Pincus — felt that it is sufficient to focus on the paternal role in discussing the history of the Pill. Hence as a chemist, I have in past autobiographical accounts illuminated primarily the equally indispensable "maternal" role of the chemist, especially since most of the historical and journalistic accounts have always focused on the subsequent biological and clinical studies. In part, this is understandable, because chemistry is largely communicated through the pictography of chemical structures, thus causing most historians and other chemically illiterate authors to jump over the chemical part of the story or even misrepresent it. But this time, rather than repeat what I have described in such excruciating detail elsewhere with extensive references to literature citations in peer-reviewed journals, I shall only describe some personal highlights as background for my subsequent focus on the "bitter" aspects of what I considered for many years a "sweet" Pill.

Just as it is clear that Ludwig Haberlandt merits identification as the paternal grandfather, having unambiguously recognized and proclaimed the role of the female sex hormone progesterone as nature's contraceptive (women do not get pregnant during pregnancy, when they secrete progesterone during the entire nine months), Gregory Pincus — despite the uncertainties of paternity generally — deserves to be called a father of the Pill. The initial rabbit experiments in the early 1950s by M. C. Chang in Pincus's laboratory, which confirmed and extended Haberlandt's work of the 1920s, clearly were metaphorically the sperm that fertilized the chemical egg. The subsequent implantation of the embryo and eventual fetal growth of what eventually became the Pill can largely, though not entirely, be ascribed to further experiments conducted in Pincus's laboratory with substances provided by us as well as other chemists. But Pincus was not only a prolific and highly experienced endocrinologist, he was also a charismatic entrepreneur. Many times, this latter quality is more difficult to find than mere scientific brilliance; it took entrepreneurship of Pincus's caliber to bring the steroids provided by the chemist to the stage where clinical trials of the Pill could be initiated and where John Rock, as leader of the clinical team, could assume the mantle of metaphoric obstetrician for the eventual birth of the Pill. While Rock's name is inexorably connected with that role, others, notably Celso-Ramon Garcia (the first professor of obstetrics and gynecology at the University of Puerto Rico Medical School) and Edith Rice-Wray (Medical Director of the Puerto Rican Family Planning Association) contributed heavily to the planning and implementation of the first clinical trials in the San Juan area. Rice-Wray subsequently directed a Family Planning clinic in Mexico City where she continued her clinical studies, this time with Syntex's norethindrone.

Some Autobiographical Observations: Cortisone and Oral Progestins

Curiously, my own involvement in the development of the Pill started with cortisone — a wonder drug for the treatment of rheumatoid inflammations, which medicinally had nothing to do with contraception except that both substances have a common basis for the chemist: they are steroids. And this in turn requires a simplified definition for which I shall have to resort to chemical shorthand in terms of Figure 1.

The word *steroid*, meaning "like a sterol," is derived from the Greek. Sterols, in turn, are solid alcohols (Gr. *stereos*, solid + *ol*) that occur widely in plants and animals — the best known being cholesterol, the most abundant sterol in humans and other vertebrates. All steroids (and all sterols) are based on a chemical skeleton that consists of carbon and hydrogen atoms arranged in four fused rings and known generically by the forbidding name "perhydrocyclopentanophenanthrene." Steroid chemists communicate among themselves not in such polysyllabic jawbreakers, but in diagrams like the one shown in Figure 1. They simplify life for themselves still more by dropping the symbols for carbon (C) and hydrogen (H). This shorthand is meaningful (to a steroid chemist) if you also assume the following: all angles are occupied by carbon atoms; all carbons are connected four ways to other atoms; any unaccounted-for connections are taken up by hydrogens. With these conventions in mind, one can read Figure 1 as showing three rings (A, B, C) each with six carbons and one ring (D) with five. Some of the carbons have three unmarked hydrogens connected to them, others two, some one, and two (numbered 10 and 13) have none,

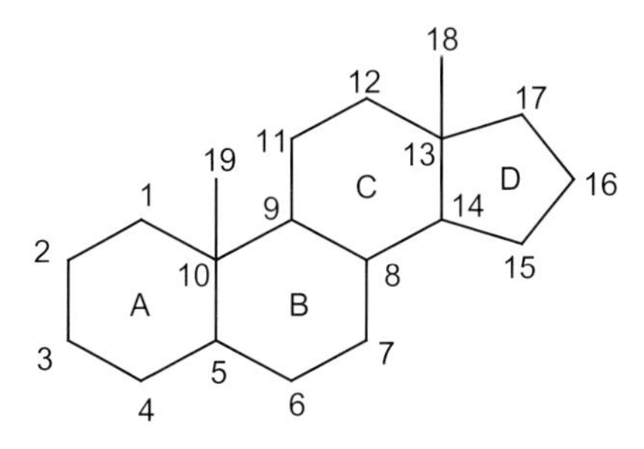

Figure 1

all their bonds being occupied by connections to other carbons. Together, these fused rings define a steroid. Note that atoms 18 and 19 are not part of a ring but are attached as methyl groups. (The full chemical notation of "methyl" is CH_3, but for shorthand purposes, it is simply written as a vertical straight line.)

Thousands upon thousands of synthetic, and many hundreds of natural, compounds are based on this fundamental steroid skeleton made up of carbon and hydrogen atoms as depicted in Figure 1, all differing slightly in chemical structure by the attachment of some additional atoms (usually oxygen) at various locations, most commonly at positions 3 and 17. The variations, however minute, produce dramatically different biological results. Many of the most important biologically active molecules in nature represent slight variations on the steroid skeleton: the male and female sex hormones, bile acids, cholesterol, vitamin D, the cardiac-active constituents of digitalis, the adrenal cortical hormones (related to cortisone and usually referred to generically as "corticosteroids") and many plant-derived products. The wide-ranging biological activity of steroids — for instance, the fact that one (the male sex hormone, testosterone) is responsible for male and another (the female sex hormone, estradiol) for female secondary sexual characteristics — is, in part, associated with the introduction of a third atom, oxygen (O), to special positions of the steroid skeleton.

By the fall of 1945, a twenty-two-year-old, newly naturalized American citizen with a Ph.D. degree from the University of Wisconsin and a wife, I returned to CIBA (the pharmaceutical firm in New Jersey where I had worked for one year after graduation from Kenyon College) for another four years, to resume work on antihistamines and other drugs. One day in the spring of 1949, I received an unsolicited employment offer from Syntex, a company I had never heard of. Although the position, as associate director of chemical research, seemed tempting to me, the location of Syntex in the scientific backwater of Mexico made the offer seem ludicrous. Fortunately, I am a tourist at heart; when I received the invitation, "Come and visit us in Mexico City, all expenses paid," I went. And as a bonus decided to include a visit to Havana in my itinerary.

The invitation came from George Rosenkranz, then technical director of Syntex and barely past thirty years old, who impressed me enormously as a sophisticated steroid chemist; he also charmed me personally. Rosenkranz showed me rather crude laboratories, but he promised lots of laboratory assistants and substantial research autonomy to devise a practical synthesis of cortisone and to pursue other aspects of steroid chemistry that might interest me. Furthermore, even though the labs were primitive, Syntex could boast of some advanced equipment such as an infrared spectrometer at a time when neither CIBA nor my *alma mater*, the University of Wisconsin, had such an instrument, which proved to be enormously useful for steroid research.

I arrived at Syntex in the late autumn of 1949, just around my twenty-sixth birthday. I have never regretted that decision, even though at that time my American colleagues considered me mad to move to a country that, although famous for *mariachi* music, bull fights and pre-Columbian ruins, had only generated the barest of blips on the radar screen of international chemical journals. Yet I was convinced that the best route to the academic job still eluding me was to establish a reputation in the scientific literature. I felt intuitively that Mexico was the right place for me. Syntex had the same objective I did: to establish a scientific reputation. Our common goal — devising a chemical synthesis of cortisone from a plant raw material — was one of the hottest scientific topics in organic chemistry at that time. I was young and willing to gamble on a few years in Mexico — partly because living in another country and learning another language appealed to me, but also because I thought that any scientific achievement from a laboratory in Mexico was likely, upon publication, to make a much bigger impression on academia than one coming from the usual elite laboratories in North America or Europe. Consequently, I really had only one requirement before I accepted the Syntex offer, and that was to publish any scientific discoveries promptly in the chemical journals. Syntex agreed to this and stuck to its bargain. From my previous industrial experience, I fully understood that discoveries have to be patented by the firm in whose laboratory the work is performed before they are written up for publication. But instead of having patent attorneys deciding whether and when to publish, at Syntex Rosenkranz and I called the shots — extraordinary for a pharmaceutical company. As a result of this policy, during my first two years (1950–1951) at Syntex we published more rapidly in the chemical literature than did any other pharmaceutical company, or even many university laboratories.

Until 1951, the only source of cortisone was through an extraordinarily complex process of thirty-six different chemical transformations

Press conference announcing the first synthesis of cortisone from a plant source at Syntex in Mexico City, 1951. Standing, left to right: Gilbert Stork (consultant), Juan Berlin, Octavio Mancera, Jesus Romo, Alexander Nussbaum. Seated, left to right: Juan Pataki, Enrique Batres, George Rosenkranz, Carl Djerassi, Rosa Yashin, Mercedes Velasco

starting from animal bile acids — a *tour de force* pioneered by Lewis Sarrett of Merck and Co. For many years, this had proved to be the longest and most complicated synthesis of any chemical on an industrial scale. Now that cortisone had emerged as a wonder drug — a discovery for which the Americans E. C. Kendall and P. S. Hench, and the Swiss chemist Tadeus Reichstein were awarded the Nobel Prize in Medicine in 1950 — developing an alternative chemical synthesis from a plant raw material became one of the most acclaimed scientific projects, with a number of powerful academic and industrial research groups in Europe and the USA competing to be first. At the outset, nobody even realized that a small research team in Mexico City had entered the race. But when we completed ahead of everyone else in June 1951 our synthesis of cortisone from diosgenin, the resulting publicity was astounding. Thus, long before Syntex sold drugs under its own name to the medical profession, its international scientific reputation in chemistry was well established. Ten years after my temporary move to Mexico, when Professor Louis F.

Fieser of Harvard analyzed in 1959 the references in the latest edition of his text *Steroids*, the recognized bible of steroid research, he found that no laboratory in the world — academic or industrial — had published as much in the steroid field as Syntex had in that time. Chemistry south of the Rio Grande had finally made the grade.

Which finally brings me to the Pill and to a summary of my personal involvement in the first synthesis of a steroid oral contraceptive. Within the confines of the present chapter, it would be superfluous to go into greater detail since I have described that story extensively in my earlier autobiographical works and have cited all relevant chemical and biological literature references in a number of review articles in the scientific literature replete with references and numerous chemical structures.

The Synthesis of Norethindrone

At the time that I became interested in the chemistry of progestins — steroids that are chemically related to the natural female sex hormone progesterone — one of the dogmas of steroid chemistry was that almost any chemical alteration of the progesterone molecule would either diminish or destroy its biological activity. This belief seemed puzzling in light of the fact, well known at the time, that estrogenic steroid hormones (e.g. the female sex hormone estradiol), which occur naturally in a variety of forms, as well as synthetic chemicals not even based on the steroid skeleton, display marked estrogenic potency. In 1944, Maximilian Ehrenstein (an emigrant from Nazi Germany), then working at the University of Pennsylvania, published a paper that was mostly overlooked, but had made a deep impression on me while still a graduate student. By an extremely laborious series of steps, Ehrenstein had transformed the naturally occurring steroid cardiac stimulant strophanthidin into a few milligrams of an impure oily substance, named 19-norprogesterone. While he had obtained only enough material for biological testing in two rabbits, in one of them his compound had displayed higher progestational activity than the parent hormone. A positive test in one animal out of two could, of course, have been just a fluke. What made Ehrenstein's results so unusual was what "19-nor" in the compound's name signified. It meant that Ehrenstein had removed carbon atom No. 19 (between rings A and B of the steroid skeleton depicted in Figure 1) from the most inaccessible site of the steroid molecule to replace it with a hydrogen atom. On paper — or in words — the change sounds trivial. Given the state of the art of organic synthesis at the time, however, this was so difficult an operation that it had required several years for completion. Moreover, if the biological results were real, Ehrenstein's observation demolished the previous assumptions about the inviolability of the progesterone structure. But there was another problem: Ehrenstein's oily product was, as I indicated, impure, a mixture of at least three "stereoisomers" — molecules that, while structurally identical, were, like mirror images, as alike — and fundamentally different — as your left hand and your right. In biochemistry, which often requires molecules to fit together like a hand in a glove, such a difference can be crucial. Which one of the components, if any, was responsible for the putative progestational activity? It took seven years for someone to come up with an answer. Our ability to do so led us almost straight to the Pill.

Part of my Ph.D. thesis at the University of Wisconsin in the early 1940s had dealt with devising a synthesis of the then-inaccessible estrogenic hormones estrone and estradiol from the more readily available androgens, such as testosterone. For years, the estrogens were only available by isolation from the urine of pregnant women (and later of pregnant mares, even today the source of one of the more frequently prescribed estrogen compositions in use for hormone replacement therapy). Hans H. Inhoffen, at Schering A.G. in Berlin, had demonstrated the practical feasibility of such a chemical conversion, but the work had been performed during World War II and experimental details were scant and had to be partly reconstructed. Syntex had started to use the Inhoffen process (which had not been patented in Mexico) for the production of modest quantities of estrone and estradiol. Upon assumption of my research position there, I suggested to Rosenkranz that Syntex examine another and potentially proprietary route to the estrogens directly from testosterone. In less than three months we succeeded in accomplishing this aim, which in chemical jargon would be described as the "aromatization of ring A of conventional steroids," and published it in the *Journal of the American Chemical Society*, one of the most prestigious journals in the field.

These aromatization studies turned into the impetus that led us in a fairly straight path to the first synthesis of an oral contraceptive. From a technical standpoint, upon my arrival in Mexico City, I felt that the time was ripe to follow up on Ehrenstein's lead of 1944. Using various chemical methods developed as part of our estrogen synthesis as well as methodology perfected by the Australian chemist, Arthur J. Birch (subsequently a long-term Syntex consultant), my Syntex colleagues and I prepared for the first time in 1951 gram quantities of pure, crystalline 19-norprogesterone, which, when assayed in rabbits by Elva G. Shipley at Endocrine Laboratories in Wisconsin, was found to be four to eight times as active as natural progesterone. In other words, Ehrenstein's observation with an oily mixture tested in one rabbit was more than confirmed: replacement of carbon atom 19 by one hydrogen (see Figure 1) had produced the most active progestational steroid known at that time. This observation was crucial, because Ehrenstein's mixture of stereoisomers had also the wrong configuration at C-17 — a change that was known to destroy progestational activity in progesterone itself) — and from an experiment in a single rabbit, it was not clear whether the beneficial effect of removing the angular methyl group between rings A and B was real.

With that lead in hand, we turned to another accidental discovery that had been made in 1939 in Germany, where chemists at Schering, again under the leadership of Inhoffen, found that if acetylene is added at position 17 (see Figure 1) of the male sex hormone testosterone, its biological activity is changed markedly: for unknown and totally unexpected reasons this androgenic compound displayed weak progestational activity. Far more important, it also proved to be orally active. On the reasonable assumption that removal of the 19-carbon atom increases progestational potency and addition of acetylene confers oral efficacy, we at Syntex put both these observations together. On October 15, 1951, Luis Miramontes, a young Mexican chemist doing his thesis work at Syntex under my tutelage, completed the synthesis of the 19-nor analogue of Inhoffen's compound — that is, 19-nor-17α-ethynyltestosterone (Figure 2) or, for short, "norethindrone" (or "norethisterone" in Europe) — which turned out to be the first oral contraceptive to be synthesized.

We immediately submitted the compound to our favorite commercial testing laboratory in Wisconsin for biological evaluation and gloated happily when Dr. Elva G. Shipley reported back that it was more active as an orally effective progestational hormone than any other steroid known at that time. In less than six months we had accomplished our

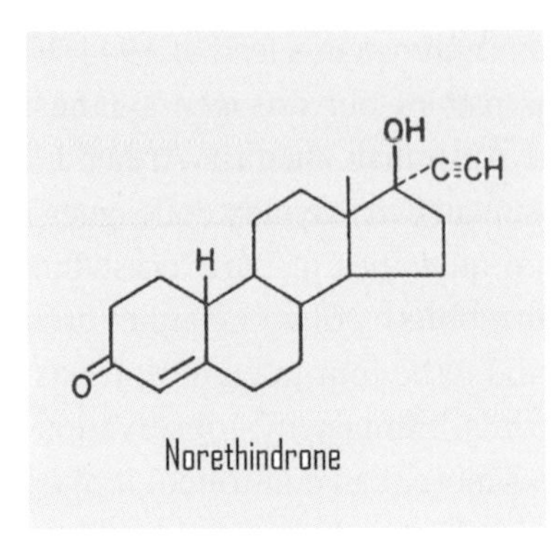

Figure 2

goal of synthesizing a superpotent, orally active progestin! Returning to my reproductive metaphor, I classify our synthesis of norethindrone as the release of the fertile egg waiting now to be fertilized. Once that is understood, the roles of Ehrenstein and Inhoffen as older maternal uncles become obvious. Since Ehrenstein lived in the US, our paths crossed in the 1950s at several scientific meetings. I know that he was pleased at our elevation of his original 19-norprogesterone work from a piece of chemical esoterica to one of seminal significance.

My encounter with Inhoffen was different. We met only once at an international scientific congress. There, his comments seemed frosty, leaving the impression that my work as a graduate student at the University of Wisconsin had constituted an intrusion into his early work on the partial synthesis of estrogens. But in 1999, our paths crossed again, twice, though on his part posthumously. Early that year, I received the Inhoffen Medal at the Technical University of Braunschweig, but a more moving event occurred later that year in Graz. I had given a typical academic talk on the History of the Pill and had done so in German, which meant that it moved slower than it would have in English. When I realized that I would be running out of time, I decided to skip some slides. One of them was a picture

Schering A.G.

Hans H. Inhoffen and Huang Minlon at Schering A.G., Berlin, ca. 1938 of Inhoffen together with the father of Chinese steroid chemistry, Huang Minlon. The room was crowded and I had to cope with many questions before the audience broke up. Suddenly a tall, serious man, probably around sixty years old, approached me to ask quietly, "Did you know Inhoffen and his work?" That's when I found out that I was speaking to Peter Inhoffen, a Catholic theologian and only son of Professor Inhoffen, from whom he had become estranged. I couldn't read his expression: was his question prompted by curiosity or by still smoldering filial pride?

Our patent application for norethindrone was filed on November 22, 1951 (it is the first patent for a drug listed in the National Inventors Hall of Fame in Akron, Ohio), and I reported the details of our chemical synthesis, together with the substance's high oral progestational activity, at the April 1952 meeting of the American Chemical Society's Division of Medicinal Chemistry in Milwaukee. The abstract of this report under the names of Djerassi, Miramontes and Rosenkranz was published in March 1952, and the full article with complete experimental details appeared in 1954 in the *Journal of the American Chemical Society*. Readers may well be irritated by such an avalanche of dates, but chronological precision is the baggage of scientists preoccupied with priority — a foible I would be disingenuous to hide.

2,744,122
Patented May 1, 1956

2,744,122

Δ^4-19-NOR-17α-ETHINYLANDROSTEN-17β-OL-3-ONE AND PROCESS

Carl Djerassi, Birmingham, Mich., and Luis Miramontes and George Rosenkranz, Mexico City, Mexico, assignors, by mesne assignments, to Syntex S. A., Mexico City, Mexico, a corporation of Mexico

No Drawing. Application November 12, 1952, Serial No. 320,154

Claims priority, application Mexico November 22, 1951

4 Claims. (Cl. 260—397.4)

The present invention relates to cyclopentanophenanthrene derivatives and to a process for the preparation thereof.

More particularly the present invention relates to Δ^4-19-nor-androsten-17β-ol-3-one compounds, having 17α-methyl or ethinyl substituents and to a process for producing these compounds.

In United States application of Djerassi, Rosenkranz and Miramontes, Serial Number 250,036, filed October 5, 1951, there is disclosed a novel process for the production of 19-norprogesterone. As set forth in this application, 19-norprogesterone has been found to be even stronger in its progestational effect than progesterone itself.

First page of US patent covering Norethindrone

A few weeks after having synthesized the substance and having received from Dr. Shipley confirmation of its anticipated oral progestational activity, we sent it to various endocrinologists and clinicians: first to Roy Hertz at the National Cancer Institute in Bethesda, Maryland and to Alexander Lipschütz in Chile; later to Gregory Pincus at the Worcester Foundation in Shrewsbury, Massachusetts, to Robert Greenblatt in Georgia, and to Edward Tyler of the Los Angeles Planned Parenthood Center. It was

Tyler who, in November 1954, presented the first clinical results of using norethindrone for the treatment of various menstrual disorders and fertility problems. All of these biological investigations can be equated to sperm that is surrounding the egg. But since this is a record of the history of the Pill, we need to address the source and origin of the particular sperm that led to the fertilization of our chemical egg and thus to the ultimate birth of an oral contraceptive.

While we were aware of Haberlandt's work and progesterone's status as "nature's contraceptive," initially, we were not focusing on birth control when we developed an oral progestational compound, because contraception was of no commercial interest in 1951 to the pharmaceutical industry. Our aim, of course, was to create a new proprietary drug for Syntex that would enter then acceptable medical practice based on the established clinical efficacy of the natural hormone progesterone for the treatment of menstrual disorders, for certain conditions of infertility, and at a research level, for the treatment of cervical cancer in women by local administration of a high dose of the hormone. Such administration was extremely painful because it involved injecting a fairly concentrated oil solution of large amounts of progesterone into the cervix. What drove us was the desire to create a more powerful progestin that would be active orally. As it happened, the progesterone treatment of cervical cancer did not pan out, but the clinical use of our norethindrone (under the trade name Norlutin and licensed to Parke, Davis and Company — at that time a large American pharmaceutical company) for the treatment of menstrual disorders was approved by the FDA in 1957 and is one of its therapeutic indications to this day.

Each of the biologists mentioned above had his own area of expertise and interest in the field of progestational activity. Gregory Pincus and his colleague Min-Chueh Chang of the Worcester Foundation for Experimental Biology in Shrewsbury, Massachusetts, were focusing on how progesterone worked to inhibit ovulation (the mechanism behind Haberlandt's "temporary hormonal sterilization" and Makepeace's confirmation). Among the many steroids tested in 1953 by the Worcester Foundation group for such activity two substances stood out: our norethindrone and another substance, norethynodrel, that had been synthesized about one and a half years following our first publication by Frank Colton at G. D. Searle, a pharmaceutical company in the Chicago area. The chemical history of norethynodrel has been told many times by me as well as others and is not a pretty one, since it illustrates one of the less attractive features of scientific research: the drive for scientific priority and the attempts to circumvent patent priority. In this instance, the stakes were higher than usual, since commercial considerations and financial returns quickly entered the equation.

For historical accuracy and appropriate credit, it is important to note (simply by comparing the relevant patent application filing dates) that even though norethynodrel was made well over a year following the publication of our successful synthesis of norethindrone, it was norethynodrel under the trade name Enovid that first entered the market as an oral contraceptive. Chang had found norethindrone and norethynodrel to have been the two most promising candidates in his initial animal studies. But Chang's boss, Gregory Pincus, who was a consultant for Searle — a company that financially supported much of his research — selected the Searle compound for further work. Syntex, not having any biological laboratories or pharmaceutical marketing outlets at that time, licensed Parke-Davis & Co. of Detroit to pursue the FDA registration and market the product in the US. It was only after 1957, when both norethindrone and norethynodrel had received FDA approval as drugs for non-contraceptive, gynecological purposes, that the paths of the two companies diverged. Chemically, norethynodrel is trivially different from Syntex's norethindrone; on exposure to acid or simply in the acidic environment of the stomach, norethynodrel has been shown by Pincus as well as others to be converted in part to norethindrone and thus can be considered a pro-drug of the latter. Is generation in the stomach of a patented compound via a pro-drug an infringement of a valid patent? I urged that we push this issue to a legal resolution, but Parke-Davis, our American licensee, did not concur, because G. D. Searle was an important customer in another field (anti-motion sickness).

In the mid-1950s, Searle actively supported clinical trials of the contraceptive efficacy of norethynodrel. The work was conducted in Puerto Rico, under the direction of Pincus and especially John Rock, a clinical endocrinologist and gynecologist from Harvard. Around the same time in Mexico City and Los Angeles, Syntex sponsored contraceptive trials with norethindrone. But fearing a possible religious backlash, Parke-Davis suddenly chose not to pursue these results through the FDA approval process, and returned the contraceptive (but not gynecological) marketing license to Syntex. Alejandro

Zaffaroni, Syntex's Executive Vice President, eventually negotiated a favorable marketing arrangement with the Ortho Division of Johnson & Johnson, a company with a long-standing commitment to the birth-control field, but the shift to a new company meant a delay of nearly two years before Syntex's norethindrone received FDA approval as a contraceptive. By 1964, three companies — Ortho, Syntex and Parke-Davis (having changed its mind after realizing that no Catholic-inspired boycott had developed) — were marketing 2.0-milligram doses of Syntex's norethindrone (or its acetate, named Anovlar and introduced by Schering in 1961 in Germany under license from Syntex as the first contraceptive in Europe), which by then had become the most widely used active ingredient of the Pill.

There is no question that Searle deserves kudos for marketing norethynodrel first — despite a possible consumer backlash by opponents of contraception. But given the extraordinary importance of these steroids, why did the Searle group never disclose in the peer-reviewed literature any of the chemical research that led them to their pill, whereas we did so within a few months of our completed synthesis? The only date supporting their claim for "independent simultaneous discovery" is Searle's patent filing date of August 31, 1953, a date that sounds "simultaneous" only without juxtaposition to the November 21, 1951 filing date of Syntex's patent application. Thus, their unaccountable shyness about the most significant product in their corporate history, norethynodrel, can only raise questions. Why, for instance, did Gregory Pincus, the person most responsible for persuading Searle to market norethynodrel, make not a single reference to any chemist (not even Frank Colton) in his 1965 book, *The Control of Fertility*? Why does his book make no mention of how the active ingredient of the Pill actually arrived in his laboratory? And even worse: why did Pincus drop Haberlandt's name into a black hole of anonymity?

Alejandro Zaffaroni with Carl Djerassi pointing at the chemical structure of Norethindrone (trade name Norlutin), Mexico City, 1959

My preoccupation with establishing unequivocally the priority — and thus the metaphoric maternal identity — is not just my admittedly strong competitive drive. I am realistic enough to acknowledge that it really does not make any difference to the world who does what first. At least that's what I thought until February 27, 2012. On that evening, an unlucky contestant in the wildly popular German TV show *Wer wird Millionär* (*Who Will Become a Millionaire?*) did not give the correct answer to the question "*Den gebürtigen Oesterreicher Carl Djerassi bezeichnet man als Vater der...?*" (The Austrian-born Carl Djerassi is considered the father of ...?), which was followed by the options: A. *Antibabypille*; B. *Nylonstrumpfhose* (nylon stockings); C. *Tiefkühlkost* (frozen food); D. *Digitalfotographie* (digital photography). Thus, she won only 125,000 Euros instead of 500,000. I wonder on what luxuries she would have splurged the extra 375,000 Euros if she had only read one of my autobiographies! Amusingly, she ascribed nylon stockings to me, an invention that was made by the late Wallace Carothers, the most famous graduate of the minute Tarkio College in Tarkio, Missouri where I had spent one semester in 1941 after receiving a scholarship.

But giving credit to Syntex as the corporate institution where it all first started is important to me (even though I severed all connections with that company in 1972), because institutional memories are so short. In the twentieth century, Syntex was the first and possibly the only significant example of important research in such a highly competitive and technically sophisticated field being conducted anywhere in the world in a developing country. Both qualitatively and quantitatively, the research output of Syntex during the 1950s has never been matched in the steroid field; the pride and self-assurance it provided to a cadre of Mexican organic chemists, virtually all of them trained at Syntex, was moving to witness. Yet that company does not exist anymore, because in 1994 it was acquired by the Swiss pharmaceutical colossus Roche and promptly swallowed and digested. In that digestive process, the entire research division of Syntex in Mexico, which had just moved into new quarters in Cuernavaca, was closed and all research personnel dismissed. To me, the cold-bloodedness of this corporate amputation seems unforgivable: I know of no other pharmaceutical company in Mexico that has currently any significant research presence.

Syntex, as a company, and Mexico, as a country, deserve full credit as the institutional site for the first chemical synthesis of an oral contraceptive steroid — a statement that is not meant in any way to denigrate Searle's successful drive to be the first on the market with the Pill. But there is a more charming end to this story. In the process of swallowing Syntex, Roche not only closed the Mexican research laboratories where norethindrone was first synthesized. Roche also decided to distance itself from any involvement in the contraceptive field, and promptly sold the entire Syntex oral contraceptive line, still based exclusively on norethindrone. Who was the purchaser? None other than G. D. Searle — the company that went to heroic lengths to circumvent the Syntex patent on norethindrone and now had to pay good money to market it as its lead oral contraceptive long after the original patent had expired. But the story does not end there. G. D. Searle itself was acquired several times — first by Monsanto, but eventually by Pfizer, now the largest pharmaceutical company in the world. Few are aware of the fact that around 1954, Pfizer had an option from Syntex to market norethindrone, an option the company had not exercised because its president, John McKeen, an active Roman Catholic layperson, felt that Pfizer should not touch any agent even potentially related to birth control. Yet half a century later, Pfizer entered the contraceptive market with norethindrone. What a closure to a historical circle!

Dismal Future Prospects in Contraception

Interestingly, Syntex-developed norethindrone is still a widely used active ingredient of oral contraceptives, whereas Searle's norethynodrel disappeared from the market many years ago, to be superseded by other 19-nor steroids, which chemically are all closely related to norethindrone (Figure 2). But this story as well as the many clinical and social ramifications of the Pill have been described by me and by many other authors in numerous books and thousands of articles. Rather than covering once more such well-trodden territory, I shall offer instead — in accord with my present emphasis on shadows — an explanation for the use "bitter" in the title of this chapter.

By the mid-1960s, I had persuaded myself that, prompted by the early success of the Pill and its rapid acceptance by millions of women, a new golden age of research would ensue with an emphasis on improved reversible birth control, where many different methods would be developed to eventually stock a contraceptive supermarket from which women and men could choose whatever suited their personal, medical, religious and other needs since no single contraceptive could ever be ideal for all people. In the most important article of my career dealing with public policy under the title "Birth Control after 1984," I outlined in 1970 in the scientific journal *Science* what it would take in terms of scientific knowledge, time and cost to develop fundamentally new contraceptives such as a Pill for men or a once-a-month female alternative, that would be based on a very different biological mechanism from the oral contraceptives that a woman has to take at least two hundred and fifty times a year. I concluded that with proper financial investments and operational incentives, it would take on the order of fourteen years before regulatory approval for any such truly new method of birth control could be reached — hence the use of "1984" in a chronological rather than solely Orwellian sense in the title of the article I published in 1970. But by 1989, thirty-eight years after our first synthesis of an orally effective progestin, twenty-nine years after its FDA approval for contraceptive usage, and nineteen years after publishing "Birth Control after 1984," I came to realize that the prognosis for a well-stocked contraceptive supermarket had become a bitterly disappointing *fata morgana*, which had little to do with science, but rather with economics, politics and the changed priorities of the pharmaceutical industry with its increasing emphasis on diseases of aging and deterioration.

In that year, again in the journal *Science*, but this time under the title "The Bitter Pill," I listed six fundamentally new methods of birth control rather than piddling improvements in existing ones that, if implemented, would vastly expand the choice for human fertility control for all constituencies: poor and affluent, pro-choice and anti-abortion, female and male. They would also make a difference in convenience, economic savings to the consumer, and possibly even safety. Note that improved efficacy is hardly an issue any more.

(1) Spermicide with antiviral properties (effective during normal coitus).
(2) Once-a-month pill for women, but effective as menses inducer.
(3) Reliable ovulation predictor ("red" and "green" light).

(4) Easily reversible and reliable male sterilization.
(5) Male contraceptive pill.
(6) Antifertility vaccine.

Now, as I write this twenty-three years later, only ovulation prediction (number 3 in the list) has been realized — in part because it required no toxicity expenses and could focus entirely on diagnostic efficacy and accuracy. But in terms of usage, this approach, for which I had coined the term "jet-age rhythm method" in still another paper in *Science*, is employed much more widely for purposes of conception rather than contraception. Work on an antiviral spermicide (number 1) is still progressing — primarily because of its applicability to the AIDS pandemic rather than for reasons of improved contraception, but so far with little success. Easily reversible and reliable male sterilization (number 4) would be of great advantage in both pediatric and geriatric countries, since vasectomy is practiced widely — notably in China and the US, though not in Germany — but mostly by men who are already fathers and wish no more children. If reversibility were guaranteed — a very expensive proposition, requiring large numbers of volunteers and many years of observation — then vasectomy, given its simplicity and safety, might well be practiced by many young men before they ever had fathered any children. For pharmaceutical companies, any approach based on vasectomy reversal would be of zero financial interest.

This leaves alternatives 2, 5 and 6, which would represent fundamental advances that would also fill enormous lacunae in our contraceptive armamentarium. But the costs for developing such agents would be enormous (each easily exceeding $1 billion), very time consuming (on the order of fifteen to twenty years, thus cannibalizing much if not all of the only incentive for pharmaceutical companies, namely patent protection) and very likely also prone to litigation. Only the biggest pharmaceutical companies would have the necessary scientific and financial resources for such an endeavor, and given their focus on diseases that afflict the ever increasing geriatric populations of the rich countries in Europe, Japan and the USA, it is not surprising that not a single of the twenty largest pharmaceutical companies is currently pursuing research and development work on a male contraceptive or a fertility vaccine. So what is left?

The few major drug companies that continued after 1975 to pursue any research in female contraception were those that had a significant market share of the Pill. The two most important ones (Schering in Germany and Organon in Holland) were both swallowed up in recent years by German (Bayer) and American (Merck-Schering-Plough) pharmaceutical giants, not unlike what happened in the 1990s with Syntex and the Swiss drug firm Roche. But even those two drug firms were mostly motivated to carve out a proprietary position by creating minor chemical modifications of the original 19-nor steroids created sixty years ago rather than launching fundamentally new methods of birth control.

Because of the huge cost of developing even minor chemical variants of the conventional female Pill, a considerable effort was also directed by these companies as well as some non-profit organizations such as the Population Council toward introducing new delivery vehicles, such as injectables, silastic implants, vaginal rings, skin patches and the like. In my opinion, such applied research is well justified in extending the use of these steroid contraceptives to a wider population. But otherwise, why continue to create "new" oral contraceptives which women would still have to take daily or tri-weekly to be protected from unwanted pregnancy during often just occasional acts of intercourse? Aside from narrow marketing considerations, is there a societal need and would the money expended on such endeavor not be better justified if it addressed the development of some fundamentally new methods of the type listed by me? Again from a societal standpoint, the answer must be a resounding affirmative, but what about commercial economic considerations on the part of a pharmaceutical company? In that regard, I am afraid that the answer must be an equally resounding negative. In fact, the market has spoken because none of the major pharmaceutical companies is spending any money on such new areas for the obvious reason that in terms of urgency in the geriatric countries, birth control cannot and possibly even should not have a high priority. Few will argue in favor of the proposition that spending a few billion dollars on a new contraceptive would be more useful — societally or commercially speaking — than a drug preventing Alzheimer's disease. It is this bitter, but realistic prognosis that justifies the inclusion of the word "bitter" in the title of my 1989 *Science* article or of the present chapter.

Where is the Pill for Men?

For me, the most unfortunate, indeed bitterest, side effect of the Pill is not medical, but gender related. The Pill's efficacy and convenience has meant that many men in monogamous relationships, who earlier carried the responsibility for birth control, simply shifted it through the Pill onto women's shoulders, who already bear all the responsibility — biologically as well as functionally — for human reproduction. Were it not for the AIDS epidemic, I maintain that many men who now use condoms would probably refrain from using them as well. Yet condoms and *coitus interruptus* are the only means of reversible birth control in the male and will remain so because the problem is not scientific, but an economic one. Scientifically, we do know how to create a male Pill. Indeed much clinical research, primarily under the auspices of the WHO, has already been carried out during the past three decades in this area in Germany, Scotland, Australia, USA, China, India, Brazil and other countries, but this has led to zero interest among the top twenty international pharmaceutical companies, who would have to be involved if such methods were ever to reach the public. Aside from big pharma's focus on medicine for the older segment (a resounding *yes* on treatment for erectile function, but an equally resounding *no* for contraception), the biggest problem is the fact that the reproductive span of a young man is two to three times longer than that of a twenty-year-old woman, who, for instance, will not ask whether continued use of her Pill would affect her fertility at age forty-five or fifty, whereas many a twenty-year-old man would require a guaranteed answer before he would reach for his Pill. To provide an epidemiologically valid answer to a young man would be exceedingly expensive, time-consuming and open to all kinds of litigious pressures since erectile dysfunction and prostate gland problems increase with advanced age and would be blamed by many men on their Pill rather than on the facts of life.

Although I am personally a firm believer in men assuming more responsibility for birth control and have demonstrated my conviction through undergoing a vasectomy decades ago, I have nevertheless been so pessimistic about the prospects of a male contraceptive Pill that I made the following brutal prediction already in 1979 in my book, *The Politics of Contraception,* which was addressed to a general public rather than just to scientists: "every postpubescent American female reading this chapter in 1979 will be past the menopause before she can depend on her sexual partner to use his Pill." I could, of course, have left it at that and dropped the subject from further writings or lectures. But that, of course, would be a cop-out that might even lead to a misunderstanding as shown when, in 1979, *The New York Times Book Review* carried a long critique of my book, where its author, a male political scientist, categorized that sentence as "nasty," and my attitude toward women as "depressing." His conclusion — that I seemed "most sympathetic to men who object to contraceptives" — promptly led me to see his name, A. Hacker, as a job description. But even a hacker's gut response of simply damning the messenger without bothering to understand the message demonstrates that the public simply does not want to hear about the ever-decreasing chances for a male Pill. Hence, to preempt another hacking, I shall now show how I dealt with this subject at two different levels, one scientific, and the other literary. But before doing so, I shall again indulge in my persistent habit of going off on a tangent, this time by recounting my own decision to undergo a vasectomy — a somewhat amusing event I had not thought about for years.

Shortly after my divorce, which occurred in the pre-AIDS days when sexual encounters were more permissive and diverse, I decided to undergo a vasectomy since I was convinced that my days of new fatherhood were gone. I asked a former student of mine, who had mentioned that he had been vasectomized, whether he could recommend a suitable physician and then called his office. The nurse on the phone pointed out that before undergoing that minor operation, I would have to visit the doctor's office for a complete explanation of the process and its doubtful reversibility to avoid any legal repercussions. I interrupted her seemingly rehearsed spiel — not unlike the standard landing instructions of flight attendants — to inform her that I knew all about vasectomies — actually teaching it in one of my courses — and that in any event I was too busy. "Just give me the latest appointment in the afternoon and I shall then sign whatever informed consent document you have" was my rather brusque response.

A few days later at five o'clock in the evening, I appeared in the doctor's office; my technical questions immediately convinced him that no further explanation was necessary. In fact, when he learned that I was a professor at Stanford University who had scientifically been active in the contraceptive field, the entire doctor–patient power relationship suddenly reversed. And when I asked whether the vasectomy could be conducted with a mirror that would make it possible for me to watch the procedure, he promptly told his nurse to take off for the day since I could assume the minimal assistant duties required for such a minor surgical intervention. So I washed my hands, put on plastic gloves, spread my thighs, and held my scrotum over the mirror while the physician proceeded to carry out the minor incision in order to lift the *vas deferens* out of the scrotum for disconnection. During all that time, we carried out a stimulating, semi-technical conversation, which ended with the doctor's sensible recommendation not to have any sexual intercourse for the next few days to avoid possible tearing of the sutures.

That, however, was the time when I had an all-too short but exciting relationship with a lover who had offered to visit me that evening in my ranch home with a restoring dinner for the "recovering patient." Needless to say, she was curious to hear the details of what she called "the operation" but which in manly fashion I simply dismissed as a "minor incision." In any event, before I knew it, she asked "may I see it?" and proceeded to unzip my fly. Her curiosity and the gentleness with which she examined "the wound" had unexpected tumescent consequences, which in turn led us to indulge in one of the slowest, gentlest acts of carnal congress I had ever experienced. Noticing postcoitally that the small stitches of my scrotum were still firm and undamaged, the following week could hardly be described as celibate in nature. I still chuckle at the shocked look on the part of the physician when I returned to his office for removal of the stitches and reported on my disregard of his medical warning. Upon pondering why at this late stage of my life I am suddenly willing to describe such a private event that nobody before — other than my companion — had heard about, I suggest that instead of labeling it an offensive display of geriatric exhibitionism, it should be accepted as a reassuring recommendation in favor of vasectomy for the many men who exaggerate or fear such a simple permanent solution to birth control and feel that if sterilization is the answer, the woman should undergo the more complicated process of tubal ligation. Obviously, I disagree.

A Swords-into-Ploughshares Proposal for the Military

Considering the rather dismal tone of my above conclusions about male contraception, perhaps the time has come to resurrect a proposal which requires no input by the pharmaceutical industry nor regulatory agencies, since toxicological considerations do not enter into the equation. In 1994, in the July 7 issue of the scientific journal *Nature*, the cryobiologist, Stanley Leibo, and I addressed the deplorable prognosis for a new male contraceptive in the foreseeable future, given the total lack of interest in that field by the large pharmaceutical companies without whose participation such a "Pill for Men" could never be introduced. This led us to propose an alternative approach based on a few simple assumptions.

Millions of men — admittedly, most of them middle-aged fathers rather than young men — have resorted to sterilization (vasectomy) and continue to do so. As I just described through my own experience, the procedure is much simpler and less invasive than tubal ligation in women. Sterilization among both sexes has become so prevalent that in the US it is now the most common method of birth control among married couples, even surpassing the Pill. (In China, the largest country in the world, it is the pre-eminent choice.) Artificial insemination is both simple and cheap. Furthermore, among fertile couples, it has almost the same success rate as ordinary sexual intercourse. But most important for our argument, fertile male sperm has already been preserved inexpensively for years at liquid nitrogen temperatures. Therefore, provided one first demonstrated that such storage is possible for several decades rather than just years, many young men might well consider early vasectomy, coupled with cryopreservation of their fertile sperm and subsequent artificial insemination, as a viable alternative to effective birth control. Shifting more of that responsibility to men, at least in monogamous, trusting relationships, appeared to Leibo and me a socially responsible suggestion. It should be noted that if our proposal were to turn into reality, contraception would be superfluous and the need for abortion eliminated, in addition to the substantial financial savings since no contraceptives would have to be bought for the rest of their lives.

As I stated before, the prognosis for a new male contraceptive by the year 2030 is nil, because the development, testing, and regulatory approval of a truly novel, systemic male contraceptive requires easily fifteen to twenty years. But given the absence of serious research and development in male contraception by major pharmaceutical companies — and only they can bring such a product to the general public — the expectation for a "male Pill" even by the middle of this century is unlikely, which would lead to the unavoidable conclusion that condoms and *coitus interruptus* remain the only feasible options for reversible male contraception.

Vasectomy cannot really be considered a reversible method of birth control, which is one reason why mostly middle-aged men, who are already fathers and do not wish to have more children, choose that option. Yet vasectomy coupled with semen storage can readily turn into "reversible birth control" if fertility of the stored sperm can be assured. For obtaining data that would support such an assurance, the first step would only require the establishment of comprehensive facilities for storage over many years of a vast number of human semen samples whose donors can be followed up in terms of their subsequent (conventional) reproductive experience. For the human guinea pigs in such a large scale experiment, only one act of masturbation would be required, but no vasectomies. And to get that started, Leibo and I proposed a swords-into-ploughshares initiative for financial, operational and motivational reasons.

The military services are the source of the largest number of young men with detailed medical records and the potential for follow-up. With little difficulty and relatively minor expenditure, tens of thousands of volunteers could collect their own semen to be cryopreserved by the military for many years. This step alone would generate an invaluable resource for studies on male fertility and for eventual spin-offs of the Human Genome Project.

Triennially, these sperm samples would be subject to laboratory analyses, providing statistically meaningful and invaluable data on the fecundity of the stored specimens. Quinquennially, sperm donors would report whether they had fathered (via ordinary intercourse) any children, or if not, whether they had attempted to do so. Matching responses with laboratory sperm sample records would establish the latter's predictive value.

Men whose occupation places them at risk from genetic damage to their sperm might view sperm cryopreservation as a form of genetic insurance. This is certainly relevant to the armed services, where procreation after death in military conflicts (or exposure to mutagenic agents such as radiation) might be an option, subject to legal definition of sperm ownership. Of course, only time will tell whether men — and women — will buy the idea of this new form of contraception. The indispensable first step is to start the clock running on a large-scale, long-term sperm cryopreservation program combined with follow-up protocols. Once confidence in the technical, operational and legal–ethical questions has been established, decentralized, entrepreneurially funded and operated programs (covered by health insurance) could be envisaged.

I would like to note that if the recommendations embodied in our 1994 *Nature* article had been implemented at that time — an effort that operationally and financially would have been exceedingly simple and cheap, we would now possess eighteen years' worth of extraordinarily valuable information that might well have been sufficient to move to the next phase of actually implementing this new option for male contraception on a practical scale without requiring the involvement of the pharmaceutical industry or drug regulatory bodies — the two costliest and most time-consuming aspects of any new form of birth control.

But that brings me to my conviction that by the middle of this century — largely through the desires and actions of women rather than men — the above outlined alternative may well turn into reality. The apparent anomaly that women would promote such a course for men is easily explained by addressing the third component of our proposal, namely the need for artificial insemination using the frozen sperm once procreation of a child is desired. In my opinion, the feasibility of such an approach will be due to the enormous amount of research that is dedicated currently on improvements in the freezing of eggs or ovarian tissue and to determine how long those gametes can be frozen and how effectively they can be artificially fertilized after thawing. In summary, while the Pill through its complete separation of contraception from the actual coital act has made sex without reproductive consequences a norm, how acceptable will be the reverse: procreation without sex? Of course, there are already five million children that were born through in vitro fertilization, but these were almost entirely babies from parents suffering from infertility that made procreation through normal intercourse impossible. Yet what I posit is the increasing future use of IVF methods by fertile people focusing on a 1.5 to 2.0 child family that is now the norm in Europe, Japan and the USA.

But why should increasing number of women be interested in following that path, which is not only much more expensive but also less pleasurable than conventional coitus? When Patrick Steptoe and Robert Edwards in England developed IVF in 1977, they did not set out deliberately to make possible the separation of sex from fertilization. They, as well as other clinicians, were focusing on the treatment of infertility. Infertility is itself an ethically charged topic. To put it bluntly and brutally: why should one treat infertility? From a global perspective, there are too many fertile parents, hence there are too many children, many of whom no one wants. The course of world history will not change if no case of infertility is ever treated, but it will change dramatically if excess human fertility is not curbed. But from a personal perspective, the drive for successful parenthood is often overwhelming. Infertile couples are prepared to undergo enormous sacrifices, financially, psychologically as well as physically, to produce a live child under conditions where nature has made it impossible. The question may well be asked whether the realization of parenthood by biologically infertile couples carries some ethical imperative — for or against.

The enormous ethical dimensions of the problem become somewhat easier to see if we consider the question of male infertility. This issue was addressed in 1992, when a group of investigators (Palermo, Joris, Devroey and van Steirteghem) in Belgium published a sensational paper announcing the birth of a normal baby boy fathered by a man with severe oligospermia (insufficient number of sperm).[12] The birth of this child was made possible through the invention of an IVF technique called intracytoplasmic sperm injection, ICSI, in which a single sperm under the microscope is injected directly into a human egg. Whereas in the original British IVF work, the egg was flooded with millions of sperm (as in ordinary sexual intercourse), with ICSI the artificial insemination was accomplished with one single sperm under the microscope. The technology that makes such fertilization possible also allows a radical revision of the definition of infertility: ICSI can be applied not only to men with low sperm counts, but to men who have no mature sperm whatsoever and thus are (often for genetic reasons) completely infertile. What ICSI has now accomplished is to make the uninheritable heritable!

[12] G. Palermo, H. Joris, P. Devroey and A. C. van Steirteghem, *Lancet* **340**, 17 (1992).

The first ICSI baby is now twenty-one years old, but in that period over one million ICSI babies have been born. I have felt that the questions this technology raises merit wider debate beyond the traditional venues of a journal article or academic lecture.

A Pill for Men: Facts Through Fiction

Around the time that Leibo and I made the above proposal in a scientific journal read primarily by scientists, I was already well on the way to become a novelist for reasons that I will recount in *"Writer"*. But the type of fiction I wrote was to a certain extent just a pretense for smuggling important information to a general public; thus, it should not come as a surprise that reproductive medicine was a recurring theme in some of my novels and plays. So let me use this opportunity to indicate how I addressed the issues of gamete storage coupled with subsequent in vitro fertilization (notably through ICSI) as well as the broader issue of male contraception, with an emphasis on women in the guise of fiction and plays.

I start with a scene from my first play, *An Immaculate Misconception*, which premiered in Edinburgh in 1998 and has since been translated into twelve languages and broadcast by the BBC World Service, NPR in the US, as well as by German, Swedish and Czech radios in addition to numerous theatrical venues. The following excerpt features a discussion between Dr. Melanie Laidlaw, a reproductive biologist and (in the play) the inventor of ICSI and her clinical colleague, the infertility specialist Dr. Felix Frankenthaler, whom she had invited into her laboratory. After she informed him that she is almost ready to perform the first ICSI injection into a human egg (without, however, volunteering that she will pick her own egg for such experimentation), they debate the possible implications of this work beyond simply treating male infertility:

> ***MELANIE*** *If your patients knew what I was up to in here... they'd be breaking down my door. Men with low sperm counts that can never become biological fathers in the usual way.*
>
> ***FELIX*** *My patients just want to fertilize an egg. They won't care if it's under a microscope or in bed... as long as it's their own sperm.*

MELANIE *You're focusing on male infertility... that's your business. But do you realize what this will mean for women?*

FELIX *Of course! I treat male infertility to get women pregnant.*

MELANIE *Felix, you haven't changed. You're a first-class doctor... but I see further than you. ICSI could become an answer to overcoming the biological clock. And if that works, it will affect many more women than there are infertile men. (Grins.) I'll even become famous.*

FELIX *Sure... you'll be famous... world-famous... if that first ICSI fertilization is successful... and if a normal baby is born. But what's that got to do with (slightly sarcastic) "the biological clock?"*

MELANIE *Felix, in your IVF practice, it's not uncommon to freeze embryos for months and years before implanting them into a woman. Now take frozen eggs.*

FELIX *I know all about frozen eggs.... When you rethaw them, artificial insemination hardly ever works.... Do you want to hear the reasons for those failures?*

MELANIE *Who cares? What I'm doing isn't ordinary artificial insemination... exposing the egg to lots of sperm and then letting them struggle on their own through the egg's natural barrier. (Pause.) We inject right into the egg.... Now, if ICSI works in humans... think of those women — right now, mostly professional ones — who postpone childbearing to their late thirties or even early forties. By then, the quality of their eggs... their own eggs... is not what it was when they were ten years younger. But with ICSI, such women could draw on a bank account of their frozen young eggs and have a much better chance of having a normal pregnancy later on in life. I'm not talking about surrogate eggs —*

FELIX *Later in life? Past the menopause?*

MELANIE *You convert men in their fifties into successful donors —*

FELIX *Then why not women? Are you serious?*

MELANIE *I see no reason why women shouldn't have that option... at least under some circumstances.*

FELIX *Well — if that works… you won't just become famous… you'll be notorious.*

MELANIE *Think beyond that… to a wider vision of ICSI. I'm sure the day will come — maybe in another thirty years or even earlier — when sex and fertilization will be separate. Sex will be for love or lust —*

FELIX *And reproduction under the microscope?*

MELANIE *And why not?*

FELIX *Reducing men to providers of a single sperm?*

MELANIE *What's wrong with that… emphasizing quality rather than quantity? I'm not talking of test tube babies or genetic manipulation. And I'm certainly not promoting ovarian promiscuity, trying different men's sperm for each egg.*

FELIX *"Ovarian promiscuity!" That's a new one. But then what?*

MELANIE *Each embryo will be screened genetically before the best one is transferred back into the woman's uterus. All we'll be doing is improving the odds over Nature's roll of the dice. Before you know it, the 21st century will be called "The Century of Art."*

FELIX *Not science? Or technology?*

MELANIE *The science of… A… R… T (Pause): assisted reproductive technologies. Young men and women will open reproductive bank accounts full of frozen sperm and eggs. And when they want a baby, they'll go to the bank to check out what they need.*

FELIX *And once they have such a bank account… get sterilized?*

MELANIE *Exactly. If my prediction is on target, contraception will become superfluous.*

FELIX *(Ironic) I see. And the pill will end up in a museum… of 20th century ART?*

MELANIE *Of course it won't happen overnight.... But A... R... T is pushing us that way... and I'm not saying it's all for the good. It will first happen among the most affluent people... and certainly not all over the world.*

FELIX *(Shakes head) The Laidlaw Brave New World. Before you know it, single women in that world may well be tempted to use ICSI to become the Amazons of the 21st century.*

MELANIE *Forget about the Amazons! Instead, think of women who haven't found the right partner... or had been stuck with a lousy guy... or women who just want a child before it's too late...in other words, Felix, think of women like me.*

ICSI raises many other ethical and social problems beyond those mentioned in the Melanie/Felix dialog. For example, now that the effective separation of Y- and X- chromosome-bearing sperm has been perfected, ICSI will enable parents to choose the sex of their offspring with 100% certainty. For a couple with three or four daughters, who keep on breeding in order to have a son, the ability to choose a child's sex may actually prove a benefit to society, but what if practiced widely in cultures (such as China or India) that greatly favor male children over girls? There it could prove disastrous by skewing the sex ratio even more in favor of many more men as is already happening in China through other means such as early ultrasound examination followed by abortion.

Or consider the capability of preserving the sperm of a recently deceased man (say twenty-four to thirty hours post mortem) in order to produce (through ICSI) a live child months or even years later — a feat that has already been accomplished. But what of the product of such a technological *tour de force*? Using the frozen sperm and egg of deceased parents would generate instant orphans under the microscope. The prospect is grotesque — yet does it take much imagination or compassion to conceive of circumstances where a widow might use the sperm of a beloved deceased husband so that she can have their only child? These issues are intrinsically gray; the technology occupies an ambiguous position, enabling us to enact our best and worst impulses, and the answers cannot be provided by scientists or technologists. The ultimate judgment must be society's, which, in the case of sex and reproduction, really means the affected individual. Ultimately, that individual is the child, yet the decision must be made before its birth by the parents — or more often than we care to admit, by just one parent.

It is the nature of such questions that they resist convenient solutions, not least because of their tendency to proliferate faster than we can solve them. Whereas reproduction has historically tended to exemplify the law of unintended consequences, the addition of technology has given that law added force. Consider: until very recently, the onset of menopause was welcomed by many women as the release from continuous pregnancies caused by unprotected and frequently unwanted intercourse. But the arrival of the Pill and other effective contraceptives, coupled with the greatly increased number of women entering demanding professions that cause them to delay childbirth until their late thirties or early forties, now raises the concern that menopause may prevent them from becoming mothers altogether. Whereas reproductive technology's focus during the latter half of the twentieth century was contraception, the technological challenge of the new millennium may well be conception (or infection, if one focuses on sexually transmitted diseases). In the long run, if the cryopreservation of gametes followed by sterilization becomes a common practice, contraception may well become superfluous.

Six plays later, in *Taboos*, I decided to focus on the social ramifications of the enormous advances in reproductive medicine that had been made in the intervening eight years. My motivation for returning once more to a play for exploring such crucial issues is best summarized in the following foreword to the play published in my book with the self-explanatory title, *Sex in an Age of Technological Reproduction.*

> *In* Taboos, *I return to the topic of sexual conduct in an age of technological reproduction, which can also be described as the complete separation of sex and reproduction. Sex — motivated by love, lust, or curiosity — will no doubt continue as usual, while reproduction will increasingly occur under the microscope or by other "alternative" means. But instead of focusing, as I did in* An Immaculate Misconception *on the technical "yang" of this theme, I now turn in my sixth play to the social "yin" with its much more subtle and complex components. As Chinese cosmology proclaims, only a combination of yin and yang produces all that comes to be, in other words the next generation of persons*

and of ideas. Terms such as "marriage," "family," and "parent" used to have firm denotations. They were the rock on which our cultural values rested. Terms such as "embryo," "baby," or "twin" were also considered unambiguous. Assumptions that marriage must be heterosexual and that a child cannot have two parents of the same sex were never even considered assumptions, because they were beyond questioning. All of these terms have become destabilized, their meanings blurred, their ranges extended.

Some would blame in vitro fertilization technology during the past three decades for these developments, but in actual fact major social and cultural changes — primarily in the United States and Europe — were even more responsible for the monumental shift that has caused so much fear and antagonism, especially among the ever increasingly strident fundamentalists in the United States. So why not write a play about a situation where "family" and "parent" have assumed disturbingly fuzzy meanings? This is why I have deliberately situated Taboos *in two of the most socially and politically polarized parts of the United States: the San Francisco Bay Area and the American Deep South.*

Even though I have spent half my life in the San Francisco Bay Area, I do not wish to be regarded as a proselytizer for either of the two extreme views I present in Taboos. *That is why I end the play on a biblical note, emphasizing the need for compromise in a situation where there can be no winners. I wrote Taboos mostly in London, but early parts also in Ireland and Germany, as an American agent* provocateur *born in Europe who has rediscovered his European roots and acquired a more distanced as well as more nuanced view of America. Unquestionably,* agent provocateur *is the role that suits me best as a late-blooming playwright, because the issues that interest me most are intrinsically provocative as well as complex. And few topics are as provocative and complex as the present questioning of the social meaning of parenthood and family, where every horror projection can be countered with a "But what if?" scenario. That is why in Taboos I have mostly taken the yin side of the argument.*

Over the course of the last three decades, in several hundred lectures and talks on advances in female contraception, I have encountered a wide range of questions from listeners. But one, enunciated especially by women, always recurs. "Where is the Pill for Men?" As I already mentioned above through the words of the anthropologist Margaret Mead in the late

1960s, the question was frequently posed in an aggrieved manner by feminist critics: "Why is there no Pill for Men instead of the Pill for Women?" That implied accusation was frequently continued rhetorically with "Is it not because all of the relevant scientific and clinical work was done by men, who have no compunction in experimenting on women, but are reluctant to experiment with any aspect of their own sexual apparatus?" While the question is not as simple as it sounds, a male feminist like myself (who believes that the ultimate definition of a liberated woman is that of a person in charge of her own fertility) would respond that under such circumstances, most of the decision-making power about a woman's pregnancy would still remain in the hands of men.

The more appropriate and less adversary question, of course, is "Why is there no Pill for Men in addition to the Pill for Women?" As shown in the preceding paragraphs, initially I have always replied to that complicated question — complicated because the reasons are mostly economic and cultural rather than just scientific — in the straightforward factual and mostly impersonal manner of the scientist. But more recently, as a professional bigamist leading the life of scientist and novelist, I chose to address that issue within my fiction as case histories. I write largely in the narrow and infrequently used genre of "science-in-fiction," which must not be confused with science fiction. My definition of "science-in-fiction" requires that everything depicted must at the very least be plausible, if it has not, in fact, already happened. Crassly speaking, I use my novels as a device to smuggle such information into the unwary reader's mind in the guise of an engaging fictional plot.

My novel, *Menachem's Seed* (the third volume in a tetralogy of "science-in-fiction" novels), and its successor, *NO*, concentrate on male reproductive biology. Thus in an early chapter of *Menachem's Seed*, the reader will encounter the following dialog:

> *"Reproductive biology? You must mean <u>female</u> reproductive biology. Why don't you men ever pay attention to <u>your</u> role in reproduction?"*
>
> *Even though the question was addressed to her neighbor, the woman's voice was meant to be heard by the rest of the people at their table. It was the annual fund-raiser for Brandeis University, where the guests felt entitled to register complaints. Invariably, they were handled politely, especially when raised by potential donors.*

The woman's neighbor, and the subject of her complaint, was Professor Felix Frankenthaler, one of Brandeis's stars, invited this evening to demonstrate to the guests the kind of value they could expect from their donations. "A fair enough question," Frankenthaler responded diplomatically. "I have to admit that I made my reputation in the fallopian tube. Even though," he raised his hand to stop any interruption, "that work dealt, in fact, with sperm motility."

"So?" The woman asked, her tone now less aggressive than amused. "What have you done for me lately?"

"Well," he announced, loudly enough that the rest of the table turned his way. "We are now hot on the trail of..."

As a scheming novelist, who would like to seduce his audience into reading his novels rather than giving away the plot, I now skip a couple of pages of dialog, before returning to a second excerpt from this novel:

"I knew it!" the woman exclaimed triumphantly. "All female reproductive biology means to you is contraception. But when you men work on your own sexual apparatus, all you worry about — "

"Now wait a moment." By now, Frankenthaler didn't give a damn that he was supposed to be buttering up prospective donors. "If you can't get it up," he hissed, "you can't get it in. Only then do we start worrying about birth control..."

This, of course, is a dialog between scientist and layperson. I now move to the more sophisticated level of scientist talking to scientist — to a dialog between the same Professor Frankenthaler and Dr. Melanie Laidlaw, the director of the fictitious REPCON Foundation, which supports research in reproduction and contraception.

"What brought you to Manhattan, Felix? It sounded urgent. Was there no one on my staff that could have helped last week?"

He shook his head. "I wanted to ask you something directly...."

Melanie waited, while Frankenthaler paused, as though collecting his thoughts — or his courage, she thought.

"Could you support a project of mine through your director's discretionary fund? You know I've never approached you before about such special handling. It isn't much," he made a deprecatory gesture with his hand, "about 75K."

"I didn't know you even dealt in such paltry sums, Felix," she said. "But why not go through our regular application process? We're not some government agency; you know we haven't got much red tape — "

"I'm in a hurry."

"I could arrange for an expedited review — "

"I'm sure you could," he said impatiently, "but I have a reason. To be quite frank, I'd rather not have the competition see what we've got up our sleeves."

"Competition?" She leaned back behind her desk, ready to listen. Personal motivation among scientists, especially superstars, always intrigued her.

"You know how small a community we are in male reproductive biology. Your panel must be full of ... "

"OK," she interrupted. "So what's so hot about your project?"

"We may have a new approach to male impotence..." he began, but Melanie Laidlaw didn't let him finish. The half-done work in front of her had made her impatient, which in turn made her prone to interrupt. With Felix, at least, she masked her restiveness with banter.

"Are you trying to cause impotence or cure it?"

"Be serious. We're trying to cure it, of course." It was his turn to display irritability.

"That's all you men ever think of. Why don't you work on prevention rather than performance, for a change? In other words, pay some attention to contraception."

"Christ," Frankenthaler started to mutter, remembering the recent Brandeis fund-raiser debacle, but then decided to swallow the rest. "I've heard that before."

"So why don't you? We get fewer and fewer applications dealing with contraceptive research. And virtually none when it comes to new approaches to male birth control. All the reproductive fraternity is interested in these days seems to be treatment of infertility."

"Hot advice; and not one to be spurned from a fountain of unrestricted money," he countered, realizing he wasn't going to get his 75K without working for it. "But it's not a fraternity any more. We're getting more and more women in the lab. So what would you work on if you were in my shoes?"

"How about ...

Dr. Melanie Laidlaw then outlines some potentially promising novel approaches to male contraception, many of which have been shown in actual research performed during the past few years to also affect libido.

Frankenthaler let out an audible sigh. "I suppose so. It would also shut off libido."

"You could offset it with testosterone administration."

"A real pain."

"You poor men: having to take a pill all the time, like women. Still — what about occasional injections of long-acting testosterone esters?"

"A pain in the butt. Sorry," he said quickly, "I just couldn't resist it."

Melanie shook her head. "I guess you just aren't interested in birth control."

"Melanie, that isn't fair!" His voice had started to rise. "I'm just being realistic. Even if you could come up with a lead — and I'll grant you," he raised his index finger for emphasis, "that theoretically you have a perfectly valid point — just think of the development times required; the years and years of clinical tests. We'd have to make sure it was reversible. Because if it isn't, why bother? Why not just have a vasectomy and be done with it?"

"What about your male impotence work? Won't that take just as long?"

"The eventual application in wide clinical practice? Perhaps — especially considering the psychological component you have to account for. But the initial research, the testing of our concept?" He thought of his contretemps with the virago at the Brandeis banquet and snapped his fingers. "Either you get an erection or you don't. The time is measured in seconds or minutes — not years."

"Is that why you picked male impotence for your research?"

"I didn't <u>pick</u> it; it picked <u>me</u>."

"Oh Felix, I'm so sorry."

"What?" Felix Frankenthaler felt a deep blush sweep over his face. "I meant the <u>solution</u>, not the problem."

"Oh," said Melanie contritely. Now why am I baiting this man? she wondered.

But Frankenthaler seemed only momentarily put off his stride, and picked up again with his sales pitch.

"I've got a marvelous postdoc from Stanford who has been working in my lab on..."

In other words, in my novel in the form of a simulated but plausible "case history," I make the point that the contemporary scientific community and the pharmaceutical industry are not interested in male contraception, but rather in the much more "glamorous" topics of male impotence or infertility. As an example of the latter, I cite a monologue on the part of Dr. Melanie Laidlaw from Chapter 18 of *Menachem's Seed*, which again returns to the topic of ICSI, which was covered in the play *An Immaculate Misconception*:

> *The day I saw the title of Van Steirteghem's grant proposal, "ICSI vs. SUZI," I was intrigued. It sounded like a lawsuit, or maybe a wrestling match. It was the first application REPCON had ever received from Belgium. By the time I'd put down the last page, I was certain of our board's approval.*
>
> *I'm not a fan of acronyms, but ICSI and SUZI appealed to me — cute children's names. Especially SUZI, until I learned that it stood for subzonal insemination, a procedure for inserting sperm just below the zona pellucida, and then abandoning it to its own devices. In the end, I favored ICSI. It wasn't because of its formal definition, intracytoplasmic sperm injection, the direct injection of sperm into the egg's cytoplasm. For successful fertilization, the spermatozoon has to get inside the cytoplasm of the oocyte. If you're going to do it, I thought, you might as well go all the way. What convinced me more than any other practical considerations was that ICSI offered the much more alluring mnemonic: I can still inseminate. Once I thought of it, it stuck. Besides, according to Van Steirteghem's proposal, when success rates were measured, ICSI won hands down.*

The reason why I chose ICSI as one of the scientific themes in two of my plays and one of my novels, and why I am raising it again in these autobiographical reflections is that it illustrates the conflict in today's society between technological breakthroughs and the ethical dilemmas engendered by them. In no area is this demonstrated more dramatically than in assisted reproduction — in other words, in the treatment of infertility — and especially ICSI, where we deal with only a single sperm.

For instance, who owns sperm? Given that over the course of their lifetimes men discard uncountable billions of sperm, is a single sperm of any value at all? Is it something that can be "stolen"? This is the key question in my novel and in my first play, where the issue is made more complicated: the man whose sperm is "stolen" knows his low sperm count makes him infertile. If such sperm is not worth anything to the man, it may well — with ICSI — be worth a lot to a woman. In such a case, where do the rights and responsibilities of parenthood lie? But the questions only begin there. Should ICSI be used for sex-predetermination? With ICSI, the sperm of a recently deceased man can be aspirated and preserved for years. To what use may such post-mortem fertilization be put? The very success of ICSI has created banks of hundreds of thousands of frozen embryos, stored away against the possible failure of a first attempt, rendered unnecessary by the procedure's significant rate of success. Are these embryos life? What should be done with them? Since ICSI results often in multiple pregnancies (e.g. triplets), with concomitant risks both to fetuses and mother, what are the ethics of "selective reduction" — even if the process is performed to assure the birth of one or two babies rather than risking the loss of all three? These and many other problems are raised in my novel and/or my play. Who should make decisions about such charged issues? Scientists or the public? And if it is the latter, who is the public and what does the public know about the technical dimensions of these reproductive technologies? And then add the politicization of this most personal of personal human aspects — sex and reproduction — which in the USA has risen to absurd extremes.

In summary, a Pill for men is still in the far distance. But other genies have already escaped the bottle of male reproductive research. The non-technical public could do much worse than reading intelligent fiction or plays to understand the background on which my following prognosis is based.

The Ultimate Demise of Contraception

This somewhat startling prediction by a person who was so deeply involved in the development of the Pill needs to start with an important definition concerning today's world. Since the last World War, the separation of the world into developed and developing countries has shifted in a much more fundamental extent to one of geriatric versus pediatric countries. The prime examples of the former are Europe and Japan which now have nearly reached or even surpassed the 20% threshold of people over sixty-five, compared to the pediatric countries of Africa and certain portions of Asia, where nearly 50% of the population is below the age of fifteen. I shall not belabor the monumental economic, political, educational and health-related consequences of such a demographic discrepancy, but rather point out that in terms of reproduction, the leitmotiv of the geriatric nations with an average family size of 1.5 children is now conception, whereas in pediatric countries with four to six children per family, contraception is still of dominant concern. I would like to coin the term "*mañana* generation" for a segment of women in in the geriatric world with its virtual divorce of sex from reproduction to explain my prediction that it will be that group where the demise of contraception will first start and eventually spread. What causes me to reach such a startling conclusion?

During the past couple of decades — a femtosecond in human history — the mean age of women at first childbirth has increased in the geriatric countries from around twenty-five to thirty years of age, but according to recent data in the UK, female university graduates have their first child around age thirty-five compared to twenty-five years of age for their non-university cohorts. No such change has occurred in the pediatric countries.

I refer to these women as the *mañana* generation, because consciously or unconsciously they ignore the fact that women are born with their life's supply of eggs and that at age thirty-five 95% of these eggs are gone and the remainder is aging rapidly. Thus postponing child-bearing, saying *mañana*, but really meaning "not now," has serious consequences. The danger of late pregnancies is well known — a typical example being the five-fold increase in Down's syndrome babies born to women in their late thirties or beyond. That, of course, is the reason why many of these women had an amniocentesis after the third month of pregnancy so that the draconic step of an abortion could still prevent such a birth if an extra chromosome is detected.

But all of this has changed during the past five years and it is this medical advance coupled with the dramatic increase of such educated women choosing to sacrifice the biologically most propitious years for reproduction on the altar for professional careers that must not be ignored. The science of gamete cryopreservation — primarily through experience in cattle where it was first practiced widely — has shown that sperm can be preserved for several decades, if not longer. The next step, embryo cryopreservation, received its big push because of the enormous increase in IVF fertilizations during the past three decades, but what was missing was a successful freezing and rethawing of unfertilized eggs. While not yet totally perfected, the advances are now so impressive that clinics already exist that offer that service on a commercial basis, and there is little doubt that substantial improvements will be achieved in a matter of a few years. In other words, women — notably members of the *mañana* generation — now can consider realistically the collection and storage of their young eggs as insurance for a long delayed pregnancy and there is little doubt in my mind that many of them will proceed along this path, given their educational background, their professional ambitions and their desire to ultimately have one or two children. The financial cost is unlikely to be the main barrier for these women, but rather the question of undertaking the true inconvenience of a hormonal drug-induced superovulation for one or several months in order to produce an adequate number of eggs for long-term storage. Yet it should be noted that during the last few decades at least twenty million women have chosen some form of superovulation.

Undergoing a superovulation and storing their young eggs, of course, does not mean that the women will actually use subsequent IVF methods (primarily via ICSI) for getting pregnant. Women in their twenties will first choose this approach as insurance, providing them the freedom — in the light of professional decisions or even absence of the right partner — of not having to worry about the inexorable clicking of the biological clock. However, I predict that many of these women will in fact eventually decide to be fertilized by IVF methods for the simple reason that concurrent with the improvements in assisted reproductive techniques, enormous advances have occurred in the area of genomics with pre-implantation genetic diagnosis the ultimate factor. Since it is now possible to accomplish genetic analyses on one single cell of an embryo prior to implantation, a process that previously involved amniocentesis in the third month of pregnancy, many women will consider this option as a major incentive to actually pursue the IVF route to their baby. And once that happens, then IVF will start to become a "normal" non-coital method of having children.

In some all-Catholic countries such as Spain, Italy or Austria, the average number of children per family is well below 1.5, meaning that the *de facto* separation of sex and reproduction has already occurred there. Couples over their life-time have sex hundreds of times after the birth of their only child, which in fertile women has been conceived through a single act of sexual intercourse. The new generation of university-educated professional women will then only proceed one step further — having one or two children through IVF rather than through one or two coital acts. I believe that this is the population where the use of IVF in fertile women will start and that over the course of the next few decades — say by the year 2050 — more IVF fertilizations will occur among "normal" (i.e. fertile) women rather than the current five million of fertility-impaired ones. This is an absolutely crucial point which has been ignored by the opponents or questioners of IVF, who cite possible post-delivery complications or low success rates requiring repeated IVF interventions. They ignore that virtually all such studies of IVF were conducted on children of parents with impaired fertility and in the majority of cases on older women with older eggs. We are now talking about IVF of fertile women with their young eggs, where the success rate is nearly equal to that of normal coitus! And once these women proceed in that direction, why not get sterilized after their young eggs are in the bank — a decision that would save considerable amounts of money as well as hassle.

Fifteen years ago, I made that prediction in my above-cited play, *An Immaculate Misconception*, before the freezing of ovarian tissue had been technically solved. But as noted, dramatic advances have been made in an incredibly short time to convert speculation into reality. If women of the *mañana* generation will be as numerous as I predict and when costs will also have decreased dramatically or, even more logically, covered by health insurance, then this scenario of preservation of young gametes, followed often by concomitant sterilization (why concern oneself with contraception when the desired child will be generated through non-coital methods?) and years later by IVF fertilization will surely spread to a wider population group beyond these educated, middle-class women. This brings me to a conclusion that has so far not been part of public discourse.

And if that happens, certain religions, such as Catholicism, will finally have to question the continuing validity of its proscription that only sexual intercourse with potential reproductive consequences is allowed, i.e. no sex solely for love or pleasure. The church's position, based on its desire to encourage human reproduction while condemning non-reproductive carnal congress, could hitherto be rationalized in terms of an exclusive requirement for reproduction-associated coitus, because there was no other way to give birth to a child. But once gamete preservation and subsequent IVF — in other words non-coital reproduction — becomes conventional reality for many people, why continue to decry sex for the sake of sex rather than exclusively for procreation? After all, the need for abortion (other than in extreme medical emergencies) will not exist in my scenario. At present, according to WHO estimates, 50% of all conceptions are unexpected and half of those unwanted, thus leading worldwide to an estimated 125,000 abortions every twenty-four hours, many of them illegal ones (notably in the all-Catholic countries of Latin America). It is a topic that a future pope will have to face realistically. *Pax vobiscum*!

A Shadowy Coda

Given that the Pill has affected directly or indirectly most of my long adult life, it is understandable that this has become one of the longest chapters in my latest autobiography. It also explains why, over a period of nearly half a century, in speeches and writings I have produced such a potpourri of thoughts, opinions, and sometimes even wavering conclusions. I shall now end on the same dour note on which I started this chapter: with a dark shadow that was thrown as late as 2011 — a year that for the first ten months seemed quite bright.

The year 2011 was the year in which the fiftieth anniversary of the Pill's introduction to Germany was celebrated by the media and the scientific community. One such event was the issuance in November of a huge 266-page special supplemental issue of the *Journal of Reproductive Medicine and Endocrinology (Journal für Reproduktionsmedizin und Endokrinologie*) under the overall heading *50 Years of Oral Hormonal Contraception* that was dedicated to me as stated in the following lovely sentences by the two editors:

> *Dear Carl, without your inventive ambition and experimental skills, we would not be able to celebrate the 50th anniversary of the Pill in Germany. We are delighted that you enjoy the best of health and unstoppable creativity and will give a celebratory lecture in Heidelberg in October at the handover ceremony of this supplement.*

I contributed to that special issue with a long article, "The Pill at 50 (in Germany): Thriving or Surviving?" whose first paragraph I reproduce herewith:

> *Which 50th Birthday?*
>
> *How many people — other than those trying to hide their age — celebrate the same birthday in successive years? And why should that happen to a drug? Yet in 2001, several people (starting with Carl Djerassi's memoir) celebrated the 50th birthday of the Pill, while 9 years later a media frenzy exploded all over the world with another 50th anniversary celebration of the Pill [referring to the date of the FDA approval]. And now, in 2011, we are doing it again by commemorating the date, 50 years ago, when the Pill was introduced in Germany. In terms of this article's title, this would clearly mean that the Pill is thriving. And in a way it is, especially in light of two overwhelming facts of the last half-century — the global population explosion and the rise of women's rights — without which oral contraceptives would just have been another medical advance and not an invention with enormous societal consequences. Yet in this article, I shall also make the contradictory argument that the Pill is only surviving, because nothing else is on the horizon in terms of fundamentally new methods of birth control.*

The day in Heidelberg where I presented my lecture and received the first issue of that journal culminated on another enjoyable note, the conferment of an honorary doctorate by the University of Heidelberg, Germany's oldest university.

Yet a very dark shadow was thrown on this bright event when five days later that pleasure dissipated upon receipt through the mail of a long article, "50 Jahre Pille in Deutschland" in other words on precisely the same topic, but written with a hodgepodge of seventy-five citations several of them bordering on the automythological which appeared in the German chemical journal *Chemie in unserer Zeit*. What the authors, two chemists named Sabine Streller and Klaus Roth, wrote simply flabbergasted me. Not just because of some of the errors, misinterpretations and out-of context quotations and lack of appropriate references, but because of the following *ad hominem* attack which I had never experienced in close to seventy years of life as a scientist:

> *The Pill, one of the "Seven Wonders of the Modern World" did not arise through a stroke of genius by one person, but represents the crowning of decades-long efforts, beginning with the first physiological experiments on the female cycle in laboratory animals, the isolation and structure determination of the sex hormones, the synthesis of orally active analogs, the performance of pre-clinical and clinical studies until government approval. Innumerable persons in many countries contributed to the successful market introduction, so that it is totally impossible to single out one person or to attribute to him or her exclusive "parentage." Yet that is precisely what Carl Djerassi has attempted to do over many years in innumerable publications and interviews by identifying himself at times as "father," and at others as "mother" of the Pill. Thus authorized, he declared the 15th of October 1951 as the birthday of the Pill. On this day, Luis E. Miramontes, George Rosenkranz and he synthesized at Synthex [sic] for the first time 19-Nor-ethynyltestosterone (Norethisteron); ignoring the fact, that this "birth" occurred unnoticed by everyone, even Djerassi, since no one at that time could imagine a Pill. When in 1960, Enovid appeared on the US market as the first Pill, Djerassi's active ingredient was not one of its constituents and only gained its importance subsequently. Nevertheless, he celebrates periodically with substantial media participation the birthday of the Pill and especially himself. As evidence, he developed an entire genealogy of the Pill, with Gregory Pincus as the other parent (at times as Father, at times as Mother). Ludwig Haberlandt as grandfather, the gynecologist, John Rock, as obstetrician, and Russell Marker as distant great-uncle. Other heavyweights, like Birch, Butenandt, Doisy, Hohlweg, Inhoffen, McCormick, Miramontes, Rosenkranz, Sanger and others are curiously not attributed any blood relation. A plain look at the history of the development of the Pill shows unambiguously that all of the people who participated in the development and market introduction of the Pill deserve our highest recognition. If one really wanted to establish a genealogy, then most appropriately they should all be considered godfathers and godmothers. Carl Djerassi clearly belongs among them, but he is not the only one. And if we are to celebrate any birthdays, then it can only be the day on which women had the Pill in their hands. In the USA, this was August 18, 1960 and in Germany and other European states June 1, 1961.*

Presentation of National Medal of Science by President Richard M. Nixon at the White House, 1973

Given what I have written throughout this long chapter, if this venomous claptrap were only partly correct, then the National Medal of Science presented to me in 1973 in the White House for the first synthesis of an oral contraceptive should be revoked as should the citation of our Norethindrone patent — the first drug patent to be so recognized by the US National Inventors Hall of Fame; not to speak of revocation of some German recognitions such as the Grosses Verdienstkreuz der Bundesrepublik Deutschland, the Lichtenberg Medaille der Akademie der Wissenschaften zu Göttingen, membership since 1968 in the German Academy of Sciences (Leopoldina), and most relevantly to the Streller/Roth article, the Inhoffen Medal and Lectureship of the TU Braunschweig. Fortunately, a recall of my other German honorary doctorate from the TU Dortmund would not be required, since it was awarded purely for my activities as a novelist and playwright during the past twenty-five years. Yet it is precisely during these past twenty-five years that I criticized in every one of my five novels and in half of my nine plays the type of self-promotional PR with which I am now painted.

A month later, I published a long rebuttal in the same German journal, which I will not bother to repeat here except for the following introductory paragraphs of my long refutation.

> *Let me rebut the personal accusations of Streller and Roth and also use this opportunity to correct at least some of the most egregious errors or misrepresentations in their article. I will do so not just by reference to my recent article [in the* Journal of Reproductive Medicine and Endocrinology*] which in any event the authors could not have seen, but by reference to a few articles and books among my zahllosen Publikationen [countless publications] of which Streller and Roth cited not a single one.... I am not a believer in automythological recollections or fantasies, but feel that only precise literature citations in refereed journals or serious (as compared to journalistic) books count. I leave it up to the*

> *interested readers whether they wish to consult any of them or simply take my word for the following rebuttals.*
>
> *In 1966, in an article, "Steroid Contraceptives," I listed every relevant article from the chemical as well as biological literature, none of which is cited in the Streller/Roth article. I did so again, with emphasis on the relevant chemistry, in the famous 1992 special issue of the journal Steroids to which all then living chemists (e.g. Rosenkranz, Zaffaroni, Colton, etc.) contributed and in which the priority of the Syntex group was clearly documented. .... Even greater detail — historical as well as personal — was contained in four books written by me over the course of three decades and I find it outrageous that the last one was not even cited by Streller and Roth, given that many of their accusations are taken totally out of context from that book. Consequently I shall start specifically with the supposed claim that I referred to myself at time as Father and at others as Mother of the Pill.*
>
> *As far as the "Father" attribution is concerned, I challenge anyone to produce a single literature citation where I personally referred to myself in that form. On the contrary, I have often made and continue to make the point that any chemist — not just Carl Djerassi — should be given the maternal role in the development of a synthetic drug. Most people, especially biologists and clinicians frequently ignore totally the role of chemists and I am completely taken aback that two chemists, Streller and Roth, object to this metaphoric analogy.*

I do not know Sabine Streller or Klaus Roth nor am I aware of any competence of theirs in the steroid field. But what would have caused two strangers to write something so utterly nasty? And why their asinine objection to my identifying October 15, 1951 as the true and only birthday of the Pill? Would one declare Mozart's birthday as the day his first symphony was heard by the public or as the day on which he was born? I suspect that the answer is most likely envy — a phenomenon that I have encountered over the years in many guises — be it among journalistic or scientific circles. Here is one that I reprint directly from my first autobiography:

> *Starting in the 1960s, I was often asked, "How do you feel about the social outcome of this work?" Depending on the circumstances, I may have grinned affectionately, shrugged my shoulders modestly, or even answered seriously that if I had to do things over again, there is little I,* as a chemist, *could or would have changed.*

But there is one question, frequently just implied or transmitted through look or voice inflection, that turns me irritably defensive. It deals with the questioner's perception of the mint of money that supposedly ended up in my pocket as a result of my name appearing first on the list of inventors of U.S. Patent No. 2,744,122. I can give two answers.

One is short and devoid of humor or even suspense. As a full-time employee of Syntex, my employment agreement contained the standard clause that every chemist working for a pharmaceutical company affirms: for $1.00 and/or "other valuable considerations," the inventor agrees to sign all patent applications and to assign to the company all rights to any issued patent. "Other valuable considerations" refers to the security of one's employment, the salary one receives, and possibly even a bonus or stock option, but never to a royalty based on a percentage of eventual sales. That would be reserved to outside inventors or other third parties.

I prefer, however, to answer the question by recounting my tussle with the Berkeley Barb, *an acerbic muckraking tabloid which bit the dust in 1980. Three years prior to its demise, the paper published a long article criticizing the financial gains that had accrued to various university professors as a result of their association with the many biotechnology firms that had started to flourish in the San Francisco Bay area and around Boston in the shadows of Harvard and MIT. Even though my own scientific research had never impinged on the biotechnology revolution, the reporter quoted an apparently uncontaminated Berkeley professor to the effect that my academic position.*

> *... hadn't kept Stanford chemist Carl Djerassi from privately patenting birth control steroids he discovered under his own name for profit, even though he had discovered them while doing NIH [National Institutes of Health]-funded research. Perhaps significantly, Djerassi ... used his own company to market such steroids.*

I was not a reader of the *Berkeley Barb,* but several copies of this particular issue promptly landed on my desk. Since their allegation — that I used government funds to feather my personal nest or that of my industrial employer — could and should have had a major impact on my academic career and on any further government funding of my academic research, I responded immediately. I pointed to the public record, showing that the patent application on the oral contraceptive was filed in November 1951, that the patent was assigned to my then-employer Syntex, that my Stanford University affiliation had started only in 1959, and that I had not filed a single patent application since that time. (While there is nothing illegal or even improper, especially under current government regulations, in filing personal patents for inventions made with governmental subsidies in universities, I have never chosen to follow that practice.) I also added that I had never received any royalties for my work on oral contraceptives or for any of the other hundred-odd patents of which I was an inventor while employed full-time by industry. Though the *Berkeley Barb* was not known as a paper likely to print retractions, in this instance they published a full-page palinode.

My reason for telling this story is that while the reporter printed an unequivocal *mea culpa* for not having checked the public record or interviewed me, he did insist that he had both quoted the Berkeley professor correctly and been given the impression "that Dr. Djerassi's alleged private patents on birth control drugs were common knowledge in the scientific community." In that respect, I believe the reporter to have been dead right. There is little I can do about that perception, which is caused by a mixture of academic naïveté and wishful thinking, often also tainted by professional jealousy. Perhaps I should have told the *Berkeley Barb* that the continuing acceptance of the Pill by millions of women all over the world is the most valuable consideration, worth all the gold in Fort Knox.

Even as recently as 2011, during a TV talk show in Austria, the question was put to me: "And how much money did you get?" To me, the tone smelled of an equal mixture of tasteless curiosity and evident envy, but this time I just shrugged my shoulders and said, "What's the difference?" I shall end this obvious bitching on my side with the observation that the question about money is almost always asked by men rather than women. Is it because the true value of the Pill has not escaped the latter?

Heimat(losigkeit)

In none of my earlier autobiographies, indeed in very little of my writings, have I ever explicitly addressed the question of *Heimat*. Even though this chapter is written in my adopted language of English, I use the word *Heimat* from my mother tongue, because it has connotations that simply do not exist in the English word "home." For me, *Heimat* emphasizes to a much fiercer extent than in English interpersonal connections, rather than some physical attachment, and I shall describe in a deliberately rambling manner why I have lacked one for over seventy years in the sense that my Duden defines *Heimat*: *oft als gefühltbetonter Ausdruck enger Verbundenheit gegenüber einer bestimmten Gegend* [a full of feeling expressed connection with a specific region]. While I may lack a *Heimat*, I do have a home, indeed I have four, which sounds like good news. The shadowy aspects will soon become evident.

San Francisco

I have a lovely spacious apartment in San Francisco with a fantastic 361 degree view (the extra degree representing a visual exclamation point) of the San Francisco Bay, the Golden Gate, Alcatraz, and the shimmering night lights of the entire city. In 2012, in a collection of admittedly bitter poems, *A Diary of Pique*,[13] I described it in the following words:

> The lights of San Francisco from his bed,
> From the shower,
> Even the toilet seat.
> "This is the place to have diarrhea in"
> He boasted.
> She agreed.

[13] C. Djerassi (translated by Sabine Hübner), *Ein Tagebuch des Grolls 1983–1984* [*A Diary of Pique: 1983–1984*]: *A Bilingual Poetry Collection*. Haymon Verlag, Innsbruck and University of Wisconsin Press, Madison (2012).

The view from his couch:
The light beacon from Alcatraz,
Like a diamond flashing in the sun.
The veil of fog dropping unto the Golden Gate
Until the bridge was draped;
The gradual lifting of the veil;
No striptease here, quite un-American,
Mere delicate exposure with a Moorish touch.

My late wife, Diane Middlebrook, and I created that home nearly a quarter of a century ago by combining four single apartments on the fifteenth floor of the highest building on the very top of Russian Hill. In addition, I own a beautiful all-redwood house in the Santa Cruz Mountains close to Stanford University, an hour south of San Francisco, a house I built over forty years ago as a weekend retreat when I was still married to my second wife, Norma. In 1972, it won a national award from the American Institute of Architects as one of the outstanding single-family homes in America, with the citation reading: "The design and simplicity of the details as well as the warmth and formal correctness of all components of the house render it one of the classical creations of its kind." Yet Norma never truly cared for it. When we got divorced a few years later, one of the few issues she did not contest in the otherwise bitter proceedings was how our jointly owned real estate would be divided. We both concurred that I would retain that architectural jewel deep in the woods, where I then led a more or less bachelor-like existence for seven years before marrying again, whereas Norma retained the suburban home which we had built very close to Stanford University when we moved in 1960 from Mexico City to California.

Having left that house during my divorce in 1976 without ever returning to it, I am tempted to skip any further reference. Yet in retrospect, this would be a mistake, because I felt more "at home" during the sixteen years I lived there than anywhere else. Hence I shall digress by quoting from my earlier autobiography *The Pill, Pygmy Chimps, and Degas' Horse*:

My wife Norma and I found an ideal spot in Portola Valley, a fifteen-minute drive from Stanford University. The lot of less than two acres was located in a hilly and heavily wooded area, the live oak, madrone, pine, and eucalyptus trees shielding us on three sides from neighbors and yet providing a superb view of the Santa Cruz Mountains in the distance. Norma remembered the former wife of one of my Wayne State

University colleagues in Detroit, who had moved to Taliesin West to study architecture under Frank Lloyd Wright, married another Wright disciple, and settled on the other side of San Francisco Bay in Sausalito. When we visited them to get architectural advice, they offered "as friends, with no further commitment," to provide us not only with a design but also with a model. So we took them to our dream lot and outlined our desiderata: decks on three sides of the house, whose floor plan should resemble that of a cross, thus providing four separate areas, and lots of wall space for pictures and books. "Forget family and dining rooms," I emphasized. "Three large bedrooms, a study, and a large living room is what we want." As I read these words now through my self-analytical glasses, I wonder whether "cross-shaped" and "forget family room" were not unconsciously precise formulations of my view of a suitable life style for an academic paterfamilias.

Six weeks later, in October 1959, we returned from Mexico City [where we were then living] to San Francisco and drove in great anticipation across the Golden Gate Bridge to Sausalito. When we entered our architect's living room, there was our model covered by a white cloth. With a flourish, he whisked off the cover to display the carefully constructed model. But where were the decks I'd been dreaming about all these weeks? I turned to my wife and saw stony New England disapproval. The architect mistook my desperate expression for timid admiration. "Lift off the roof,*" he encouraged me, "and look at the inside." It was lovingly constructed, like a doll house, but all wrong. I hardly knew where to start. "Are the bookshelves adjustable?" I finally stammered, picking something simple but to me, the stickler for details, important.*

"Of course not." He looked startled, as if I had asked an inane question. "That would look messy."

"But I told you I wanted adjustable shelves," I almost whined. "We have lots of art books."

"You can always stack them horizontally." His tone was that of an adult instructing a child how to stow a tricycle.

"There is hardly space for any pictures," I observed, pointing at the Lilliputian living room. "I told you — "

My complaint was stopped with a policeman's raised hand. "Look," he pointed to a spot near the mini-fireplace, "this is the display area. It will take a good-sized picture."

"Picture?" I practically levitated from my sofa seat. "I said we needed space for pictures." I hissed the s as if there were several attached to the word. "Lots of them."

"No problem," the architect smiled. "It's all taken care of. Here is a storage area for the art. You display one item at a time, while you store the others. You'll like this rotation."

I wondered how the director of the Guggenheim Museum felt when Frank Lloyd Wright first described the curved walls and slanted floors in his plans for the projected building in Manhattan. Let the curators worry about trivia like hanging large canvases on curved walls! But this was not Frank Lloyd Wright, I reminded myself, nor was our future home a public edifice. I was being talked down to by one of his many apprentices, who had all too well learned the master's manners. By now, I didn't even worry about the absence of the decks — "too fussy," I was told later; I knew the project was beyond repair, and that I should have known better than to mix business with friendship.

The next evening, dining at the home of the Nobel laureate Joshua Lederberg (chairman of Stanford's new genetics department) and lamenting the time we had lost in planning our house, we were given the name of his architect, William Hempel, whom he much admired. Two days later, back in Mexico City, I called Hempel to ask whether he had ever been to Mexico. Hearing he hadn't, I invited him to spend three days in our house to learn how we lived and to see what we wanted in our California home. Within one month, Hempel had shipped to us three different drawings of cross-shaped houses surrounded with outdoor decks and replete with display space for art. Adjustable bookshelves, I decided, would be mentioned later. In January 1960, we flew once more to California to approve the final plans. The architect and the Finnish builder–carpenter were nonplussed that we were ready to make all detailed decisions right there and then, down to doorknobs and bathroom fixtures.

"The house has to be ready on September eighth in the evening,' I announced, "because we are flying up from Mexico City in time for our children's school opening. We want to sleep here the very first night." We did indeed sleep there that night, and the house thereafter proved an unmitigated joy. I was deeply saddened to leave it sixteen years later, when my wife retained it as part of our divorce settlement.

SMIP Ranch

But back to that ranch house and how it really got started. In 1965, when my daughter Pamela was fifteen and my son Dale twelve, we had agreed to spend a goodly portion of my Syntex-generated affluence on some spectacularly beautiful land in the Santa Cruz Mountains, before developers discovered and ruined it. There were redwood forests, deep canyons, sweeping views of the Pacific, deer, coyote, bobcats — even the occasional mountain lion; it was only a few miles from Stanford University and an easy commute from a metropolitan area inhabited by several million people. In less than an hour, one could drive from the San Francisco Opera House into the magic solitude I had named SMIP. At first the letters stood for *Syntex Made It Possible;* later they came to signify four more consequential words. The first ninety-five acres I purchased were deep in the forest; over the next few years, I acquired additional parcels that extended from the redwoods, through clumps of madrone and oak trees, to the undulating meadows pushing toward the coast — a truly feminine landscape of breasts, thighs, bellies and buttocks, the grassy skin tanned golden-brown in the summer and colored lusciously green in the winter and spring. By 1970, SMIP had become a 1,200-acre spread — two-thirds of it held in the names of my children — on either side of Bear Gulch Road, a winding county road ending at the property of our neighbor, the rock musician Neil Young. On the eastern half of the property, which rises to over 2,000 feet and drops to 800 feet within our own confines, we erected a twelve-sided barn and a ranch manager's residence, centered on open grazing land, which became the site of a purebred, polled shorthorn cattle operation. Twelve-sided barns are rare, but the impetus here was chemical. Because the barn site was visible from various hilltops, I wanted the roof to be a hexagon, the organic chemist's favorite six-membered ring so prominent among the steroids. But since the barn's sides were mostly open to allow

Distant and closer views of twelve-sided barn, SMIP Ranch and Pacific Ocean

Outside and inside views of house at SMIP ranch

cattle to enter and leave *ad libitum*, the architect reasoned that Pacific winter storms might rip off the roof if it were only mounted on six columns. Consequently, he doubled their number to create a dodecagon — a cyclododecane ring system that a chemist finds much more difficult to synthesize. Around that difficult ring my family began to crystallize a cluster of three homes that was to be the scene of events that changed my life forever.

On the edge of the redwood forest, Norma and I built a small second home, exquisitely designed by Gerald McCue, then chairman of the architecture department at the University of California and soon thereafter of Harvard University. The setting was so private that one could not even drive to it, but only reach it by descending, through a screen of bay laurel, live oak and fir, some seventy-five irregular steps made from railroad ties. (A dozen years later, in my post-divorce bachelor days, when I ended up in a full leg cast and crutches for over eight months as a consequence of a horrendous hiking accident, the seventy-five steps created an insurmountable barrier that drove me out of the house. I never resumed full-time residency in it thereafter.)

Son Dale ages 27 and 55

Around the same time, my son Dale, soon to graduate from Stanford University, was the beneficiary of a trust fund based on an early gift in Syntex shares, which had multiplied many times in value. He requested that the bulk of the trust be used to construct his own home — built in the shape of a hawk — near a pond on the western portion of the ranch, an hour's hike from my ranch abode. In 1974, my daughter Pamela — then living in La Jolla in Southern California while her husband was finishing medical school in San Diego — followed suit. Her home and studio went up on the west side of Bear Gulch Road, half an hour's walk from her brother's home. Sometimes, when she sat on one of her hills overlooking the Pacific, and the ocean wind blew the right way, she could hear Neil Young rehearse.

Pami and her husband had moved into their ranch home when the latter started his radiology residency at Stanford, where they had met as undergraduates. In July 1978, I'd been divorced for nearly two years and was also living at SMIP in the redwood house. For well over a decade, we had hiked weekends, and frequently on weekday evenings, all over the property. Still, there were many areas we'd never explored; some sections were simply too rugged or otherwise inaccessible. In 1983, on one such hike with a group of my students from Stanford, I fell while climbing over some fallen logs in a creek bed at the bottom of a steep canyon. By the time I hit the ground my stiff leg was badly broken. It took seven hours for the paramedics and state foresters to winch me up from the bottom of the canyon to eventually get me to the Stanford hospital, which was only a thirty-minute drive away, showing how inaccessible that accident site was. During that interval I received enough morphine to sedate a couple of adult elephants. If my students had not been with me, who knows how long I would have lain there before my body had been found? But I survived, whereas my daughter was already dead by then. How I managed to cope with this most traumatic event of my life will be described in *What If?*

Dale and Pamela, ages twenty and twenty-three respectively

Ingeborg Kuhn

Entrance to SMIP Ranch

People had often asked me what SMIP meant. Instead of giving the simple, if corny, answer, "Syntex made it possible," I usually challenged them with the reply, "guess." Depending on the respondent as well as the occasion, the guesses were extraordinarily wide ranging, convincing me that S M I P is clearly a wonderfully promiscuous array of acronymic letters. "See me in private" or "sexy man invents pill" were one extreme of a continuum ending with "surely mankind is precious" on the elegant side, with "spend my investment portfolio" or "some men ignore promises" on the dismal extreme. At one stage, I had counted over forty variants. But the most a ppropriate — and in the end the one that stuck and now greets visitors on a moss-covered sign by the entrance — was "*Sic manebimus in pace*" [Thus we shall remain in peace]. Its originator was the physicist Felix Bloch, Stanford's first Nobel laureate, who produced it within thirty seconds after my challenge.

For a long time, I thought that nothing could surpass that acronym until I recently made the mistake of Googling under SMIP and discovered to my horror acronymic monstrosities such as *Senior Management Institute for Police*, *Sorghum and Millet Improvement Program Switch Modularity Interface Platform*, and *Small Modular Immunopharmaceutical*. Given my own orthopedic problems, starting with my 1957 knee fusion from a skiing accident, *Sports Medicine and Injury Prevention* seemed relevant, but www.smip.com — a German site with the English acronym *Short Message In Picture* — looked even more promising until I encountered the Canadian Parliament's maxi-acronym for SMIP — *Special Committee on the Modernization and Improvement of the Procedures of the House of Commons* — which clearly took the cake as an unsurpassable mouthful. In any event, following my marriage to Diane and my move to San Francisco, the house at SMIP (still retaining the name *Sic manebimus in pace*) turned only into a lovely but mostly empty abode that I visit a few times a year. A home it is no more, because my third marriage coincided with my increasing Eurocentricity.

Oxford and London

David Loveall

Carl and Diane, 1985 (year of marriage) and 2006 (one year before Diane's death)

My wife Diane, though a true West Coast American — she was born in Idaho, raised and educated in the state of Washington, except for a few years at Yale University, and taught in California — was, as bchooves an English literature professor, an Anglophile. In 1986, the year after our marriage, she served as director of the Stanford University program at Oxford. For the first time in my life, I became a faculty spouse. Every day I would drive her to the Stanford site at Magdalen College, take my morning swim, buy some groceries and then settle down for the rest of the day to write short stories — an activity that within a few years seduced me to assume a new professional life as a fiction writer. We were staying in a two-story house on Woodstock Road, the guests of my son's in-laws, who owned Headington Hill Hall, a sumptuous estate whose round swimming pool I used. (For the first time in my many years of aquatic exercise, I now had to switch to swimming in circles rather than the accustomed laps.) The upstairs occupant was another recipient of their hospitality, the former British Prime Minister Harold Wilson. One afternoon, he dropped in to regale us for about an hour with recollections of prime-ministerial minutiae in the 1960s that my wife found maudlin. The Anglophile in me was charmed.

As I write these sentences, instead of simply describing my four homes, I have already twice gone off on a tangent. But rather than starting over once more, I realize that only by continually going off on tangents will I be able to account in this chapter what *Heimat* means and does not mean

Balcony and common garden of London home

in my life. So, dear Reader, consider yourself warned: if tangential disclosures are not your cup of tea, stop here and skip this chapter!

During that summer of 1986 in Oxford, we devoted so many evenings to the theater in London that the following year, we experimented with spending the entire academic summer break there. Both of us found London to be a very livable city, ideal for our work and lifestyle. As we did not want to be disturbed by anyone, the eight-hour time difference from our California stamping grounds coupled with our daytime writing schedule (Diane on her famous Anne Sexton biography, I on my first novel) provided an ideal buffer. We repeated the London summer sojourn in 1988 by renting a flat owned jointly by the novelists Alison Lurie and Diane Johnson in the Little Venice section of Maida Vale. We promptly fell in love with that area and purchased a modestly sized flat of our own, which for the next three years became our summer home and principal writing nest. Writing seven hours a day, seven days a week; sifting only once every ten days during those pre-e-mail days through our snail mail, pre-culled in California by our secretary; going to the theater or concerts or opera almost every evening; and meeting a new circle of friends — most of them writers of one sort or another — proved to be the ideal setting for a new intellectual life. One day in 1991, while passing a real estate office in the neighborhood and looking purely out of curiosity at the posted pictures, I was struck by one showing a spectacular balcony overlooking a large garden. To our surprise, the house was located across the street from our small apartment and a week later we had purchased what became our favorite home until the death of my wife in

Dining room, London

2007. I call it "favorite" not only because of its innate attractiveness — the high ceilings and tall French windows of a well-preserved building from the 1860s — but the fact that we furnished it jointly as a true testimony of togetherness. As my wife decided to assume early emeritus status at Stanford, turning into a professional biographer, and I had closed my laboratory to concentrate more and more on writing, we spent increasingly more months each year in London.

By then, I was in my late seventies, but had not intended to retire from my academic position at Stanford. I had stopped chemical research, but continued to teach, although in fields that were very different from those my other departmental colleagues taught. They were sufficiently unusual that I will describe them in some detail in *"Professor of Professional Deformation"*. But one day in late 2001, when my wife — sixteen years younger than I — had already decided to retire from Stanford to become a full-time biographer, she turned to me and said, "Chemist (she always called me "chemist" — never "Carl" or "darling" — though with a non-reproducible affectionate tone) why don't you also become an emeritus and just write?"

I was taken aback and yet flattered. Flattered, because it meant that she had accepted that I was not just a dabbler in literature — as so many scientific colleagues continued to assume, especially those who had never read any of my books — but that she now recognized my having crossed the fence to her side of a jealously guarded professional turf. But I was also taken aback, because I had regaled my wife for years with macho pride that I intended to become the Strom Thurmond of academia. If Senator Strom Thurmond of South Carolina could serve as the first hundred-year old member of the US Senate, why should I not aspire to become the first non-retired centenarian professor at Stanford? I am not sure whether I really meant it, because I always allowed the barest of smiles to cross my face whenever I bragged in that fashion. But in late 2001, still twenty-two years shy of this Thurmondian ambition, I realized that Diane was right. Our chemistry department at that time had a chairman — a first-class chemist but culturally also one of the most circumscribed ones I had ever met — who did not disguise his disapprobation for the direction in which my professorial and intellectual life had gone by writing that "in recent years your interests have been far removed from those most highly valued by your department, even if they relate to broader matters of science and culture." Frankly, I was not even certain whether the majority of my chemistry colleagues did not also share his disdain, though none of them ever expressed it quite that bluntly. Within an hour after my wife's offhanded comment, I surprised her by showing her the e-mail I was about to send to our chairman indicating that I would move to emeritus status effective April 1, 2002. Nobody seemed to have noticed that I had picked April Fool's day as my formal abandonment date from my ambition to equal Senator Thurmond in academia. While this sounds amusing, the following tangential remarks are far from it. They are sad, and I would be lying if I did not admit that they have bothered me to this day. *In Retrospect*, I am beginning to realize, is starting to turn dangerously into a waterfall of complaints. Hence, I shall try from time to time to dam the water upstream in order for only trickles to splash on the reader. Yet I cannot guarantee that on some occasions most of the water can be retained by the dam and the following interlude may be one such example. I encourage the reader to don an imaginary raincoat for the rest of this chapter.

Stanford University

Institutions, especially universities, have generally short memories. Nevertheless, it is customary — and Stanford University is no exception — that before dropping a new emeritus professor into the black hole of anonymity there be held some sort of official departure celebration ranging from a celebratory symposium or possibly the presentation of an inscribed silver platter or even more maudlin gesture to at least a dinner with toasts and corny jokes, the actual choice and planning generally being left to the chair of the department. In over forty years of active professorial service at Stanford, I had attended my share of such events. They were generally considered *de rigueur* and extended also to departing secretaries, occasionally even long-serving janitors, although their exodus was usually limited to doughnuts, ice cream and soft drinks. In my instance, it was nothing, not even stale potato chips and lukewarm Cokes. Not long thereafter, I turned eighty — again a milestone that is often celebrated in similar fashion. While I did receive several hundred birthday cards and best wishes from students and colleagues all over the world, including a truly charming surprise party at Christ's College of Cambridge University — a university with which I had no formal connections other than subsequently receiving an honorary doctorate — Stanford University was the origin of only a single (albeit lovely) birthday card sent by my secretary but no one else!

That this hurt me should not come as a surprise. But the ultimate affront occurred in October 2010, a few days before my eighty-seventh birthday and the fiftieth anniversary of the day when I arrived as a new professor at Stanford together with my friend and colleague from the University of Wisconsin, the late William S. Johnson, who was to assume the position of head of the Stanford chemistry department. When in October 2010 the annual Johnson Symposium on Organic Chemistry was held — an event that I had initiated twenty-five years earlier as homage to Johnson as the best

Jeff Seeman

Lynette Cegelski

William S. Johnson, Head of Stanford's chemistry department, 1986 and view of the Chemistry Gazebo, 2012

department head we ever had, while Johnson was still alive to appreciate that tribute — I suggested to two of my senior colleagues that it might be appropriate for me to give the after-dinner speech to reminisce about our joint arrival half a century earlier at Stanford, an event for which there were no other survivors. The suggestion was neither accepted nor declined; it was simply ignored. What in retrospect seems to be a display of embarrassing masochism on my part, I persevered by proposing that I host a dinner in my home at my expense for the symposium speakers and members of the department faculty to reminisce what had brought Johnson and me to Stanford so long ago. My invitation was politely declined on the grounds that it was unlikely that many people would be interested in attending. It is only this humiliating declination that prompts me now to disclose publicly what I simply could not tell these colleagues in private, namely that my decades' long service at Stanford was not piddling and did not merit such gross dismissal. After all, for the first thirty years, I had taught at least as many chemistry courses as my colleagues; I had published more scientific papers in prestige chemical journals than any of them; I had received more honorary doctorates than all faculty members combined; I had probably won more awards than all but one or two Nobel laureates in our department; and I am only one of two American chemists to have been awarded both the National Medal of Science and the National Medal of Technology from two American presidents in the White House. In addition I had created and ran for the first ten years our Industrial Affiliates Program, which had brought hundreds of thousands of dollars into the departmental kitty, and perhaps most relevantly, had raised completely alone the funds for a small hexagonal building — the "Chemistry Gazebo" — which became and still is the favorite seminar and meeting site of the department. (In fact, I contributed all of the royalties of a book to purchase the furniture.)

But considering that fifty-year jubilees, like golden fifty-year wedding anniversaries, cannot be celebrated *post facto*, I shall at least present my portion of the story why I moved to Stanford in a later chapter. If the Stanford chemistry faculty is not interested, others might be.

Readers may well wonder why I use a chapter with the heading *Heimat* to go off on this particular tangent to reflect on my estrangement with the University where I had spent half a century. But while the traditional

Heimat is generally inherited, there is also a professional one in which one is not born, but which is only acquired as an adult. As a person, who turned "*heimatlos*" [homeless] as a teenager, the professional alternative seemed even more precious. Clearly fifty years at Stanford should qualify that university as my academic *Heimat*, but as I stated at the outset, a *Heimat* involves personal relations in which the other inhabitants voluntarily accept you or at least make you feel welcome. That even my decades-long conviction about Stanford having been my academic *Heimat* had ended up as an illusion hurts deeply as a wound that has not healed. Yet as in so many other disasters in my life, my shift to a literary career and thus a departure from my earlier academic life has helped me cope as I shall describe in more detail elsewhere in this book, in the chapter "*Writer*".

London

And now, back to London. By the time of my official retirement from Stanford, London had become the focus of a new social life that was very different from what I experienced during my twenty-five-year-long second marriage to Norma where most of it revolved around my academic or industrial colleagues. The deterioration of that marriage was gradual, but inevitable: we were both changing greatly during that time, but unfortunately in directions that moved us apart rather than keeping us together. Norma was a highly educated, intelligent woman, a professional until she became pregnant when she voluntarily turned into a home maker. Eventually, that grated on her, but instead of accepting that this had been her decision, she ascribed most of the blame to me. Neither one of us openly addressed that problem; we silently accepted it and that also reflected itself in an increasingly separated social life until the sad end became inevitable.

My third wife Diane was a professional in her own right, which immediately put our joint life on a very different footing. In addition, she was not only a fabulous hostess, but since that London period coincided with my own move from science toward literature, our common friends covered an enormous range from journalists and authors to theater people with scientists definitely in the minority. Our annual summer party for nearly one-hundred-and-fifty guests packed mostly on the balcony overlooking a huge garden became a truly sought-after event. In addition, Diane established a women's salon that included some of London's most distinguished female authors and intellectuals.

The observant reader will notice that I enumerate mostly people rather than the physical features that made that home so attractive: the art, the carefully chosen furniture, the books. I do so because for me, the essential feature of *Heimat* are the people with whom you have something in common and by whom you feel appreciated. Yet within weeks of Diane's death, even that balloon of delusion was pricked in a brutal manner.

In 2001, when Diane was suddenly diagnosed with a rare form of cancer, a retroperitoneal liposarcoma, her seemingly successful four-hour operation in San Francisco gave her confidence that she would be able to continue her newly chosen career as a full-time biographer. Liposarcomas generally do not metastasize, but they do grow promiscuously unless eradicated completely. Her first surgery had eliminated most, but not the entire tumor, and in 2004 an explosive regrowth occurred, which involved a second surgery with a poor prognosis: not all of the tumor could be removed and brutal chemotherapy was proscribed. My super-elegant wife, who had always paid so much attention to intriguing hairstyling, became bald and had to wear wigs. The gastric side effects were horrendous. The oncologist's frank answer to her direct question, "How long have I got?" was "Six months to a couple of years." Her second question, "How will I die?" led to a hardly more assuring response, "You will probably starve to death with not too much pain."

Most people would have given up at this stage, but not Diane nor I. I helped her search for alternative treatment in 2005 — first an experimental immunological approach (dendritic cell therapy) in Germany that had not yet been approved in the USA and subsequently finding a superb German surgeon, Dr. Rainer Engemann, in the regional hospital of Aschaffenburg who was willing to attempt one more operation which the American surgeons had judged to be impossible. The operation in January 2006 lasted over eleven hours and resulted in the removal of over five kilograms of liposarcoma. To heap insult on injury, I had broken my hip the preceding month in Oxford and was myself unable to travel. My stepdaughter, Leah Middlebrook, managed to get leave from her professorial position at the

University of Oregon to be with her mother in Aschaffenburg. Two days after Diane left the intense care unit of the hospital, Leah e-mailed me:

> *By this evening Mom was in pretty great shape. She was reading, and being taken care of by a nurse who had read her Billy Tipton book in German! She also ate a real dinner, involving pickles. She was feeling very happy and composed, though tired. She actually made friendly, sweet overtures towards sending me away after about an hour. I think she's tired of being the center of attention, and just wants to read until she's well enough to get back to her life! Like everything else, that's one more excellent sign ... you two will have a lot to catch up on.*

Within a couple of weeks, we were together again in our London home — I on crutches and my wife regaining her energy. Both of us decided to follow Goethe's prescription from Wilhelm Meisters Wanderjahre: "Healing the soul's pain is never achieved by understanding, rarely by reasoning, more frequently by time, but always through committed activity." We did so with a vengeance, becoming total workaholics except for evening ventures to the theater, accepting some dinner invitations, and, in Diane's case, continuing with her women's salon. We both realized that we were living under the Damocles sword of her limited life expectancy, because the eleven-hour long operation, though heroic, was not a cure but only a gift of limited life extension. Not all of the liposarcoma could be removed and even the surgical wizard of Aschaffenburg had to concede that a further operation would be impossible.

Diane resumed her daily writing routine until late October 2007 when she gave the last public lecture of her life at the Centre for Gender Studies at Cambridge University. She had been one of the greatest and most popular lecturers I have ever met, the recipient of all major teaching awards at Stanford University, but this one brought tears to my eyes as well as to many others in the audience. To me, because I knew that the Cambridge audience was listening to a dying lecturer, and to the others, because her stunning — almost ethereal — presentation was so moving. A few weeks later, she knew that the end had arrived and flew to San Francisco to die in a hospital on December 15, 2007.

For the second time in my life, I had to scatter the ashes of a deceased beloved in the small waterfall on our property which decades earlier I had

Norio Sugano

Scattering Diane's ashes in Harrington Creek, SMIP Ranch, 2007 and gathering with my son, Dale, and grandson, Alexander

Anne Pantelick

Memorial service (DRAP, January 27, 2008): grandson, Alexander, with parents, and stepdaughter, Leah Middlebrook, with husband and aunts, listening to cellist Joan Jeanreynaud

selected with my daughter as the site where I wished to have my ashes dispersed. Instead, in 1978 I performed that duty with my daughter's and nearly thirty years later with my wife's ashes. A few days later, with my stepdaughter, I arranged a truly touching memorial at the Djerassi Resident Artists Program's site where several hundred friends and admirers of Diane's came. I asked Joan Jeanrenaud, the great former cellist from the

Paul Dyer

Outside and inside views of Diane Middlebrook Memorial Writers' Residence at Djerassi Resident Artists Program

Kronos Quartet, whom both Diane and I had so admired over the years, to start and end the memorial with two solo cello pieces, Maurice Ravel's *Kaddish* and Max Bruch's *Kol Nidrei*. We also announced that in memory of my wife, who had played such a crucial role in the founding of the Djerassi Resident Artists Program and served for years on its Board of Trustees, a fund had been established for the construction of the Diane Middlebrook Residence for Writers so as to increase the annual capacity of the Artists Program by forty additional writers. (Some three years later the project was completed.)

All the important London newspapers printed long obituaries (mostly authored by her women friends), which was not surprising, given that her last published biography, *Her Husband* (which Diane had dedicated to me) was that of the British poet laureate Ted Hughes and his wife Sylvia Plath; furthermore, Diane was one of the few foreigners to have been elected to fellowship in the Royal Society of Literature. I felt that London, our adopted home, should surely be the site of another memorial for Diane and tried to enlist the help of Diane's closest American friend who had always helped and participated in our big summer parties. She turned me down, but pointed out that Diane's other women friends and salon members intended to hold one. That they did, but without me. I was simply not invited. In fact, with just a couple of exceptions, none of these supposed friends, whom I had received and entertained so often in our home, ever contacted me again. Needless to say, I was stunned. But that shock was of lesser consequence compared to what followed. A few months later, the director of the Artist's Program sent me a list of the donors — some one-hundred-and-fifty of them — to the Diane Middlebrook Writer's Residence fund.

A number of generous contributions came from England, but not a single one from any of her women friends. Even though the solicitations did not come from me, I decided to write the following note to Diane's closest British woman friend:

> *The director's office forwarded to me a list of contributors and to my utter amazement, I discovered that there were only five contributors from the UK — all of them men — and not a single one from Diane's London women salon friends. I would like to believe that this only happened in that the message of the Diane Middlebrook building fund somehow never crossed the Atlantic, even though this would not explain the quite generous contributions of the five British male friends and their spouses.*
>
> *When L. wrote to me on June 23 to explain why I was not invited to the memorial gathering, she justified it by the statement: "The group had bonded around her in a particular way and wanted a time to have an intimate gathering." But it seems to me that aside from a coffee klatch and reading, Diane's memory and in fact any bonding with her would be served in a much longer lasting and meaningful way by contributions to the Diane Middlebrook Writers Residence Fund.*

The reply was so sickeningly hypocritical that I reproduce it herewith:

> *In answer to your letter perhaps I should point out that virtually all writers who write for a living are dirt poor (including many of the best-known). That is surely why you set up a foundation to help them. It was immensely generous but asking poor authors to help build a retreat designed for people like them is a bit like soliciting contributions from the blind to a charity for the blind, something only those who happened to be well off could do on any significant scale.*

The "dirt poor" author writing this apologia was a distinguished best-selling British writer who two years earlier had won a major literary prize valued in excess of $50,000 with which she bought a Greek vacation house.

Thus, I could not help but reply once more:

> *I still feel so wounded about the whole Diane memorial affair that I really should not answer your letter. But your belief that the Diane Middlebrook Building fund is equivalent to the blind being asked to contribute to the blind is in such bad taste, given that we are speaking about a*

memorial for Diane and not some charity for dog lovers or saving the whales, that I simply must respond. (But if I were a blind writer, that is exactly the type of charity to which I would contribute.)

Of the couple hundred or more people who so far donated to the Diane Memorial fund, the vast majority is far poorer than the "dirt poor" London writer–friends of Diane. Many of those American donors contributed in the $100–500 range and knew Diane much less than her British women friends. Indeed many did not know her at all. But as I had learned inadvertently (I am not the collector of that money) not a single of Diane's British "dirt poor" friends contributed as a gesture of affection in spite of the fact that their putative poverty does not preclude them from taking many more holidays abroad than any of their American counterparts or going to the theater, etc. etc. Many of the American dollar contributions amounted to the cost (in pounds) of a single London West End theater ticket or two round trips to Cambridge or Oxford in which many of you indulge repeatedly. And at least one of the "dirt poor" Brits was left a sum of money in Diane's will (which I disbursed as Diane's executor) where 1% of it would have constituted a contribution at the very top range of the American donations I just mentioned.

In the last few years, both Diane and I have been quite embarrassed about many aspects of American politics and even culture. But in this instance, I am embarrassed about the solipsistic manner in which Diane's local women friends want to commemorate her: by writing something and then spending money on privately publishing their own contributions. Do you really think that this would be more meaningful than helping create a place bearing Diane's name that will help hundreds of writers in the future?

I am truly sad that I have to write such a letter to you whom Diane considered one of her two closest British friends.

Sugar Daddy

My possibly paranoid interpretation for this truly unbelievably cheap and tasteless affair is simple. The Djerassi Resident Artists Program with which Diane had such an intimate connection (as will be described in detail in *What If?*) bears my name and to that extent is all too often interpreted by strangers as simply a display of a wealthy man's philanthropic generosity, a gesture that might be commendable but to which surely others need not contribute. In actual fact, the Djerassi name refers to my daughter, whose suicide that program commemorated. I sold virtually my entire art collection (other than the works of Paul Klee, which I donated to two museums) in order to fund it. Not long thereafter, it had become a public charity with frequent subsidies from national and state government entities as well as many private donations. In general, the Brits are well known for responding to all kinds of do-good projects, be it for the protection of endangered local bat species or the elimination of female genital mutilation in Africa. It is inconceivable to me that there would have been such a total women's boycott in support of a writer's colony if it had been located in the UK and not under the name of a man, whose affluence — however non-ostentatious — they envied. This subtle envy is a fact of life that often bothered Diane. The envy was really directed at her, whose marriage allowed her a living style to which her supposedly "dirt-poor" academic literature colleagues or other literati could only aspire. I still recall one such manifestation by a distinguished Stanford English department colleague (if the term "colleague" can even be used in this instance), the author of sixteen books, one of whose novels supposedly featured a female faculty member and her sugar daddy — according to Diane, a thinly disguised description of her and me. Deeply wounded, she made me promise never to read that novel. Less crude but still rather obvious displays of envy by her friends were the all too frequent comments, "of course, we can't afford this … while you…" It grated on Diane and has had such an impact on my current life as a widower that I cannot resist the temptation of going off on another, though short tangent.

I ended the earlier chapter, *Suicide*, with an advertisement I once encountered in the German weekly, *Die Zeit*, which contained the following lines: *I search for a marvelous young lady, music, theater, much humor, good English, intelligent, uncomplicated, natural, slim, non-smoker. I am a very old man, Jewish, German/US/UK background, minor walking restrictions, entrepreneur, very clever, fancies much travel as well as life abroad. Ready to travel?* I now repeat them, because with minor amendments (substituting *German* by *Austrian* or really *Viennese*) the preferences listed there also apply to me (chronologically clearly "a very old man"), although quite frankly I would not have included "slim" among the adjectives, since I do prefer a modest lipid layer between a woman's bones and skin. But all the other indicated preferences imply that in his opinion the man (in this case me) is in fact intellectually (I write and lecture all the time), physically (I go to a tough gym five times a week and travel all the time), sexually (no explanation needed, if the broader definition of the term as cited in the OED is understood: *Relative to the physical intercourse between the sexes or the gratification of sexual appetites*) and esthetically (be it in art, music or clothing) only barely past his prime and hence interested in female companions who are decades younger. But *interest* and *realization* are not synonymous if the sugar daddy syndrome is recognized. On a number of occasions since my widowerhood, I have met women in their forties and fifties where the enormous age discrepancy simply disappeared and, in fact, still has not surfaced. But in each instance, that evaporation of decades applied to private and not public intercourse, because as we appear in public as a "pair," the inevitable and completely understandable question arises, "but why is she hanging around with this old man…?" Therefore, any current or future intimate companionship can realistically only be pursued in total privacy, which eliminates a great number of pleasures that really require public acceptance.

I shall never be able to overcome the sugar daddy perception because I am obviously well off; yet in terms of wealth, I am orders of magnitude away from the inflated misperceptions of the envious, who naïvely believe that every contraceptive pill swallowed by the hundred million women who annually use the Pill globally leaves some financial residue in my pocket. Not only have I given away most of my art collection, the bulk of my real estate (for example the SMIP Ranch to establish the Djerassi Resident Artists Program), but I am a person who disdains open displays of wealth. Perhaps that explains my almost exclusive use of the Underground in London or the U-Bahn in Vienna. As I read what I am here disclosing in such an unattractively complaining and exculpatory fashion flavored with straightforward realism, I am reminded of my feelings when I read the autobiography of another Jewish Austrian refugee scientist in America, Erwin Chargaff's 1978 autobiography, *Heraclitean Fire.* Elegantly written by one of the most cultured scientists I know, I was struck by its uniformly bitter tone, and yet how deeply it attracted me at a time when my own discontent had barely surfaced. But now I seem to be finding discontent in the form of a continuously appearing shadow created by the dimming light of advanced age. Perhaps it is inherited: the congenital fault of so many Viennese — belly-aching *ad libitum* and *ad nauseam*.

In retrospect, the sugar daddy syndrome may not be that recent. The following stanza from a long poem, "Hairshirt", in my 2012 poetry book, *A Diary of Pique 1983–1984*, describes how it affected my relation with the great love of my life:

> During their seventh year
>   (sabbatical from concubinage?),
> During her academic sabbatical
>   (At Harvard — where else?)
> Reliving the pleasures of genteel academic poverty,
> She discovers the danger of his golden net
> And cuts it.
> A younger lover
> (Jew, not survivor nor scientist)
> Helped.

Adieu London

Needless to say, after the utterly distasteful episode associated with Diane's women friends, any past *Heimat* feelings about that city quickly dissipated into thin air. I still occupy the London flat from time to time, since London is an exciting place to visit and I do have some friends there who have maintained contact with me. But "home" it is not any more — neither in the emotional nor proprietary sense. Shortly after Diane's death, I donated my London flat to Cambridge University (with the proviso that I could occupy it during my remaining years) so that the resulting funds could be used for an annual Diane Middlebrook visiting professorship in Cambridge's Centre for Gender Studies — the venue where she gave the last lecture of her life. By now, five distinguished professors from abroad have honored Diane's memory, with many more to follow.

The bitter mood from this London episode coupled with my growing realization of the severe limits to San Francisco's cultural scene caused me to look for another "home" that would suit me as a widower. During my long academic life, when during the day I was surrounded by people, hence leaving little time for undisturbed reflection, I welcomed evenings spent in the near solitude of my ranch home deep in the woods. But now, as a writer working totally alone during the day, I want to be in a city that at night is vibrant with events, with people (even strangers), and with theater, music and opera. San Francisco, though blessed by superb scenery and climate, simply does not measure up. Thus, when I returned from London in mid-December and wished to go out in the evening, I found that the three main theaters, ACT (American Conservatory Theater), Berkeley Rep and Magic Theater were all closed, not to open again until early January. There was no opera, since the San Francisco Opera, though of high quality, is only open for three months in the autumn and another month in the summer. The San Francisco Symphony — the only serious musical venue — was sold out. London, of course was totally different in that regard, but I was searching for another city uncontaminated by the emotional fiascos I had experienced so recently there. I also felt that if I wished to create a new social life, it should be done in a location that would not remind me continually of the lost companionship of the preceding two decades.

Given that my European roots from childhood, which I long thought to be petrified, had in fact started to sprout again, I concluded that it would have to be in Europe, rather than New York — the one American city that would have offered the cultural diversions I craved. If a female partner had been involved in that choice, I could have visualized Paris, Madrid or even Rome, since I probably would have become linguistically competent in a relatively short period of time. But since the choice was solely my own, I reached a decision which on the face of it seemed counterintuitive: to pick a city where German, my mother tongue, is spoken — the language in which I thought and dreamed until my teenage years, but which I then abandoned for half a century after my forced emigration from the Vienna of my youth.

My short list consisted of Zurich, Berlin and Vienna. Zurich headed the list, because it is the only one of the three cities that I had visited on numerous occasions during my American life as a scientist; furthermore, it carried no baggage from Nazi times. But Zurich also became the first city I struck from the list, because for me it ran up against the same problem as San Francisco: culturally too small. Furthermore it seemed to suffer from a disadvantage that I expressed, only partly tongue-in-cheek, in my play,

Calculus: when describing one of the main characters, a Swiss named Louis Frederick Bonet:

> ***CIBBER*** *For heaven's sake, why bring in the King of Prussia's Minister to England... hardly a scandalous occupation?*
>
> ***VANBRUGH*** *None of my characters have scandalous occupations... least of all Bonet. It's their scandalous behavior I wish to unmask.*
>
> ***CIBBER*** *Germans are never scandalous. Learned? Yes... Hard working? Always... Dull? Often... Cruel? Perhaps... But scandalous?*
>
> ***VANBRUGH*** *Our Bonet is not German. Some would call him Swiss —*
>
> ***CIBBER*** *Swiss? Good heavens, John! Even worse than German! For the Swiss... what is not prohibited is proscribed. I advise eliminating him from the play.*
>
> ***VANBRUGH*** *This Bonet is from Geneva.*
>
> ***CIBBER*** *Why didn't you say so? That is a mitigating fact... possibly even promising. French scandals are the best... and Geneva is right at the border.*

Berlin, a city I had only visited a few times prior to the 1990s — and then only for professional reasons — had become for me since that time the most exciting and lively metropolis in continental Europe. I even looked at some furnished apartments for a trial period and found a suitable one in terms of location, size and furnishings on a street in the center of the city, close to the Holocaust Memorial, with the enticing name "Hannah Arendt Strasse;" enticing, because Hannah Arendt was in the process of becoming one of the main figures in my play, *Foreplay*. I had visions of her looking over my shoulder and kibitzing as I described her intellectual and personal combat with another of my characters, Theodor W. Adorno.

Vienna

In the end, I picked Vienna for complicated reasons. In terms of baggage, both positively and negatively, Vienna clearly carried the heaviest load.

Here I was born and in retrospect, through the blurred glasses of long-passed childhood memory, I mostly remember it as a glorious place in which to be brought up, including four years of first-class education in an elite central European *Gymnasium*, much more advanced than contemporary American public high schools. For instance, by age fourteen, given that there was only soccer and skiing, but no TV to distract me, I had already indulged in much significant belletristic reading; I had already studied Latin for four years through Ovid's *Metamorphoses* where the lines *Aurea prima sata est aetas, quae vindice nullo, sponte sua, sine lege fidem rectumque colebat* still today ring in my ears; where Shakespeare, though all in German translation, had become familiar through his collected works and my Burgtheater attendances; where books by Charles Dickens, Mark Twain, Jack London, Edgar Allan Poe and other English-language authors were avidly read — but all in German. But then I was suddenly ejected as a Jew. (No wonder, then, that the chapter following this carries the short but pregnant title "Jew" with the indispensable quotation marks.) In retrospect, the lack of forewarning about the Nazi Anschluss seems astonishing, but I lived with my mother and grandmother in a naïvely apolitical matriarchal setting that now seems incredible. Once I fled from Vienna via Bulgaria to the USA to become, at age sixteen, a new American immigrant, my mother tongue atrophied for fifty years. First as a college student wanting to become instantly Americanized; then as the successive husband of three American wives who spoke not a word of German; and most importantly, as a workaholic scientist during the post-War decades when English rapidly became the *lingua franca* of the realm.

Between 1938, the year of my emigration, and 1992, the year when I was first invited back for a lecture in Vienna, a half century had elapsed during which I felt no pangs of homesickness nor any wish to return to the city of my birth. Not only had I been kicked out, but after decades of successful scientific accomplishments, I had not once been invited for a lecture by any post-War Austrian academic institution. On more than one occasion, I reminded listeners that until 1992, there had been only four countries in Europe that had never invited me to lecture: Albania, Malta, Portugal and… Austria (I am not counting Monaco, San Marino or Andorra). I do not mean to imply that this was a personal vendetta, because many Viennese refugees of my age group who had established international scientific reputations found themselves similarly ignored. Many people, especially in America, responded that if the Viennese of that immediate post-Nazi generation were not interested, why should you care? Yet underneath, it is now clear to me that I did care.

I can trace an event in Germany in 1991 as the seed that eventually did bring me to the decision to establish another part-time home in Vienna and thus proceed in my current mode of triangular commuting between San Francisco, London and Vienna. When in that year the German translation of my first novel *Cantor's Dilemma* was serialized in its entirety in the *Frankfurter Allgemeine Zeitung* over a period of two months prior to the publication of the book, my publisher arranged a reading tour, starting in Berlin. Reading a text aloud in German presented no difficulties and my slight Austrian accent scented by whiffs of occasional Americanisms, certainly did not put off the audiences. Initially, I was worried how I would cope with extemporaneous answers in German to *ad hoc* questions. But as I proceeded from Berlin to Hamburg, then Cologne, Frankfurt, Braunschweig, and onwards, my German loosened almost hourly. Even impromptu live TV and radio interviews in German did not faze me, especially since in my experience the German interviewers and journalists are generally better prepared through much more extensive background reading than their American counterparts back home; furthermore, the interviews are generally longer, thus reducing the pressure to produce sound bites in response to questions that deserve considerate commentary. Twenty years later, much of my German has returned. On at least one occasion, I even dreamed in German — to me the ultimate testimony that a reviving mother tongue is poking holes in a psychic barrier that, prior to the late 1980s, I had assumed had become impenetrable. Yet to this day, my German chemical terminology is hopelessly limited. Thus, during a live radio quiz show over an East German radio station, I — who had just been introduced pompously as a world-famous chemist and "Father of the Pill" — could not think of the German word for HCN. To my total embarrassment, all I could mouth was a German pronunciation of "hydrogen cyanide" which bore not the vaguest resemblance to the correct *Blausäure*. Even the idiomatically English "prussic acid" — rather appropriate, given the radio station's location — had escaped me.

It is strange that this initial recovery of my Central European roots did not occur in Austria, the country of my birth and early education, but rather in Germany. This was the country, after all that had severed those roots in the first place when Hitler's legions had driven me from Europe. All my relatively infrequent visits to Germany had been as an American adult, to scientific lectures or congresses, and these had been held entirely in English. Except for a smattering of small talk with waiters and taxi drivers, and the occasional skimming of some German newspapers, I had always considered myself, and acted like, a visiting American. All that has changed since the publication of my fiction in Germany. It took a modern German voice — that of my translator Ully Mössner — and a woman's voice at that, to bring me to terms with my European origins.

In 1992, the year after my most successful novel, *Cantor's Dilemma*, was published in German translation, which caused me to give readings from it in a couple of dozen German cities, I was finally invited to Vienna. Not by the chemists, but by the literati, the Österreichische Gesellschaft für Literatur, which had suddenly discovered me as a novelist and born Austrian writing in the rare genre of "science-in-fiction." Subsequent and ever more frequent invitations then came from medical, not scientific circles, notably the Austrian Society of Gynaecology and Obstetrics, and subsequently from the Austrian media — the Austrian Radio and TV (ORF) and many newspapers — all during the time when Austria, some three decades after Germany, had officially started to re-evaluate what had occurred during the Nazi and immediate post-Nazi period in Austria. It has to be emphasized that in this respect, the Austrian government started much earlier than the academic community. One important gesture was the offer in 1995 to all former refugees from the Nazi period to again receive Austrian passports without having to relinquish their American citizenship. I was one of the recipients of this mass mailing and after long deliberation accepted that offer and completed the necessary documentation. Nothing happened for a year, which initially I ascribed to bureaucratic inertia associated with *Schlamperei*, an untranslatable semi-pejorative equivalent of sloppiness but, on further inquiry, I discovered that this time the cause was not the all too well-known Viennese *Schlamperei* but the perseverance of the Austrian bureaucracy. Somewhat apologetically, I was informed that they had located the divorce proceedings of my parents — a document I had never seen before — which to their seeming regret demonstrated that I, who had been born in Vienna, had never been an Austrian citizen in the first place, thus making the entire citizenship reinstatement moot. To this day, I cannot imagine what had caused some eager bureaucrat to even look for the divorce documents. But before explaining what my parent's divorce had to do with my citizenship, I must digress once more for reasons that the following excerpt from an earlier autobiography will explain.

Divorces

If you asked most of my friends, they'd say I had been married twice; in fact, I have had three wives. I married Virginia before I reached the age of twenty. "I see," you might nod, jumping to the obvious conclusion, but you'd be wrong. I married, not because my bride was pregnant, but because I thought I was old enough to marry, even though — or especially because — I was still a virgin. I had already graduated from college (where I had met Virginia on a blind date at a neighboring college) and had worked for a year as a research chemist at CIBA in Summit, New Jersey. On my way to start a doctoral program at the University of Wisconsin in Madison, I stopped in Dayton, Ohio for the wedding at my twenty-four-year-old bride's home. Our bridal night was spent on the train in a Pullman compartment — a locale quite in keeping with fantasies I'd had as a teenager traveling on the glamorous Orient Express between Vienna and Sofia on my annual summer visits to my father.

Six years later, still childless, my wife and I moved to Mexico City, where I had accepted the position with Syntex as associate director of chemical research. Early in 1950, I asked Virginia for a divorce to marry the woman who was pregnant with my first child. Virginia could have been nasty or at least recalcitrant, but she was neither. Since there was no argument about money — apart from my salary we didn't own much — we decided to retain one lawyer for both of us and to get the quickest Mexican divorce possible. The drive to Cuernavaca (where Mexican residents could be divorced in one day) and our dejeuner à trois *were so civil, that our* licenciado *asked whether we were really certain about a divorce; he had never encountered a more* simpático *couple who were about to get unmarried. But two hours later, I was an ex-husband and, within weeks, a first-time father with a new wife. Soon thereafter, Virginia also married someone in Mexico and promptly became a mother. My relief on hearing*

With my first wife, Virginia, in Madison, WI in 1943 and with my second wife, Norma, in 1950 in Hidalgo, Mexico

that news was, of course, partly attributable to its dilution of residual guilt about my extramarital affair; now, I felt, our respective family portraits were being drawn on similar canvases.

When I was asked to fill out a biographical inquiry for Who's Who in America *or some such compendium, I listed my second spouse in response to the question "Name of Wife." "Date of Marriage," I left blank. These were not lies; but once I'd denied the existence of my first wife in print, it wasn't easy to resurrect her, even for those close to me. Should I have announced, as soon as my two children were old enough to understand, "By the way, I've been married once before"? It seemed cavalier, in the absence of a good reason, to bring up the topic, which their mother and I never discussed. During one of our wedding anniversaries, when my nine-year-old daughter asked, "Papa, when were you and Mom married?" I fudged and moved the date back a year.*

Although it took me years to see the parallel, my own parents divorced when I was six years old and kept the fact from me until I was nearly thirteen. People to whom I tell this are usually shocked. How come I didn't notice? And, more to the point of my own history, why did my parents keep it a secret?

My mother and my father had met when both were attending medical school at the University of Vienna; after graduation they had settled in a house on Ulitza Marin Drinov in Sofia. Eight months pregnant, my mother returned to the Viennese hospital where she had trained and which, in her opinion, was the only institution suitable for delivery of her first child. Two months after my birth on 29 October 1923, we returned to Sofia, on a day so cold that all water pipes had frozen. It was not a good omen for my mother, who disliked living in Sofia, never really learned Bulgarian, and, not surprisingly, found it difficult to establish a practice there. Besides, to any sophisticated Viennese, Sofia was the backwater of Europe and no place for the education of an only son. Thus, when I was old enough to enter school, my mother and I moved back to Vienna; and thereafter it didn't seem odd to me to see my father only when he visited us on holidays or during the summers when I traveled to Sofia. I suppose I was simply too young and generally too happy to wonder why my parents didn't live together.

With my father, Samuel Djerassi, in Sofia and my mother, Alice Friedmann Djerassi, in New York City just before and after my emigration from Europe in 1939

The extreme possessiveness of my mother for her only son — a comfort during my childhood, but an ever-growing discomfort in my late teens and early twenties — finally resulted in a complete break between us by the time of my real manhood. Attempting to maintain her place as the dominant woman in my life, she behaved toward my first wife with increasing intrusiveness. Repeated threats of suicide, coupled with the presence of bottles full of pills — both before and after my marriage — became nerve-wracking. Even though Virginia was remarkably tolerant, my mother was clearly a contributing factor to the failure of that marriage, which ended when I was twenty-six. Adjustment to the new Mexican environment, to a second wife, to my first child, and nearly total commitment to exciting research left me with a low tolerance for maternal pressure. At the first suicide threat during my second marriage, I said, "Enough!" and asked my mother to leave me alone. Only then did she resume medical practice as an attending physician in a New York hospital. Except for rare letters, and my financial support during her last years, our rupture was complete. When I encountered my mother again, she was suffering from advanced dementia and did not recognize me.

My father, however, gave me a glimpse of his private life the summer I was thirteen. His specialty was venereal disease. In those days, before penicillin, syphilitic patients had to be treated with arsenicals for several years. They were, moreover, embarrassed to be seen in such a specialist's waiting rooms, and appointments had to be scheduled so that patients wouldn't meet. I hardly ever encountered any of my father's patients, even though his office and living quarters were in the same apartment. One day, however, I saw a handsome woman reading on the sofa, who not only wasn't embarrassed but greeted me by name. The following Sunday, she joined my father and me on our weekly hike to the Vitosha mountain, at whose foot Sofia was located. When we got home that evening, my father — whom I recognized even then as a dashingly confident and eloquent man — become tongue-tied. At last, he stammered that my mother and he had been divorced, but that the time had come for me to understand why there was a woman friend — "not a patient," he hastened to add. To his apparent surprise, the revelation of my parents' divorce made no particular impression on me. I had just fallen in love for the first time and was wondering when and where I'd get my first kiss. Somehow, the fact that my father also had a girlfriend just made everything more intriguing.

Years later, when my own daughter had reached about the same age, she walked into my study after dinner and sat on my desk. A contented look on her face, she mentioned that one of her classmates, already living with a third father, had complimented her that morning in school on the evident stability of our domestic life. On the spur of the moment, I decided to bite the bullet. "Actually, I've been married once before," I announced in an offhand sort of way, as if I had just remembered that fact, and quickly went on, "but not your mother. It's her only marriage."

Like my father, I ended up surprised. "Papa, you too?" she exclaimed, breaking into giggles, rather than being shocked. "Tell me more." It didn't take long for my daughter to get the whole story: my first wife's name, the circumstances of our meeting, what she looked like and the absurdly young age at which I married. I even had to produce the group of pictures from my former connubial life, which I'd so carefully guarded from everyone among my chemical files. After studying them, she asked, "How long were you married?"

"Six years," I replied.

My daughter looked at me distantly, saying nothing for a few seconds. "Six years? But I was born — "

When I had finally confessed it all — I, a synthesizer of the first oral contraceptive — she broke into a whoop. "Wait till I tell Dale!" she exclaimed, and ran out to look for her brother.

My father never knew about that episode with my daughter. Perhaps I should have told him. For there was also at least one tale he should have told me.

My father was dashing and unconventional for prewar Bulgaria — his predilection for women extending even to occasional flaunting of mistresses. (Concubinage he used to call it, pronouncing the word in the French manner.) But he faced a real dilemma: he did believe in the institution of marriage, and no one in his large family or wide social circle had ever been divorced. His only son, he thought, would be embarrassed to be the child of divorced parents. And when I asked him why he had kept his divorce from me for so long, he always answered, "We did it for your sake." Only when he arrived in America, nearly sixty years old, did he remarry, and he remained married to my stepmother, Sarina, who was two decades younger, for the rest of his life.

Until the age of ninety-five, my father was judged by most persons to be, mentally and physically, twenty years younger. He took up skiing at fifty, when other men become cautious and stick to golf; he learned to drive in his sixties, and continued to do so until he was nearly ninety-five. Although he had been afraid of water most of his life, shortly after his eighty-fifth birthday he somehow learned to swim the backstroke. After that, forty minutes a day in the pool became one of his routines until the last year of his life, when he broke his hip stepping off the scale in the exercise room.

On the last day of his life, he lay unconscious in the hospital, all life-support systems disconnected. I sat by his bed holding his hand. My son, Dale, and his cousin, Ilan, from the paternal side of the family were close by. Suddenly my son's whisper startled me. "Papa, why didn't you tell me Grandpa was married before?" "Before?" I echoed stupidly, thinking he was referring to my mother. "Yes, in Bulgaria. Before he came to this country." He motioned toward his cousin. "Ilan just told me. His mother knew her in Bulgaria." I looked at my father, wanting to beg "Papa, please don't die yet. Who was she? Why didn't you tell me?" But it was too late. He was no longer breathing.

Until a year or so after my second divorce, and years after telling my children about Virginia, I still had not acknowledged her existence to anyone else. Then one day, I received out of the blue a letter from her, sent from her home in the Midwest. She had seen me on a television program and, in spite of my silver hair and the partial mask of a beard, recognized me immediately. She asked whether we could meet again, since she was about to visit California on a vacation. I agreed to meet, though wondering whether I would recognize her.

One's imaginings about long-delayed disclosures usually prove to be exaggerated. I recognized Virginia right away. The bare outlines of our separate lives we told each other in just a few hours. Yet how does one reconstitute a quarter-century of absence? She could not know, nor I tell, that I had denied the existence of our marriage — all six years of it. If she guessed, she was discreet about it as she had been discreet during so much of our married life.

After her return to her home, she sent me a thank-you note together with a gift, whose box indicated that it was an electric yogurt maker. It was the last thing I needed. Modern laboratory chemist that I am, I make my yogurt in the old-fashioned, large-scale Bulgarian way: I bring the milk to a boil, let it cool until I can dip my finger in it without discomfort, stir in a couple of spoonfuls of yogurt, and keep the mixture overnight in a wide-mouthed thermos. I thanked Virginia for her gift and then, without unpacking it, put it away and forgot it. Several months later, I had occasion to move the box. Hearing something rattling, I thought the yogurt maker had probably broken and that it served me right for being so thoughtless. I ripped open the box, only to discover inside thirty daffodil bulbs — one for each of the years since our divorce.

According to the accompanying note, Virginia had dug them up in her own garden and intended them for my ranch house in northern California. I planted them within the week; and the following spring, all thirty bloomed.

So what did my parent's divorce have to do with being turned down for a restitution of Austrian citizenship? At the time of their marriage, it was still obligatory for the wife and any offspring, irrespective of the place of birth, to assume the citizenship of the husband. Consequently, my mother had willy-nilly turned overnight into a Bulgarian, but at the time of her divorce, I now learned that she had petitioned to have her Austrian citizenship restored, a request that was granted. She also asked that her son, born in Vienna and by then in school in Vienna, be given Austrian citizenship, but as the document showed, this request was denied on the logical grounds that something could not be given to the Vienna-born son that he never had possessed. I never knew these facts, having assumed that I always had been an Austrian citizen, but this historical fact ended this particular form of restitution. Nevertheless, in 1999 the Austrian government presented me with the Austrian Cross for Science and Art (*Österreichisches Ehrenkreuz für Wissenschaft und Kunst*) followed shortly thereafter by similar decorations from the City of Vienna and from the Government of Lower Austria, the latter leading ultimately to my now possessing an Austrian passport. The logical question, "How come?" deserves the unexpected answer, "Because of Paul Klee."

In 2001, Carl Aigner, then director of the Kunsthalle in Krems, Lower Austria, approached me with the request to borrow eight-five works of Paul Klee from my personal collection of which the majority had already been promised by me to the San Francisco Museum of Modern Art upon my death. The exhibition was only the third major Klee exhibition that had ever been organized in Austria and the lovely Kunsthalle in Krems, so close to Vienna that many visitors came from the city, was held for a period of over three months in rooms that were specially painted a deep red. This unexpected color appealed to me so much that upon the return of these works to San Francisco, I used a similar red as the background color where to this day I have some Klee works hanging in my home. The success of the Krems exhibition did not only lead to the aforementioned honor from the government of Lower Austria, but it caused Carl Aigner to urge the powers to be in the Foreign Ministry to arrange for conferment of the Austrian citizenship through a special act of the government. This actually came to pass on the day before my eightieth birthday with the statement "*dass die Verleihung der Staatsbürgerschaft wegen der von Ihnen bereits erbrachten und von Ihnen noch zu erwartenden ausserordentlichen Leistungen im besonderen Interesse der Republik liegt.*" [The bestowal of citizenship is based on the already realized as well as still anticipated extraordinary achievements in the special interest of the Republic.] I found these words charmingly presumptuous, since as justification for their action, the authorities did not just cite my past accomplishments, but also those that the Austrian Republic expected from me in the future. A truly surprising and amusing addition was the further question whether my wife also wished to accept such citizenship, evidence that the patriarchal grounds under which my mother automatically lost her citizenship now presented my wife with the non-obligatory choice of acquiring mine. I still recall my surprise in the Vienna City Hall when I was presented with that question. I picked up the telephone and called my wife in London, "Vocalissima (my synonym for *darling)*, do you want to get an Austrian passport?" After a short pause and laugh, my wife responded "Why not?" and that curt reply I transmitted to the official sitting next to me with the additional comment that my wife did not speak a word of German. While this did not faze him, what did cause a problem was that the space in Austrian passports for "place of birth" is simply much too short for "Pocatello, Idaho." I suspect that Diane was the only Austrian citizen from Pocatello, but the restricted space only listed "Idaho" as her place of birth. The greatest fringe benefit of our new passports was that whenever we entered London and passed through Heathrow Airport's notoriously inefficient immigration lines, we always displayed our Austrian rather than American passports so as to pass through the much faster EU and UK queue.

These honors and benefits were clearly expressions of apology for the crimes committed during Nazi times, and I accepted them as such. But these were followed by one that I truly did not expect. One day after my eightieth birthday, I received the following e-mail, while sitting in Bangkok where I had just arrived.

Dear esteemed Mr. Djerassi,

I am responsible for special issue stamps of the Austrian Post Office.

Yesterday, I heard over the radio that you have again been awarded Austrian citizenship. While I was aware of your scientific and literary accomplishments, it was only after the radio broadcast that I occupied myself with your impressive and tragic biography.

For me, it would be a great honor if the Austrian Postal Service could issue a special "Dr. Carl Djerassi" stamp in recognition of your personality, your accomplishments and your fate. While we have missed the ideal time (your 80th birthday), it is after all only a date. We can find another one. I would, therefore, like to ask you politely to inform me whether you are in agreement with this idea. If affirmative, then I shall make to you several suggestions how we could proceed. I am looking forward to your (hopefully positive) reply.

With friendly greetings,

Dr. Erich Haas
Austrian Postal Service

The rest of the story is summarized in the following article in the Spring 2005 issue of the journal *Philatelia Chimica et Physica*, whose existence, quite frankly, had hitherto escaped me. I wrote it at the instigation of that journal's editor and reproduce it herewith verbatim to convey the pleasure this stamp has given me. In my Viennese youth, I was an avid stamp collector, but I never expected to receive an envelope through the mail that could only be delivered because my face on the stamp had made it legal tender.

Autobiography of a Stamp

Since March 8, 2005, people in Austria — by now in the thousands — have been licking the back of my head and presumably will continue doing so for quite a while. Unless, of course, there are 400,000 stamp collectors in the world who would not want this complicated stamp image to be marred by an impersonal post office franking machine in which case I shall remain unlicked. So how did I end up in that enviable position, which could never have happened to me in the USA where I would have had to be dead before people would have been prompted to stick out their tongue? Or did the Austrians know something about my life expectancy of which I was unaware?

It all started on October 30, 2003, the day after my 80th birthday, when I received an e-mail message from a total stranger, Dr. Erich Haas, chief of the special issues department of the Austrian Post Office. He asked whether I had any objection to a special stamp being issued in an edition of 400,000, albeit belatedly, in my honor since the Austrian

Government had offered me Austrian citizenship on the occasion of that birthday. The message had reached me in Bangkok, where I was presenting a chemical plenary lecture. Kudo-happy scientists — an enormous group to which I also belong — usually are willing to accept honors at the drop of a hat, but this one seemed a bit complicated. After all, the offer came from the country that had driven me out as a Jewish teenager born in Vienna and educated there until the Nazi Anschluss of 1938. I proposed meeting in person in Vienna in late January 2004 prior to the day on which the Albertina — one of Austria's great museums — was inaugurating the installation of an important kinetic outdoor sculpture by the American sculptor George Rickey that I had donated to the Albertina as a sign of reconciliation with the city of my birth.

By the time I arrived in Vienna in January, I had already committed the first faux pas, which was deflected diplomatically with Viennese politeness: instead of waiting for their proposed design, I had sent them an amusing but also slightly mad photograph of me that showed my face looking through glasses made out of contraceptive Pills. On the day of our January appointment, I was met by Dr. Haas as well as his colleague, Dr. Reinhart Gausterer, who, I discovered to my surprise, was not just in charge of all stamp printing operations, but had also been a chemist. And not just an ordinary chemist, but one who had actually read some of my papers. He took me completely by surprise by suggesting that my stamp (I already started to feel proprietary about it) ought to feature a three-dimensional conformational depiction of a steroid and its mirror image, since he was familiar with my earlier research on optical rotatory dispersion and circular dichroism. Clearly, I had underestimated the intellectual height to which the Austrian Post Office aspired. From then on, I was putty in their hands.

Austrian postage stamp, 2005

I promised to deliver a camera-ready design of the two steroid mirror image representations, which eventually found their place in the upper right corner of the stamp block, by enlisting the help of one of my oldest chemical friends from the early 1950s, Prof. Andre Dreiding, now

Professor Emeritus at the University of Zurich. In addition, Dr. Gausterer suggested as background some excerpt from a chemical paper of mine. But which one? As someone who has been accused of having published much too much in my chemical career, selecting a paragraph or two out of twelve-hundred articles seemed hopeless. But as I had turned in my sixties into a novelist and in my mid-seventies into a playwright, I proceeded on the slippery slope of literary self-promotion by suggesting that we pick a brief excerpt from my latest play, Ego — itself an appropriately self-critical title. Dr. Haas countered with the suggestion that I provide an image from some favorite Paul Klee work from my large Klee collection, which had been shown for some months in Austria. In the end, the only reference to my current "artistic" persona was the word "romancier" on the stamp that I found to possess phonetic euphony as well as the proper sugary Viennese touch compared to the brusque "novelist" or "author."

The inscription on the stamp's left upper corner is almost a verbatim translation of the English text that can be found on the base of the Rickey sculpture that used to be sited just outside the Albertina Museum, but is now on the campus of the University of Vienna. When I made that donation to the museum, I had proposed that instead of German the plaque be written in the language of the country that had accepted me after my forced emigration from Austria, but I had used the somewhat euphemistic expression "1938 exiled." I would like to give credit to the Austrian Post Office for substituting the more honest "1938 vertrieben" on the stamp as that German word means "expelled."

*A few months later, I was presented by e-mail with the first complete design of the stamp block. It resembled the final version, except that my face within the perforated stamp was superimposed on a mountain landscape. When I inquired about its significance, I was told that the stamp designer had read my autobiography (*The Pill, Pygmy Chimps, and Degas' Horse*) and had noticed that as a child I had skied in the Rax and Schneeberg mountains that supposedly were reproduced on that stamp. I countered that this was a piece of esoterica that nobody would recognize without a lengthy geographical explanation. I proposed instead the current image, which many Viennese would recognize immediately and which also had a deep personal meaning. The view is based on an 1897 drawing by one of Austria's most important Jugendstil architects, Otto Wagner, that I had first seen at a major Jugendstil exhibition at the New York Museum of Modern Art. It represented his proposal for the* *reconstruction of the quay around the "Aspernbrücke," a bridge crossing the Donau Canal, at whose northeast corner my Viennese home was situated: the spot now covered by the crossed fingers of my hand on the stamp.*

Originally, Dr. Gausterer had speculated about printing the two steroid images on the right upper hand corner holographically. But that proved to be fatal for the budget allocated to the production of the stamp. Instead, he came up with an extraordinary alternative — according to him unique in world stamp production — that the second larger background image of my face would be composed entirely of microscopic steroid structures! Considering that I had spent decades of my life as a chemist synthesizing steroids, seeing my face now synthesized out of steroids was the ultimate homage of the Austrian Post Office. Even though one needs a good magnifying glass to really appreciate the complicated chemical designs created by these enantiomeric images of the cyclopentanoperhydrophenanthrene skeleton, it should now be obvious to stamp collectors or stamp lickers why I am loath to see any franking mar that chemical construction. In other words, collect the stamp, but don't use it to send letters — not even love letters — unless they are addressed to me.

It was this gesture and the ever increasing-invitations I received from Austrian individuals and institutions that overcame my initial decision to select Berlin as a home in Central Europe. In Berlin, where I had few acquaintances, I would have had to begin socially almost from scratch. Instead, I settled on Vienna where in the past ten years I had gotten to know more people and thus could count on some modicum of social life right from the start. Or underneath it all, was loneliness driving me to a return of sorts to the place of my birth?

Vienna: Home or Interlude?

In the autumn of 2008, nearly a year after my wife's death, I decided to bite the bullet to extend my San Francisco–London commute to a triangular one involving Vienna. The first question was, where would I live? At my age, purchasing another place made no sense, so the question arose where and what to rent. In Vienna, the answer to the first was easy. I did not plan to have a car there, considering that I planned to spend most of the time in a city, which had excellent public transportation but practically no reasonable parking. Furthermore, since most of the cultural venues — opera, theater, museums and the like — were centered around the First District, I wanted to live in it or nearby. Examining my appointment diary from that period, I am startled to note what a preposterously hectic time I picked for what should sensibly have been done at leisure and with careful scrutiny.

My therapy for the problem of advanced-age widowerhood has been a workaholic existence superimposed on or combined with almost continuous travel; what happened in 2008 was no exception. According to my calendar, I started the month of October in Bielefeld, where a friend of mine, Andrea Frank, had arranged for three lectures at the university; then a week in London, followed by participation in a dramatic reading of my play *Calculus* in Munich; then a three-day stop in Austria for lectures at the Austrian Academy of Sciences in Vienna and at the Medical University of Graz (dealing with the provocative topic "Retrospective after 70 years: What if?); and finally three days in Bulgaria for premieres of my play *Taboos* in Sofia, Plovdiv and the ancient royal capital, Veliko Tarnovo, where an impressive Light and Sound event illuminating the city and surrounding hills was staged in my honor. Amazing, I now realize, that I passed parts of a single week first in the country that in 1938 had kicked me out followed by the remaining days in the other, which had then received me with open arms. These thoughts immediately lead me to another tangential observation.

To many Central Europeans, and most certainly the type of Viennese to which my mother belonged, Bulgaria represented the primitive Balkans incarnate, a poor and rough country in the lower right corner of Europe. This judgment, which probably still persists, ignores an overwhelmingly mitigating fact for which I have always been extremely grateful: other than Denmark, it was the only country in German-occupied Europe where the indigenous Jews were not deported during Nazi times. Essentially all survived. This was no coincidence and not solely the decision of Czar Boris, but a reflection of the inherent decency of the Bulgarian population among whom the Sephardic Jews had lived in peace for some centuries. As I have mentioned, shortly after the Anschluss, my Bulgarian father came to Vienna to quickly remarry my mother so that she could again receive the Bulgarian passport she had been so anxious to discard just a few years earlier in order to permit her and me to leave Austria in July of 1938 — in other words before the horrible period of Kristallnacht and the eventual beginning of the holocaust. Never in all my time in Bulgaria did I experience any form of anti-Semitic behavior. The subsequent teenage year I spent in Bulgaria — from a personal, familial and educational standpoint (in the American College of Sofia) — remains one of my best memories of pre-war Europe. Even Bulgaria's acknowledgment of my partial Bulgarian roots was not so long ago commemorated in an amusingly original manner. One day, an American journalist asked me "How does it feel to have a glacier named after you by Bulgaria?" I clearly took this as a joke until some Googling uncovered numerous references to the Djerassi Glacier on Brabant Island in Antarctica. For doubting Thomases, I hereby add the precise coordinates of "my" glacier: 64°13 S 62°27 W, 6 km : 7 km wide, in case they would like to check it out in person.

I had always thought that only the Danes and Bulgarians had managed to protect their Jewish compatriots during World War II. But in 2011, in Vienna of all places, I was disabused of that assumption. I had an appointment in the Hamakom Theater in the Nestroyhof, the original location of a cinema very close to my original Viennese home in the Second District where I had seen movies in my youth. Now I was visiting it to explore the possibility of having one of my plays, *Foreplay*, staged there. I was early and the theater director had not yet arrived. The receptionist asked whether in the meanwhile I would be interested to look at an exhibition they had in the cellar. "What exhibition?" I asked, whereupon I received the answer "*Besa*," which led me on an unexpected emotional detour.

Besa, it turns out is the name of the Muslim Code of Honor in Albania, which also extends to taking care of those in need and to being hospitable, a concept that is worth reflecting upon in these islamophobic times. *Besa* was practiced so widely in that small country that not only did all Albanian Jews survive the Nazi period, including the period under German occupation, but Albania also turned into a refuge for Jews fleeing from neighboring countries. At the end of the war, as demonstrated in a moving series of photographs in that exhibition, ten times as many Jews resided in Albania as were there at the beginning of the war. The apparent total absence of anti-Semitism in Albania was unknown to me and I was struck that Albania and Bulgaria, two of the poorest countries in Europe that are often held in such disdain by the central Europeans and especially the Germans and Austrians, had behaved so differently. In that exhibition, I even learned that Albert Einstein, after giving up his German citizenship, had traveled on an Albanian passport through Europe to the US.

By now, seventy years had passed since my emigration from Austria and two days later, October 28, I headed for a fabulous week in Berlin the city that I had nearly chosen for my new third home. Let me explain why I refer to that week as "fabulous."

October 29, 2008 was my eighty-fifth birthday — a discouragingly high number that I wanted to ignore, but was not allowed to. As a compromise, I invited my minute family, consisting of my son, Dale, my grandson, Alexander, and my stepdaughter, Leah, to spend a week with me in a city they did not know and whose language they did not speak. The high point of that visit *en famille* was the opening of a huge Paul Klee exhibition at the Neue Nationalgalerie during which a superb dramatic performance of excerpts from my book *Four Jews on Parnassus*, directed by the Viennese director Isabella Gregor and embellished by live music by the Icelandic pop composer Egill Olafsson and the German minimalist composer Klaus-Steffen Mahnkopf was held. The exhibition contained a number of Klee works that I had lent to the museum. The reason for combining the dramatic reading of my book with the exhibition, probably the largest Klee exhibition ever held in Germany, was that Paul Klee was one of my book's main characters. This book — my most original, and psychologically also my most important one — will be a crucial component in the chapter *"Jew"*.

Jan Riephoff

Celebrating my eighty-fifth birthday in Berlin with son, Dale, grandson, Alexander, and stepdaughter Leah Middlebrook

My return from Berlin via three days in London offered me only a five-day window in Vienna during which I could seriously start and end the search for an apartment. But since these five days also included three diverse talks, one at the instigation of the University of Vienna with the pregnant title "After 70 years: Viennese American or American

Viennese?" I had very little time to make one of the most important decisions of my recent years because by mid-November I had to depart via San Francisco for a lecture trip to Hong Kong and Guangzhou. Fortunately, a kind Viennese friend provided me with a list of suitable two-bedroom apartments which she had screened and which allowed me within hours after a whirlwind realtor-led tour to condense to a short list of three suitable ones out of eight. None of these three was located in the First District, but each in an adjacent one: the Second District, where I had lived as a child, the Ninth and the Third. The visits were made so rapidly by car that I did not pay much attention to the surrounding houses, focusing instead on the lay-out of the flats and the easy availability of public transportation. Thus, I had not noticed the large Norwegian flag waving in the breeze from the balcony of the Norwegian Embassy on the second floor, nor that the high wall across the street was enclosing the Palais Metternich, now housing the Italian Embassy or that half a block down the street stood the buildings of the Iranian, Russian, German, and British embassies. I had innocently passed into the Embassy neighborhood of the Third District, where I now reside for several months each year, but where I probably would not have ended up, had it not been for the Austrian stamp graced by my smiling face.

Right from the beginning, I learned that renting an unfurnished apartment in Vienna was different from one in the States. First of all, between obligatory deposits and realtor's fees, I was expected to shell out the equivalent of six months' rent and thereafter immediately embark on paying the rather stiff monthly rent even though I was not yet ready to arrive there for over four months until the middle of March, 2009. Surely the initial non-refundable six-month outlay should be sufficient, I argued, until I actually moved into the premises, but the agent simply replied, "We have a line of Russian *mafiosi* [actually my descriptive term, not his] willing to pay cash with no further haggling," which seemed to end negotiations right then and there. But that was when my companion interjected by pointing out that instead of cash-rich nouveau riche Russians, they would in me have a tenant represented on an Austrian postage stamp. At that point, the mien as well as demeanor of the agent changed abruptly. "Of course, Herr Professor," he replied and conceded that rental payments needed to commence only after my actual move.

Having solved that problem, I was taken aback by the local meaning of "unfurnished," which in this instance really meant "absolutely and totally unfurnished." Except for a couple of bare bulbs hanging from the wall, there were no closets in the bedrooms, no medicine cabinets in the bathrooms, no shades or curtains or other accoutrements that I would have expected. I had never bought a free-standing wardrobe in my life — not in the USA nor in London — but now I had to start from scratch. Here again, my Austrian friend came up with what seemed a superb and quick remedy. She would drive me all the way across the border to Italy where in Tricesimo, a small town near Udine, one could find a large number of Italian home furnishings stores — in fact so many that operationally it was almost equivalent to shopping in a giant IKEA store. As an aficionado of upscale Italian design, which was well represented in my San Francisco apartment, I immediately accepted because I wanted to furnish my Vienna abode in a style suitable for a daytime workaholic like myself. The eight-hour long drive to Italy proved well worth the effort when I saw one emporium after another overflowing with everything I required — from sofas, chairs, desk, bathroom accessories to light fixtures, and the ubiquitous wardrobes needed to make up for the absence of the built-in closets of all American habitations. In one single day, I ordered everything but door knobs, with the expectation that within days I would live in a furnished place. That's when I received another rude shock.

In view of the size of my purchases, the Tricesimo stores offered free transport to Vienna, but it turned out that with the exception of some small items, every piece of furniture had delivery times ranging from four to twelve weeks. I was horrified, but since there was no other reasonable alternative, I yielded and essentially led a hobo-like existence for some months except for the art works that were promptly hung on the walls and some of the books I had shipped from San Francisco. I do not wish to whine about the other annoying minutiae of getting settled, because once I was entrenched in the apartment with its two balconies and the extreme quiet of Embassy Row at night, it was clear that I had made the right decision. I was finally ready to write in a pleasant study with a lovely view beyond the balcony unto the tops of large old trees growing in the adjacent garden of a church–with leaves that turned into a glorious multi-colored tapestry in the autumn.

Dumplings

Joking in my study in Vienna with a Turkish visitor (Birol Kilic)

Aside from the cultural amenities of Vienna, which certainly put San Francisco to shame, there was one unexpected pleasure; the Viennese cuisine with which I had grown up and then had to abandon for decades. I never truly appreciated the extent of my dumpling deficiency, although an indication surfaced in 1988, when I made a rare trip to Austria to accompany my wife to a conference in Kirchberg am Wechsel on literature and psychoanalysis where she was lecturing. I was simply the faculty spouse and here is what I wrote:

> *Arriving at our Gasthof, we sit down for lunch. The choices in our small inn are limited to various Schnitzel. At the sight of my Naturschnitzel mit Champignons, drowned in cream sauce, the Viennese in me salivates, even as the weight-conscious, lipophobic Californian draws back in horror. I assuage my calorie guilt complex by deciding to behave abstemiously at dinner. After a few hours of deep sleep, we stroll to the restaurant where the other visiting academics are assembled, to face the second Austrian menu of the day. "No main dish," I proclaim to my wife in a voice full of virtuousness, "just soup and a modest dessert." I should have known better, but I had not been back to Austria for a long time: "just" and "modest" — at least in matters culinary — have a very different meaning in this country. The soup is Leberknödel Suppe, which I, born with a soupspoon in my mouth, have not tasted for decades. Attacking the huge liver dumpling, I rediscover Archimedes's principle: I have consumed hardly half the Knödel when I find the remaining broth barely covering the bottom of the soup plate. Culinary symmetry and gustatory nostalgia lead me to choose Germknödel for dessert. Knödel, the German word for "dumpling," has no plural, like sheep in English. Since my departure from Vienna in 1938, I have tasted on occasion Marillenknödel or Zwetschkenknödel — small dessert dumplings stuffed*

> *with apricots or plums; but my last Germknödel dates back to the pre-Anschluss days. I have forgotten that this comes as one giant Knödel, squatting over the entire plate, generously freckled with poppy seeds, drenched in butter, and, most delectable, stuffed with Powidl, the Austrian plum jam. In California, I would have gone on a four-day fast to make up for this megacaloric sin, but here, in Kirchberg am Wechsel, instead of sinking like a stone to the bottom of my stomach, the first bite of the Germknödel immediately penetrates my blood–brain barrier. Like a crack smoker after the first puff, I experience an instant high.*

But now living in Vienna for weeks at a time, I found myself binging repeatedly in restaurants on liver dumplings in the soup, on savory dumplings as the indispensable companion of most main dishes, and on my favorite dessert dumplings, although these were replaced in the apricot off-season by *Salzburger Nockerln*, *Kaiserschmarren*, *Mohnnudeln* or other Austrian culinary enticements, which in terms of megacaloric content were equally pernicious. The consequent impact on my midriff was only partly assuaged by vigorous exercise during my daily visits to a nearby gym and by the automatic dumpling abstinence during my sojourns in London or San Francisco. Yet putting aside for a moment my California-induced caloriephobia, I readily admit that Vienna is my gustatory *Heimat*, something that even the Nazis were not able to destroy. Furthermore, since any of my current homes are really just places where I buy the plane tickets for my numerous travels, almost all professional in nature, I soon appreciated another advantage to residing partly in Vienna, this time a geographical rather than culinary advantage. Not only was its airport much smaller than the monstrosities of Heathrow, Charles de Gaulle or Frankfurt; even more importantly, its location in the middle of Europe meant that almost every one of my trips, be it to London, Madrid, Copenhagen, Hamburg, Paris, Berlin or Prague to just name the destinations I noticed in my 2010 calendar, were never much longer than two hours.

Returning to School

There were other events that encouraged a still ongoing rapprochement with Vienna. During the past decade, various media — notably the ORF radio and various Austrian and German TV channels — became interested in documenting aspects of my earlier life in Europe and in the process introduced me to events and locations I had essentially forgotten. One of the most charming ones merits recounting.

Eberhard Büssem, a film maker from Munich, had been filming me for over a year in various locations — San Francisco, London, Sofia, Vienna — in preparation for a documentary to be shown on Bavarian and Austrian TV. During his shooting schedule in Vienna, he surprised me one day by taking me to the Czerninplatz Volksschule in the Second District, the primary school that I had attended from age seven to ten (I had attended the first grade in the "German School" in Sofia) and had not visited since 1933! When I attended that school, there were two entrances since boys and girls were segregated in adjacent wings. But in 1930 I had arrived (from Sofia) a few days too late and, since the boy's section was full, I and a few other male stragglers were placed among the girls. I can proudly prove through the original school documents (*Schulnachricht*) under the signature of the director, Albine Nagel, and the classroom teacher, Emma Starkel, that I was a student in the Girl (*Mädchen*) school, thus justifying my claim that having been brought up in an all-female household, having even been delivered by the first Viennese female obstetrician at the hospital, and then having started my primary education surrounded by females, where even the teachers were women, I was imprinted for life in terms of my preference for feminine company. (*Se no è vero, è ben trovato!*)

Apparently, the school had been heavily damaged during the war and the reconstructed building now had only one entrance since it had turned coeducational. For its one-hundredth anniversary, the school officials had searched their archives for distinguished alumni and found a number of interesting ones, whose pictures and biographies they mounted on the walls for the students' edification. Since the Second District — formally named the Leopoldstadt, but at times also referred to with an implied sneer as the *Matzeinsel* (Matzo Island) — had been heavily populated by Jews, it was not surprising that all of the former alumni pasted on the walls proved to be Jewish, of which the co-discoverer of nuclear fission, the physicist Lise Meitner, was clearly the most renowned. But as the camera team brought me to the school where a teacher, Renate Gründler, met me at the entrance to lead me up the stairs, I had a triple surprise. The walls along the staircase were covered with drawings by the pupils and I immediately realized that they were all based on or at least influenced by Paul Klee. When I arrived on the second floor where alumni such as Lise Meitner; Alfred Bader, the chemist, philanthropist, and art collector; and Viktor Frankl, the neurologist and founder of logotherapy, were highlighted, I also noticed my picture. But at that moment all the assembled students broke into a collective "Happy Birthday, Herr Djerassi," which was all dutifully filmed. While charming, it was nothing compared to what followed.

I was led into a smaller room where the twenty First Graders were waiting for me, standing at attention around a long table carrying a birthday cake, which on closer inspection proved to be an edible replica of a famous Paul Klee water color. Half a century ago, I may have been at one of the first-grade classes of my children, but if so, I remember no details. But this encounter was touching as well as startling. Touching, in that Frau Gründler asked each six-year-old child to stand up and wish me happy birthday in its mother tongue; yet startling, in that I heard these birthday wishes in thirteen languages, covering the alphabetical range from Albanian to Ukrainian. Fewer than five members of that class were native-born Austrians who had been brought up in a German-speaking home. Even more remarkable was the Paul Klee connection. The cake was the masterwork of another teacher, Elisabeth Schmid, whom I eventually got to know much better as demonstrated by a picture taken at a private lunch in her home where I faced my second Klee cake. Other than the cake, the Klee ambiance, which I had already noticed upon entrance to the school was not in homage to me,

Birthday celebration with "Klee" cake and first-graders led by their teacher, Renate Gründler, at my former Viennese grammar school, Czerninplatz

but rather due to the fact that each year, Renate Gründler used one famous painter as the *leitmotiv* for all the instruction — from art to simple math. I just happened to have arrived during the year of Klee, but soon encountered also vestiges of earlier Matisse and Hundertwasser years.

Herbert Schmid

Herbert Schmid

Poppy seed cake adorned with the edible image of Paul Klee's "Und schämt sich nicht" created in 2010 by Elisabeth Schmid to whom I am expressing my gratitude after a slice had already been cut and tasted

I was truly struck by the quality and personal attention this particular class got from their teacher and the sophistication of the art motive as it was woven into everyday instruction. The language diversity was truly remarkable as was the speed with which most of the children were learning German or even spoke it fairly fluently. But then my next question suddenly threw a dark shadow over this apparently rosy picture. Remembering that in that school of my time, at least 40% of the students were Jewish, I asked how many Jews there were now among the 400 odd students. "One," was the answer. Did this mean that the Nazi attempt to make Vienna *Juden-frei* had succeeded, decades after the end of the holocaust? The answer, of course, is more complicated: the vast majority of the current Jewish population of Vienna — now probably fewer than 10% of what it amounted to at my time in the 1930s — does not consist of the former typically super-Viennese secular Jews, but of post-war immigrants from the East, who are much more religious. Their children now go mostly to Jewish schools — another example of total, albeit voluntary, segregation.

Since that visit with a film crew, Renate Gründler and Elisabeth Schmid have invited me each year for a birthday celebration with first-graders at my old *Volksschule*, each celebration featuring another Klee cake, which I was always hesitant to cut, considering it a form of artistic mutilation performed on my favorite artist's works, yet accepting it as another manifestation of Vienna's unexcelled pastry artistry. On each visit, some small event makes me think of what *Heimat* might mean to those children. Were they gaining a true one through that school or was Vienna only a cultural way station in their life? At my last visit, a Turkish Muslim girl approached me, asking "Tell me, is it true that Muslims kill Jews?" What could I say but acknowledge that fact? But then I added, "The reverse is also true. And neither one should do that." She seemed satisfied and thanked me. Frankly, neither Paul Klee nor religious killings were subjects that I had expected to encounter among first-graders.

Until now, I seem to have been circling around the term *Heimat* and its deeper meaning, at times approaching it from one side or another, but never truly head on. That only occurred on November 9, 2010 when I appeared on a TV talk show "Talk im Hangar 7" dealing with the question "*Gibt es Identität ohne Heimat?*" [Is there an identity without home?]. What caused me to accept the invitation in the first place was the thoughtful overall approach as outlined in the initial letter from Servus-TV, a channel owned by Dietrich Mateschitz, the Austrian tycoon owner of Red Bull: "In the first place, since the discussion should not be political but rather philosophical, I wish that the emotional meaning of Home — also in terms of a psychic Home — be analyzed." The live interview was held in Salzburg in front of a substantial audience in a spectacular airplane hangar containing the collection of Mateschitz's airplanes and race cars, and moderated by a sophisticated Hamburger, Ruprecht Eser, with four other participants — academics, one each from Switzerland and Austria, and two from Germany. All of them were philosophically competent as well as philosophically oriented, none of which applied to me. I have always regretfully conceded that the biggest lacuna in my formal education was philosophy. In addition, I wanted to focus on the question of *Heimat* in a totally personal manner, which turned out to be fitting, especially since I was the only member of that quintet who had lost his home through involuntary emigration.

The topics were listed in a series of questions to which brief answers were solicited from each panelist. The questions, together with my answers are reproduced below and are so self-explanatory that only a few deserve further comment.

Can Identity exist without *Heimat*?

1. What does *Heimat* mean to you?

What I lost in 1938 and can never truly regain.

2. Does *Heimat* have a prospective function?

For a baby surely. But for me, a refugee in 1938 while still a teenager, it seems impossible except if one refers to a second pseudo-*Heimat*.

3. Can Identity exist without *Heimat*?

Indeed. At least personally, I found this possible. In many respects, a homeless identity is to be preferred.

4. Is Death the end of *Heimat*?

Death is the end of everything except for renown (both in the sense of fame or notoriety).

5. Why can the concept of *Heimat* be so easily misused?

Because the word is subjective and not objective, even though many persons — perhaps even including you — consider it objective.

6. Does Globalization strengthen *Heimat*?

Not for me, but most likely for many others.

7. How much *Heimat* does a person need?

That depends on the person. I have learned that one can live without a real *Heimat*.

8. Is *Heimat* defined through the existence of foreignness?

On the whole, I can accept that definition, except if one is expelled from one's *Heimat*.

9. Can one love a *Heimat* in which one encountered pain? (e.g. Expulsion, Migration).

A masochist certainly can. With me, it is a much more complicated feeling: a bitter-sweet mixture, which can never be pure love, but which may have some positive emotional component.

10. Can *Heimat* be equated with fate?

No.

11. Why do people let themselves be manipulated so easily by *Heimat*?

Because certain persons — especially the very "right" (for instance in a political sense in Austria) or fundamentalists (for instance in a religious sense in the USA) — are prepared to be manipulated.

12. Is there a virtual *Heimat*?

Indeed, a personal one, which one only creates as an adult, but which has nothing to do with birthplace.

13. Is religion the only globalizing *Heimat*?

I would deny this, since I am totally irreligious. But if the answer were affirmative, then I see all kinds of disastrous consequences of which the fundamentalist religious extremists (be they Muslims, Christians, or ultra-orthodox Jews) are the worst examples.

In addition, I was asked a question that pertained only to me, since it was based on the statement, "Homelessness is only noticed later in life," that I had supposedly made in some earlier newspaper interview. Instead of simply replying that this certainly was true of me, I gave the following third-person response: "He lost his Viennese home in 1938, but sixty-seven years later, an Austrian stamp with Djerassi's image was issued. Does that now make him a Viennese? He still doesn't know." But since that time, two humorous stamp-related events occurred that merit telling.

One day, upon my return from a trip abroad, I found a note in my post box requesting that I pick up from the post office a registered letter that required my signature. When I arrived, I was asked to produce some identification with my picture. The one I always carry with me is my California driver's license which even the ultra-paranoid security personnel at American airports accepts as valid proof of identity. But not the snippy *Beamter* in Vienna manning the counter. After suspiciously eying the proffered driver's license, he pulled out a booklet, skimmed it briefly to then announce that it was not on the list of approved Austrian identifications. Truly pissed off, I returned home and instead of picking my Austrian passport, I returned half an hour later brandishing the Austrian postage stamp bearing my face and name. I had the bad luck to end up in front of the same officious man, who took one look at "my" beautiful huge stamp and just dismissively waved his hand. "What is this?" he asked. "An Austrian postage stamp," I replied, "of me. Just look at the picture." "Not on the official list," he countered tapping his booklet. When I pointed out that even though the booklet may not have included Austrian postage stamps as standard identification, surely in an Austrian post office a stamp printed by them should be acceptable proof for just handing over a registered letter. By now, the line of people behind me had turned restless, but I persisted as immovably as the jerk behind the counter. I demanded to see his boss, who eventually appeared. After hearing my story and my threat that if I had to return once more I would do so in the company of a muckraking journalist, he compromised. He asked his subordinate to go to another counter, not wanting to embarrass him since he was properly sticking to the letter of the law, but then gave me my envelope without further question.

A few weeks later, I again had to go to the same post office to pick up a package of books where I knew I would be asked to show some identification. Deliberately, I reappeared with my stamp, but this time made sure not to end up in front of the earlier postal nemesis. A youngish clerk, not looking too bureaucratic, was about to release the package when he remembered to ask for some identification. Looking at the stamp I offered and then at my face, he started to laugh. "How and where did you have this made?" he asked admiringly. "You made it," I replied. He refused to believe my answer, even after I made it clear that by "you" I really meant his employer. But he was so impressed by my supposed effort to construct a fake stamp of myself that he waved me away with a big grin.

I could let the subject of the TV questionnaire rest at this point, but I won't, because three of the questions touched me so personally. The first was No. 4, "Is Death the end of *Heimat*?", which I cannot ignore at my age. If I had a choice, where would I wish to die: San Francisco, London or Vienna? It is a topic, I confess, that I have considered now and then and for which I have reached a surprising compromise, one that I would not have expected just a few years ago. *Heimat* invariably starts with a birth. So let me end it with death. Where I die, of course, is unlikely to be entirely my choice, but the disposal question clearly is. I would prefer that my death occurred in Vienna — the city where I was born, the city from where I was ejected, but also the city where at least late in life I have received a measure of respect that I do not find in San Francisco. But choosing a place for death does not automatically also include disposal of the remains. In that regard, I insist on cremation and on a scattering of my ashes in the Californian waterfall of my ranch in precisely the manner I have already described earlier on in this chapter. There is, however, one caveat. By death I mean a sudden and thus enviable death — a heart attack or an accident — not a lingering one that demands lengthy hospitalization or home care. In that event, I prefer San Francisco where some options for expediting the process exist.

The second is a combination of Nos. 3 "Can Identity exist without *Heimat*?" and 7 "How much *Heimat* does a person need?", about which I spoke at some length during the TV interview. Fundamentally, I do not believe that one can ever regain a true *Heimat* once one has been thrown out of it, especially under the conditions that pertained in 1938 at the time of the Nazi Anschluss. But rather than commiserating about a never-healing wound, I find that *Heimatlosigkeit* for me is not the same as homelessness. I live in three beautiful locations and find them physically comfortable and well suited to my present state of productive restlessness. Instead of retiring in one place, I find diversification in homes just as wise and assuring as in financial investments, where diversification is always the most sensible move. It has offered me a freedom that, for instance, made it relatively easy to establish a new home in Vienna at age eighty-five, because in theory I could still be enticed (presumably by someone, rather than solely by myself) to move to still another location. I find such freedom refreshing rather than burdensome.

I have left question No. 9 purposely for the end, because my more or less affirmative answer is reflected in some substantial gifts I have made over the years. The three cities where I have homes have each in one way or another, at one time or another, offered me the temporary warmth of a home, even though not a true *Heimat*. I have left some evidence of my appreciation in each of these cities through gifts of art; each of them bearing a different plaque that explains my specific feeling towards that city.

In San Francisco, it is a huge kinetic stainless steel sculpture, *Double L Gyratory,* by George Rickey, one of my favorite American sculptors, whose work I have collected since 1970 and who also became a good friend. It is a huge sculpture in an edition of two. One of them used to be on Fifth Avenue in New York City, at the entrance to Central Park. The

Gabriele Seethaler

George Rickey, *Double L Gyratory*, San Francisco

other one danced in the wind for years on my ranch, until I decided in 1997 to donate it to the city of San Francisco where it stands in front of the Main Library, across City Hall bearing the inscription "A gift from an immigrant, Carl Djerassi, to his adopted city." The best way of indicating my emotional attachment to this sculpture is to show how I smuggled it into a scene from my last novel *NO*:

> *Neither one spoke as they leaned, side by side, on the balcony railing looking down on a well-kept lawn, in the center of which stood a concrete column carrying two giant Ls mounted on ball bearings — a stainless-steel kinetic sculpture by George Rickey. The two Ls moved slowly in the light breeze, each in its own cylindrical path, seemingly ready to crash into each other but, of course, never doing so. The simplicity of the construction coupled with the complexity of the separate motions of each giant letter L were the secret behind its magic.*
>
> *Finally, Melanie broke the silence. She pointed to the moving Rickey sculpture: "Try to catch them with your eyes when they are in a mirror image relation: one L facing right, the other to the left... There they are!" she exclaimed. "Did you see it?" For the first time, she turned to face him directly.*
>
> *"The letter L," he said in a low voice, without meeting her glance, "is probably the best letter in the English alphabet: 'liberty, light, laughter, leisure — "*
>
> *"Don't forget love and libido," she laughed.*
>
> *"I wasn't finished yet. Also luck and lust; lips and legs — "*
>
> *"Yes," she murmured. "And liaison, logic, and loss."*
>
> *He looked at her. "You're right. It's not a perfect letter. There's also lie, larceny — "*
>
> *"Don't, Menachem," she pleaded. "What about leniency?"*
>
> *He nodded. "'Lama lo'" — why not? Two more Ls, but they don't count: they're in Hebrew."*

The gift to London was a spontaneous choice made jointly with my wife. We were both great fans of the new British Library on Euston Road — as users and admirers. In the late 1990s, on one of our annual visits to the Cass Sculpture Park in Goodwood, Sussex, we saw Bill Woodrow's bronze sculpture, *Sitting On History* — a huge open book chained to a heavy ball, implying that a book is often a source of information from which we cannot escape. As we sat on that book-bench, we seriously contemplated acquiring it for our London balcony, but soon realized that it was too large and too heavy. But not too long thereafter, shortly after the formal opening of the new British Library, the then chairman, John Ashworth, mentioned to us that the Library was still looking for art objects for its open spaces. I looked at Diane, silently mouthed "Bill Woodrow", and she just nodded. A few minutes later, I broke the news to Ashworth, who was quite overwhelmed since he himself had seen that sculpture and had wished to secure it for the Library. Our gift was predicated on one condition: that Diane and I would select an appropriate site in consultation with the sculptor and the art-savvy architect, Colin St. John Wilson, who had also been smitten by that sculpture. As a result, for the past fifteen years, the open bronze book is visible through the big glass window at the main entrance of the British Library with a small plaque simply stating "A gift from Carl Djerassi and Diane Middlebrook, readers and writers, 1998." Since then, countless visitors, while sitting on that bench, have polished the open page of the bronze book by bodily rubbings with their collective buttocks — in my opinion a very intimate sign of approval.

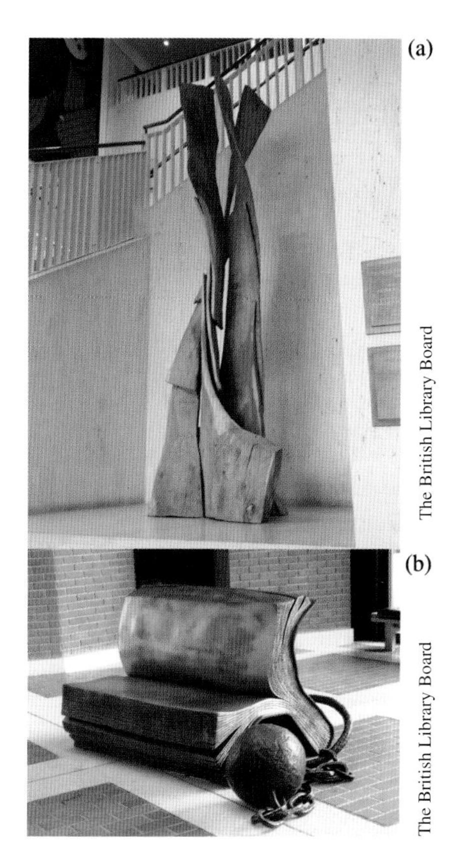

Two gifts to the British Library, London.

(a) David Nash, *Spiral Sheaves*

(b) Bill Woodrow, *Sitting on History*

Three years later, just as Ashworth was handing over the chairmanship of the Library to Lord Eatwell, he approached us once more, this time to ask whether we would be prepared to pay for the towering wooden sculpture *Spiral Sheaves* by David Nash, which had been displayed as a loan from the artist at the entrance to the Conference Center of the British Library and which was about to be returned. How could I say no? A few months later, a plaque similar to the first one, reading, "*Spiral Sheaves*, David Nash 1991, A gift of Carl Djerassi and Diane Middlebrook, Readers and Writers" was attached. David Nash was the most distinguished British artist who ever worked at the Djerassi Resident Artists Program; the deeply moving origin of three of his sculptures at our ranch in California, which I describe in the coda to the chapter of my daughter's suicide, testifies to my regard for him.

Rickey in Vienna: A Melodrama

The story of my gifts to Vienna are more complicated, which *a priori* should not be surprising given the nature of my past relation to my birthplace. But the emotional complications were nothing in comparison to the melodramatic bureaucratic ones.

In 2003, I visited Vienna no fewer than six times, always for short periods and always associated with some lecture or professional event. Since each visit was prompted by an unsolicited invitation, I took it as a gesture of welcome, an omen of increasing comfort with the city of my birth. The first visit that year occurred in mid-March and was associated with a lecture at the Leopold Museum on our Artists Program in California, followed by a very imaginative dramatic reading of my play *Calculus* under the direction of Isabella Gregor in one of the exhibition spaces of the Museum Quartier, diagonally across from the Leopold Museum. Completely by accident, an Austrian friend, the dermatologist Elisabeth Wolff, who was well connected with Viennese music and art circles, asked whether I would like to join her at the grand opening of the restored Albertina Museum since she had an extra ticket. It was truly a grand event, coupled with the opening of a huge Edvard Munch exhibition, and garnished with opening speeches by the President and the Chancellor of Austria as well as by the Albertina's director, Klaus Albrecht Schröder. The speeches were elegant and erudite and what impressed me most, delivered largely extemporaneously. I could not imagine that President Bush could have given or even read such a speech without the aid of a teleprompter if the event had been held at the National Gallery in Washington. 2003 was also the year of my eightieth birthday and it suddenly struck me that a city that spends millions of Euros during the height of the Bosnian war on the grandiose restoration of such a gem of a museum did deserve approbation. At the conclusion of the opening, while wandering around the Albertina, it occurred to me that another kinetic Rickey sculpture, which had been standing thirty years at my ranch home might look superb if mounted there as a token of my ever-increasing reconciliation with the city I was forced to leave as a youth. In my usual *ad hoc* manner, I turned to Elisabeth Wolff and asked whether she would contact the director of the Albertina, whom she knew well but whom I had never met before, to ask whether he would be receptive to installing a sculpture gift of mine outside the Albertina. Her e-mail answer came a few days later:

> *I had a long conference with Schröder yesterday afternoon. He was enthusiastic about the Rickey, about which, surprisingly, he knew quite a bit. This art work would look wonderful on the Terrace which we had considered for its location.*

I was sufficiently pleased by this rapid response that I ignored — at my peril, as it turned out subsequently — the following sentence in her e-mail:

> *Before Schröder can accept your spectcularly generous gift, he has to receive the permission of the Burghauptmannschaft (Office for the Maintenance of Historical Buildings), the Denkmalamt (Office of Monuments), and the Ministry. The Terrace belongs to the government, while the property of the Albertina ends with its bulding walls. Hence he has only limited possibilities to accept your gift, which makes him very unhappy.*

At that early stage of my rapprochement with Vienna, I had not yet realized that *Burghauptmannschaft* and *Denkmalamt* were misnomers for bureaucracies, whose exclusive function it was to find reasons why a given project should be shown not to be feasible, rather than to facilitate it. Neither aesthetics nor reason entered into their equation. Their *modus operandi* was to create delayed-release mines through which only the most sophisticated mine-detecting operative could proceed. Schröder, whom I subsequently got to know well and certainly now count among my friends,

(a) Gabriele Seethaler

(b) Isabella Gregor

(c) Karl Schwaha

Three successive Viennese locations of George Rickey's *Four Lines Oblique*: (a) Albertina Museum Bastei; (b) Stadtpark; (c) University campus, AKH

was such an expert able to cope with such bureaucratic mine fields. Yet even he needed nearly a year before the sculpture, *Four Lines Oblique*, could be set up on the terrace just behind the Albertina, facing the Burggarten and the Imperial Palace. The vernissage was finally held in January 2004, a delay that inadvertently proved to be providential, since two weeks earlier I had been presented with my Austrian citizenship, an act of governmental good will that I had not anticipated at the time I had decided upon making this gift. Visually, the location of the sculpture was superb, but climatically it turned out to be fatal. Vienna is a city notorious for its shifting winds — and not just conversationally; the particular juxtaposition of the Albertina Museum and the Imperial Palace produced occasional whirlwinds, which the sculpture had never experienced in thirty years of Pacific storms on my ranch. In the event, after two years, the elegant swaying metal arms were sufficiently damaged that it became clear that another location would have to be found once it had been repaired in Berlin by a disciple of George Rickey.

The question now arose where to relocate the sculpture and over a period of three years I found myself transformed into a begging surrogate for an elegant and expensive but homeless sculpture by a world-famous sculptor, whose works are in museums and public collections worldwide, but not so far in Austria. During that period, I was once invited to meet Austria's Chancellor at his initiative for a completely different matter. When at the conclusion of our meeting he politely offered to help in any future plans of mine, I grasped the opportunity to ask for help with the Rickey. To him, my request appeared trivial and he immediately gave instructions to his staff to help solve what seemed to be simply a non-problem. Two years later, I realized that even the Chancellor of Austria is helpless when confronted with the faceless and all-powerful Viennese bureaucracy, which was not prepared to be dictated by a federal official. After all, this was autonomous Vienna, but even its mayor, whom I approached indirectly through acquaintances only proceeded in the typical Viennese fashion of handing the problem over to an underling, who either was a member of the bureaucratic mafia or incapable of coping with it — in other words not a facilitator. In the end, through manipulations not worth recounting, I managed to get permission to mount the sculpture near the entrance of the Stadtpark, Vienna's elegant inner city park. As shown in figure (b) in the accompanying triptych the location was not only visually stunning, but also practical. It was at the edge of an intersection of four busy streets controlled by several stoplights, which meant that cars were sometimes waiting for nearly two minutes for the light to turn green while the sculpture was dancingly diverting the stalled drivers. Yet within two months, the sculpture was dismounted at the order of the Stadtpark administration, which had plans for a temporary kitschy Christmas installation. By now, the sculpture's odyssey, which had started in 2004, had reached the year 2011 when an apparent permanent solution was found on the grounds of the University of Vienna campus which since 1988 is located within the complex of the old Allgemeines Krankenhaus, Vienna's most famous central hospital of the turn of the nineteenth century and at that time probably the largest hospital in the world. It was also the hospital where I was born. As shown in figure (c) in the accompanying triptych the sculpture with the same inscription as displayed on "my" Austrian stamp (1923 born; 1938 expelled; 2003 reconciled) is now installed within a few hundred meters of my true birthplace.

I have always made it plain that "reconcile" does not mean "forgive" or "forget," but it does mean moving forward. To reaffirm this personal belief, a few years after my Rickey gift, I decided to transfer to the Albertina after my death one half of my large Paul Klee collection with the

Leah Middlebrook

George Rickey, *Oblique Column of Twelve Open Squares*, Vienna

other half passing to the San Francisco Museum of Modern Art. I have made one more gift to the Albertina, a second kinetic sculpture by George Rickey, which rotates 360 degrees rather than in just one plane and thus should prove resistant to the whirlwinds of the museum's terrace. Somehow, my very first gesture of reconciliation made around my eightieth birth year in 2003 through Rickey's *Four Lines Oblique*, is now restored through Rickey's *Oblique Column of Twelve Open Squares* near my ninetieth anniversary.

Vamos a ver what will happen at my centenary. If I should still be alive and living even part time in Vienna, I am sure to check up on these two sculptures — if necessary, even in a wheelchair. If they should still be standing there, then perhaps the question of *Heimat* will finally be resolved.

"Jew"

I insist that the title of this chapter be considered a five-letter word, because the quotation marks form an indispensable component of what I have to say about my Jewish identity — an all-too-often shadowy subject that can run the gamut from proud acknowledgment of one's "Jew" label to tacit admission or even devious denial. Off and on, two of these alternatives have also applied to me over the course of my life.

Family Connections

Both my parents were "Jewish" — my Viennese mother arrogantly Ashkenazi, my Bulgarian father aggressively Sephardic. The arrogance and aggressiveness manifested itself in each spouse's opinion of the other's origin. But my home life was virtually nonreligious (hence the quotation marks) with two exceptions, one public and the other private. In the Viennese public schools, religious instruction was obligatory. Three times weekly, the Jews were separated from the Catholics, and each group drilled in matters religious. At home, each evening, my mother came to my bed to listen to my recital of a prayer ("*Müde bin ich, geh zur Ruh...*"), a German version of the child's prayer: "Now I lay me down to sleep..." — quite consistent with the fact that my maternal Viennese home could fairly be categorized as one of Christmas-tree Jews. Like so many secular Viennese or German "Jews," the December holidays revolved around a Christmas tree (without a cross or other Christian religious signifiers) and not Hanukkah candles. Christmas trees were absent in my Bulgarian paternal home, but since my parents divorced very early in my life, I spent summers rather than winters with my father and these periods were totally devoid of any religious activities or even instruction. Yet by name (Bulgarian names usually end with "ov" or "ev"), by openly displayed identity, and even by their home language — the Ladino derived from their Spanish origin during Inquisition times — the paternal Djerassis were overtly "Jewish" as were virtually all Bulgarian Jews. Open admission rather than devious denial had reigned there for centuries, probably a reflection of the general absence of anti-Semitism in that country.

There is one aspect of the name "Djerassi" — actually a side issue — that I have never openly discussed, but in retrospect also says something about my late-in-life concern with "Jewishness." As is well known, many of the early Zionists, especially those from Eastern Europe, assumed Hebrew names that in one way or other also made religious or even political statements. Typical examples are Israel's first Prime Minister, David Ben-Gurion, born in Poland, whose original name was David Grün or the Israeli author Amos Oz, who changed his family name Klausner to Oz (meaning "strength"). Indeed, one of my own cousins was enticed at one stage to undertake a "Djerassi" to "Dvir" ("Holy of Holies") transformation which I will shortly detail.

In 1948, Bulgaria's Jewish population (which, in spite of the country's official participation on the Nazi Axis side, had survived World War II in its entirety) was permitted to immigrate to Israel. Over 80% of the 50,000 person Jewish community took advantage of that window of opportunity to leave for Israel, including nearly all members of my father's family. While my father (after first serving as a medical officer during the Arab–Israeli War of 1948) took this step only to eventually join me in the USA, virtually all other Djerassis remained in Israel. One of them, my first cousin Eliyahu (the Hebrew form of Elijah, meaning the Lord is my God), whom we always called Liko, became police chief of Kiryat Shmona in Northern Galilee, at which time he changed his name at the instigation of Prime Minister David Ben-Gurion, who was presiding over the graduation ceremony, awarding the commanding officer's ranks to each of the cadets. When the Prime Minister read the name Eliyahu Djerassi, he felt that Djerassi was not Hebrew enough for an officer of the Israeli state, and suggested the name Dvir, which combined with Eliyahu seemed contextually quite logical.

I had last seen Liko in 1955 in Kiryat Shmona during my first visit to Israel for the dedication of the chemistry building of the Technion in Haifa, but then completely lost touch with him, although his two sons, Yaakov and Ilan have remained good friends to this day. Liko was killed on February 22, 1961 by Tzvi Doctorman, a deranged Auschwitz survivor, who then committed suicide. I had forgotten our last meeting until the mid-1990s when I was working on the third novel, *Menachem's Seed*, of my "science-in-fiction" tetralogy, which in part dealt with Middle Eastern political problems of the 1970s and 1980s, including the bombing of the Iraqi Ozirak nuclear reactor. In memory of my late cousin, I named the protagonist of this third novel "Dvir," even staging his reappearance in my first play, *An Immaculate Misconception.*

But back to Vienna and the high point of my acting as a Jew without quotation marks. The occupant of the other apartment on the floor of my maternal Viennese home, Herr Hassan, an elder in Vienna's only Sephardic synagogue (the Türkische Tempel), had sensed the above-mentioned ecumenical prayer evidence of assimilation, so typical of the Austrian Ashkenazi Jews, and decided to put a stop to it. As I approached my Jewish manhood, my thirteenth birthday, Herr Hassan proposed to my mother that I have a real bar mitzvah; all arrangements could safely be left in his hands. My mother, who did not catch the significance of real, did not say no, and that was how I ended up with a super bar mitzvah: for one day, I became the central character of the entire service in the "Turkish" synagogue. (Even the term *Turkish* — this generic oversimplification for Sephardic — was an indication of Ashkenazi disdain.) After I had spent weeks learning by heart (since I did not read Hebrew) the Hebrew prayers and melodies for a particular day of the Jewish calendar, a telegram arrived from my father in Sofia, informing my mother that my rite of passage had to be postponed by one week because he had a new syphilitic patient. My father's specialty was venereal diseases, and in those pre-penicillin days, a new and affluent syphilitic guaranteed a significant contribution to a doctor's income for at least three years. As a result, I had to take crash lessons to prepare myself for an entirely new service for another day, the rabbi singing to me and I singing back until everything had sunk in, including the look heavenward at every mention of God. When the lessons were finally over, the rabbi asked what I would wear. I had never worn long trousers; *lederhosen* was the usual garb, and knickerbockers the longest pants in my possession. As the most religious day of my life arrived, I effectively entered manhood, I remember, in new knickerbockers and a man's hat. A few years later, the Nazis burned the "Turkish" temple to the ground, and there have not been enough Sephardim in Vienna since to build a new one. As far as I recall, my next visit to a Synagogue occurred sixty years later for literary research rather than religious reasons. Yet as I will show, it was my move to literature that raised my "Jewish" conscience.

When the Nazis took over in the spring of 1938, I, like all my Jewish classmates, was labeled a "Jew;" and it was as a "Jew" that I was driven from Vienna. Religious observance played no role whatsoever since even prior conversion to Catholicism or Protestantism did not count. In this instance, even the Spanish Inquisition five hundred years earlier was more flexible. The label "Jewish" (actually *mosaisch*) followed me throughout my childhood: my birth certificate was issued by the "Türkisch Israelitische Gemeinde zu Wien," and every school certificate of mine identified me as "*mosaisch*." In some respects, some variant of labeling has continued to this day. Even in 2012, when my name was used for the 500,000 Euro segment of the German blockbuster TV show "Wer wird Millionär?" the moderator referred to me as "*Carl Djerassi, mittlerweile US-amerikanischer Staatsbürger österreichisch-jüdischer Herkunft*." [Carl Djerassi, meanwhile, an American citizen of Austrian–Jewish descent.] If I had been a Protestant, would he have added such a denominational appendage? The answer is clearly no, although the reason for marking me as of "*jüdischer Herkunft*" should not be ascribed to anti-Semitic motives but to a tagging process that has a great deal to do with the quotation marks I employ for the title of this chapter.

In 1938, identification as "Jew," of course was pernicious and ultimately often fatal. It started with the yellow star, which I did not wear. Instead, my lapel button bore the horizontal white, green and red stripes of the Bulgarian flag, indicating that I was a foreigner. This was the label under which I left Vienna in July 1938 and officially also the one under which I arrived in America shortly after the outbreak of World War II.

Jew in America

The first decade of my life in America was a period of seemingly total assimilation and of nearly total immersion into chemistry. But I was not just a chemist; I was a "Jewish" chemist. At that time, I would have made that pronouncement with much hesitation, because advertising (as distinguished from admitting) my Jewish origin was the last thing I was prepared to do. In the 1940s, many chemistry departments of major American universities had not a single Jewish faculty member, a fact I was ready to ascribe to active discrimination. Even in 1960 when I moved to Stanford University, I was the first Jewish professor of chemistry, although this has changed dramatically since then. This should not be attributed to intentional discrimination by the university, since there were Jewish faculty members in many other departments, notably Physics and the Medical School, where all three Stanford Nobel laureates were Jewish. To this day, I know several Jewish colleagues of my generation — faculty members in some of our most distinguished universities — who refuse to broach the subject of their Jewish origin.

For years on end, I lived with the self-imposed, but rarely admitted, burden of wondering how to dodge the question "Are you Jewish?" without lying; of anticipating that question with typical Jewish paranoia, and attempting to change the direction of the conversation, when the other person may never have even thought about the topic; and yet on numerous occasions displaying from my side the same inquisitiveness, "Is *he* Jewish?" I never seemed to ask whether *she* was Jewish, because my psyche recognized only male anti-Semites. Clearly there are plenty of female anti-Semites. Did I ignore that reality because I never refrained from entering into a relationship with gentile women? In fact, all my three wives were gentiles. While I have not stopped wondering about other people's Jewishness, I find that nowadays, among professional peers, I practically flaunt my own. Is it because I have finally turned impervious to overt anti-Semitism, or because of the recognition that I would not have become a chemist if I hadn't been born a Jew in Vienna? I didn't have any childhood chemistry sets; I never blew up our basement; prior to my sixteenth birthday, I never had any chemical "hero," not even Madame Curie. If I hadn't been born a Jew, I wouldn't have left Vienna and would doubtless have ended up as an Austrian physician — possibly even one voting for Kurt Waldheim — since both my parents were doctors. My acute awareness in those years of possible American anti-Semitism involved me in a curious insecurity, even though the Americans who eased my path in those early years couldn't have been more benevolent — a psychologically significant insecurity as shown in the following excerpt from my earlier, now out-of-print autobiography.

Although half a century has passed, I still remember every detail: the big ears projecting straight out like a wary deer's; the tooth missing just above the thick lower lip, its gross thickness accentuated by the virtual absence of its upper partner; the eyes, big yet hooded; the tousled black hair; the grin moronic but also devious; and finally the nose — after the ears, the boy's most prominent feature.

His image occupied the center of a filthy poster plastered on the walls in our neighborhood in Vienna, just after the Nazis had taken over in 1938. The head was attached to a gangly neck, protruding from an absurdly adult suit, its black vest buttoned almost to the sternum, leaving visible only the knot of a black tie. In a remarkably succinct way, his attire managed to stigmatize the boy as a sly street peddler. The poster's brutal message consisted of just three words: Tod den Juden — *Death to the Jews!*

I encountered the face the second time in a newsvendor's stall in the Midwest during the early 1940s. Still extraordinarily sensitive to every real or imaginary anti-Semitic innuendo, I did not touch the picture. I knew exactly what it stood for. And in my state of shock, I overlooked the fact that this grinning boy's nose was somewhere between triangular and bulbous, rather than sharply Semitic.

At the time, I did not tell anyone what I had seen, just as I hardly disclosed anything about my past life. It was my way of attempting to "pass," which even without my accent would not have been easy in this small Midwestern college town where I was the only Hitler refugee; many of the locals had never even met a Jew.

"Where're you from?" they'd ask as soon as I'd finished a sentence or two.

"My mother lives in upstate New York," I'd reply, sometimes mentioning the hamlet near the Canadian border where she worked.

"Yes, but where're you from?" they'd insist. "What kind of an accent is that?"

"Bulgaria," I'd lie, hoping the remoteness of that country would deflect the inquisition.

Of course, some of my inquisitors were more persistent. (Was it my paternal Sephardic background that invariably made me attribute to innocent Midwestern curiosity fifteenth-century Spanish inquisitorial motives?) "Why didn't you stay in Bulgaria?" ("Idiot," I wanted to retort but didn't, because that would have taken explanations incompatible with "passing.") "Were you born there?" Once I owned up to having been born in Vienna, the questions tended to become more precise and, worst of all, more intrusive. Still, I equivocated. Only when asked point-blank, "Are you Jewish?" did I acknowledge the fact, and then I promptly changed the subject.

Some years later in the 1950s — probably in Michigan where I taught and where rabid anti-Semites like Gerald L. K. Smith and Father Coughlin operated — I again came upon that face: on the cover of a publication with the implausible title of MAD. *But having by this time become an American citizen, I felt more secure. I picked up the magazine and flipped it open. I was stunned to find it filled with comics — an American infatuation to which, along with football and peanut butter, I had never succumbed.*

I was too preoccupied with other matters, and also too impatient, to delve into the contents of MAD, *but I did make some discreet inquiries about the nature of its cover picture. To my surprise, virtually every person I asked knew the identity of that boy: Alfred E. Neuman.*

"Where does he come from?" It was my turn to ask that pointed question, only to be told that nobody knew or even cared. He had just existed as long as my informants could remember.

"N E W M A N?" I spelled the name.

"No," I was corrected. "N E U M A N."

"Aha!" I cried out triumphantly. "I knew it! It isn't 'Nooman.' It's 'Noyman.' German, of course."

Decades passed, and the boy's face receded again from my conscious memory. Then one day, on a visit to Yad Vashem, the Holocaust Memorial in Jerusalem, as I stared at some of the enlarged photographs from that most despicable and horrible period in European history, Alfred E. Neuman's face seemed to surface here and there. I decided that the time had come to uncover the origin of the face whose memory had never quite left me.

As soon as I returned to California, I went to a local newsagent. "Do you carry MAD*?" I inquired, not even knowing whether the magazine still existed. "Over there," the man pointed. I turned and saw a grinning Alfred E. Neuman dressed in a snow rabbit's outfit stepping out of a chimney, holiday cheer practically oozing from the January 1988 cover of the latest* MAD. *I handed over $1.35 and walked to a corner of the store, where for the first time in my life, I read a comic book from cover to cover. In spite of my ingrained suspicion, it became clear to me that no Nazi had ever had his hands on that issue. In fact, it was* *not even obvious to me why kids would read it: the political cartoon on the last page featuring Gary Hart and Ronald Reagan was clever and biting. I would not have been surprised to find it on the cover of a leftist magazine like Mother Jones.*

I was puzzled: how could I reconcile my memory of that taunting face of over forty years before with this benign comic? My first American vision of Alfred E. Neuman's face had been around 1942, give or take a few months. Yet on telephoning the editorial office of MAD *to inquire when the first issue had appeared and how I could secure a copy, I received a preposterous reply: number 1 of* MAD *had hit the newsstands only in October 1952. Even more absurd was the claim that Alfred E. Neuman — face as well as name — had not graced the cover of* MAD *until 1956. Had the Nazis sold the original magazine to some innocent purchaser with the proviso that the origin of the publication be disguised? Everyone knows of notorious examples of the falsification of historical facts. If* MAD *was just another such victim, it was time for me to correct the record — if not for the public's sake, then at least for mine. Two weeks later, I flew to New York and headed for 485 Madison Avenue, then the perch of* MAD.

The bemused tolerance with which the small editorial staff received me was reflected in the genial disarray of their offices, in which, after a short search, they located the bound volumes of the magazine starting with the first issue. Its cover featured a terrified family, the man yelping, "That thing! That slithering blob coming toward us!"; the woman screaming, "What is it?"; and the small child at their feet exclaiming, "It's Melvin!" Melvin Coznowski, I was told, was Alfred E. Neuman's predecessor.

The face I'd remembered — the face that had remained with me for decades and brought me to MAD*'s New York office — first surfaced in* MAD *in November 1955. It appeared above the masthead in number 26 (surrounded by Socrates, Napoleon, Freud, and Marilyn Monroe), but so small that it occupied less than half the space of the central letter A in the title. The next issue, number 27 of April 1956, had a somewhat larger boy crouching at General Eisenhower's feet amid a bewildering crowd of at least sixty characters ranging from Thomas E. Dewey, Adlai Stevenson, and Richard Nixon to Churchill, King Farouk, and Khrushchev. Not until the December 1956 issue did the likeness of Alfred E. Neuman — the famous Norman Mingo portrait apparently familiar to all Americans but me — fill the cover in lonely splendor. He was featured as a write-in candidate for president under the slogan "What — Me Worry?"*

I was totally perplexed by the discrepancy between these facts and my memory until I read an early Letters to the Editor section, where an amusing collection of feisty and succinct missives presented no fewer than eleven different images of Alfred alias who-knows-who, sent in by readers claiming to have known the Ur-Alfred. In three pictures, the hair was actually slicked down; he could have been a neighborhood school kid. In the three craziest ones, he wore various kinds of hat. The other five approached my image from Nazi days.

These letters and many other fascinating exhibits were in a huge binder containing background material from a copyright suit that had been filed against MAD *in the 1950s. I found myself rooting for* MAD *— my belated and, by now, favorite introduction to American comics. I was relieved to find that the magazine had won by demonstrating an abundance of prior art with that face and with legends such as "Me worry?" or "Da-a-h ... Me worry?" There were references to a publication of that face by Gertrude Breton Park of Los Angeles around 1914; to a 1936 advertisement from Brotman Dental Lab in Winnipeg; to a somewhat corny book, Hall of Fame, published in 1943 in Toronto by one J. J. Carrick. There was no question that, at least in terms of chronology, that face existed when I was a teenager in the Midwest.*

I had almost forgotten my role as a Nazi hunter, but then I got warmer. Not hot, not quite there, but warm enough: a postcard with the Nazi version of the face, except for the hooked nose, and the legend "Sure — I'm for Roosevelt." The reverse side read: "If you are opposed to the Third Term send these to your friends. 15 cards for 25c. Send coin or stamps. Low, quantity prices on request. Send to Bob Howdale, Box 625, Oak Park, Ill."

I suppose I could have flown to Chicago, searched the old phone books, and tracked down Bob Howdale. Maybe he was one of Father Coughlin's followers. But I had lost my taste for chasing down the real Alfred E. Neuman. I was certain that neither MAD *nor Bob Howdale could make me forget the specters of my youth. As to my own memory of Alfred's face, a line in a poem by Bruce Bawer says it all: "The past cannot move into the present uncorrupted."*

By the time I reached my fifties, Jewish identity was not anymore a subject of concern to me. The paranoia of the young refugee from Nazi Vienna, compounded by the horrors of the holocaust as it eventually unfolded in the media, had gradually dissipated, perhaps because of the security and self-assurance associated with professional success and resulting affluence. Nearly fifty years of total irreligiosity, in fact convinced atheistic secularism, caused the subject of religion to barely arise in my home. Interestingly, both my second and third wife were brought up as practicing Christians, Norma as a Protestant and Diane as a Catholic. Intellectually and educationally, they were well informed in matters biblical and knew much more about Jewish religion than I ever remembered from my obligatory Viennese religion classes. Yet both, while still in college, had turned secular and this reflected itself in the total absence of any religious raising of our children. I do not mention this as some form of braggadocio or even conviction about the benefits of such secular upbringing, but rather as a factual description that makes what follows even more surprising.

Jewish Identification through Fiction

As I describe in more detail in the "Writer" chapter, I started to turn to fiction writing in my sixties. My autodidactic *modus operandi* was that of an intuitive novelist, who refrains from first outlining the story line and delineating most of the characters. Instead, as I become intrigued by an event or a person, I start free-associating and let the characters lead me. Or at least that is what I pretended, when, in fact, I simply allowed my unconscious to run the show. No wonder that I keep referring to this process as auto-psychoanalysis. Yet because autobiography preceded my fiction writing, I soon realized that many sensitive topics I wanted to cover in the actual autobiography — notably those dealing with women — could not be disclosed, primarily for reasons of privacy or even shame. Hence, first unconsciously and soon thereafter deliberately, I chose to confront such non-divulged issues under the protection of the anonymity provided by fiction — a form of ongoing public confession without the public's knowledge. I have made this point before and I shall make it again, since it has played such a large role in my personal development during the past two decades. A relevant example is the surprising fact — initially entirely unplanned — that every main male character in my five novels turned out to be Jewish as well as of direct or indirect European origin, and never religious, whereas all but one of the female principals were non-Jews.

How much closer could I hew to my real life? So who were these Jews in my novels and what did they do there? In the first three novels, they were simply described as Jews, a fact that added nothing to the overall structure of those books. Yet without realizing it, it showed the path I would be pursuing in the subsequent years.

Ostensibly, the main theme of my first novel, *Cantor's Dilemma,* was the question of trust, without which the scientific enterprise could not function and the resulting plot revolved around that issue through a tale of academic intrigue. Yet while trust in science is crucial, it is unrelated to religious, ethnic or geographical factors. But since Professor Isidore Cantor also drives the plot, let us examine him, starting with his name, which was obviously Jewish though not quite as Jewish as I initially wanted it to be. For a starter, he never liked his forename, Isidore, insisting that people refer to him as Professor I. Cantor or more intimately as I.C. — an abbreviation that phonetically could sound as "I see" or "icy" depending upon the circumstances. Did he want to disguise his Jewish forename? If so, was this an allusion to the fact that so many assimilated Jews have done so? And what about his family name, Cantor? This last question is worth exploring, because my unconscious clearly knew into what swamp I was stepping.

It all started with a short story, initially called *Cohen's Dilemma*, my first entry in what has turned to out be the genre for most of my novels: "science-in-fiction" — not to be confused with science fiction — in which I describe the tribal culture of contemporary science. In *Cohen's Dilemma* I wrote about a scientist (obviously Jewish, considering his name) awarded the Nobel Prize for work that turns out to be flawed — a story that pleases me to this day because of its plausible premise and its ambiguous ending. It shows a scientific superstar falling in love with a spectacular hypothesis (how many of us have not yielded at one time to such dangerous temptation?) and then convincing his favorite collaborator that the experiment, designed to confirm the validity of his hypothesis, had to work. But when the lab's Prince of Wales delivered the hoped-for experimental goods and the work was published, a key competitor could not reproduce them. Was the hypothesis false? Was the collaborator's work in error? Or was it...?

Cohen's Dilemma met the usual fate of a first submission by a fledgling author: a printed rejection slip from *The New Yorker*. But the next attempt succeeded: Fred Morgan of the *Hudson Review* accepted what became my first published short story. Morgan requested that I change the name *Cohen* on the advice of an editorial consultant of Nobel Prize stature (I subsequently learned that it was Joshua Lederberg, at that time President of Rockefeller University), who felt that *Cohen* was too common a name for a fictional Nobelist. After all, two Cohens had already won Nobel Prizes and a third one was at that time a perennial candidate. Rather than arguing, I proposed *Cantor*, since for reasons obvious to any reader of my story, the name had to start with a C; and since *Cohen* was a Jew, I felt that his replacement's name might as well have a Jewish connotation. But the editorial consultant was too well informed. "What about Charlie Cantor?" he inquired, referring to the then-chairman of the genetics department of Columbia's College of Physicians and Surgeons. I was so anxious to see my first short story in print, that instead of arguing that Columbia's *Cantor* surely carried no trademark on that name, I offered an "s" for the "n." *Castor's Dilemma* was the title of the published short story and proved to be the vehicle that eventually converted me into a novelist.

A year later, I embarked on an expansion of *Castor's Dilemma* into a full-length novel. Within weeks of its appearance in the *Hudson Review*, I had started to receive letters about *Castor*: "Had the postdoc cheated?" "Was Professor Castor thinking of suicide?" And "Do you really find it necessary to wash dirty lab coats in public?" This last complaint is one I continue to hear from time to time about my "science-in-fiction," and my response is always the same: people working in white coats are bound to get dirty — a fact of life that white-coated scientists cannot afford to ignore. But as long I was going to be doing such laundry in public, it behooved me to be careful myself, so before submitting the manuscript to a publisher, I headed for the Royal Society Library in London to confirm the non-existence of any tumor cell biologist named *Castor*. To my chagrin, I discovered that such a biologist did indeed exist in Pennsylvania, whereupon I decided without further dilly-dallying to revert to *Cantor*, the name I had originally chosen to replace *Cohen*. *Cantor's Dilemma* was the title under which Doubleday published the hardcover edition, and while the novel was reviewed very widely, both in the lay and scientific press, neither Charles Cantor of Columbia University nor any other Cantor ever complained to me. (A couple of years later, Charles Cantor and I were successive plenary lecturers at a scientific congress. The organizers decided to announce my talk as "Cantor's Dilemma" and his as "Djerassi's Dilemma." The audience enjoyed it and so did we.) Some readers of the novel assumed that the initials C. D. of the title were a masked reference to my own name, but strangely enough this obvious connection had escaped my attention for the simple reason that to me, as chemist, those initials always stood for "circular dichroism," a field of esoteric research in which I had been active for nearly two decades.

Initially, the Princeton professor Max Weiss, the chief male character of my second novel, *The Bourbaki Gambit*, just happened to be Jewish without any apparent psychological or historical relevance to the overall plot of the novel, similar to the point I just made with *Cantor's Dilemma*. Yet to be sure that his Jewish identity was unambiguous and not just based on a Jewish-sounding name, I specifically had him describe himself as of "Hungarian Jewish ancestry." In addition, in the same chapter I had him walk through the streets of Vienna, where he eventually "rang the bell at Aspernbrückengasse No. 5, a turn-of-the-century apartment building at the foot of the Aspernbrücke — the bridge crossing the Danube Canal at the juncture of the fashionable First District and the Second District, in pre-Anschluss days the predominantly Jewish quarter of Vienna." By now, at least some readers

UNIQA A.G.

Aspernbrückengasse 5, the house (right corner) where I was brought up in Vienna, 1935.

will ask why I had felt it necessary to insert such seemingly irrelevant bits of esoterica into the novel. Simple: Because the address of my childhood home in Vienna was Aspernbrückengasse No. 5.

The lead personage in my next novel, *Marx, Deceased* (entitled *Ego* in the most recent German version), a famous writer with the Jewish-sounding name Stephen Marx, is identified only in passing as a Jew ("If an Irish ex-Catholic [referring to James Joyce], who didn't even ski, could flourish in Zurich, why not a secular Jew from New York?"), although I based him on a fictitious composite of Norman Mailer and Philip Roth, two of the most famous American Jewish authors, who were notoriously sensitive to negative book reviews as was my Stephen Marx. Where matters become much more personal is that Stephen Marx is not only, *mirabile dictu*, the author of a novel entitled *Cohen's Dilemma,* but also writes about a man, named Nicholas Kahnweiler, who in turn is a very thinly disguised Carl Djerassi. Thus, my own little ego appears here in two characters, the writer (Marx) as well as his object (Kahnweiler). I shall illustrate the likeness of Kahnweiler with myself through two quoted passages from that novel — one short and the other longer. In each instance, I picked the name Kahnweiler as a fictional counterpart to my own name. But what made me pick the name of Picasso's art dealer rather than one that phonetically or historically would have some relation to Djerassi? To this day, I cannot provide a truly plausible answer, other than to ascribe it to a desire to illustrate the life-long spelling or explanatory circuitry I always indulged in with my surname.

In the first excerpt, Kahnweiler, responds as follows to questions by a new acquaintance;

> *"I know practically nothing about you other than that you are a scientist. Why don't you tell me something about yourself?"*
>
> *"All right," replied Nicholas. "I'm a scientist, divorced, one child; but what is occupying me right now is a novel."*
>
> *"What's it called? By whom?" asked Gerald Bogen.*
>
> *"I haven't yet picked a title."*
>
> *"You mean you're writing a novel?"*
>
> *Nicholas wasn't sure whether admiration or amazement was behind Gerald's question. He was hoping for both.*
>
> *"It's my first attempt."*

Gerald had turned completely towards Nicholas, who could almost feel the searching, speculative — and, yes, admiring — look. "What made you do it?"

Nicholas decided this was the time to be honest. Not completely open, but honest. He would tell him why he wanted to publish it, not why he had written it.

The question, "what made you do it?", posed by Gerald, is one that I have been asked numerous times, and the answer will appear in a later chapter, *"Writer"*. But what does that have to do with Jewish identity, as any justifiably impatient reader will ask? For that, one only needs to turn to the second excerpt, which is so autobiographical that the slight changes (e.g. Germany for Austria, Romania for Bulgaria) seem almost unnecessary:

"It's the standard question I get every time I introduce myself: 'What kind of name is this?' 'Where are you from?' and frequently, 'What kind of accent do you have?' Depending on my mood, I give different answers. Usually I just reply, 'It's of German origin but I don't really know what it means.'

"The real story is more complicated, and you may be interested in it. My family is Jewish and comes from Germany. You'll find immediately that for German Jews 'Kahn' is simply a germanicized — a laundered — version of Cohen. When the emancipation of the Jews began in Germany in the eighteenth century, they were ordered to assume last names, preferably German ones. Remember, Jews traditionally had no last names. The rich ones picked allusions to wealth — gold, silver — and others names of flowers, like roses. No wonder you now find many names like Goldberg, Silberstein, Rosenkranz. But the Cohens, the priestly class, were more subtle and found names that retained a phonetic connection with Hebrew but would still 'pass.' As I mentioned earlier, the ending Weiler *is fairly common in German; hence 'Kahnweiler' sounded plausible, in that it might be derived from some hypothetical place like 'Kahnweil' and did not even sound Jewish. A more honest translation ought to be something like 'Cohenburg.'*

"To the second question, 'Where are you from?' I frequently say 'I live in San Francisco,' even though I know perfectly well that this is not what they mean. If the questioner is persistent, he'll say, 'You don't have an American accent, where were you born?' and I'll reply that where I was born won't tell him anything: I was born in Bucharest, but that doesn't make me, or my name, or my accent, Romanian. I was two months old when my parents moved back to Munich, where they were originally from.

> *"The most common question is, 'How do you spell your name?' The very first time I was asked this, shortly after my arrival in the United States, I, the sixteen-year old Nicholas, got flustered. When I tried to spell out my name the way I had heard others in America spell theirs with reference to geographical or personal names, all the words starting with a 'K' in my native tongue that came quickly to mind were 'Kalifornien' and 'Kairo,' which in English start with a 'C.' Kalamazoo was as yet unknown to me — I, the Central-European urban kid, would probably not have believed that a place could be called Kalamazoo — and neither Kansas nor Kentucky was as yet part of my spelling subconscious. Offhand, I couldn't think of a single English word starting with 'K,' so in desperation I came up with 'Kitsch,' and I have used this ever since. I got again stuck with 'W,' which of course I pronounced as a 'V.' The damn German confusion between 'V' and 'W' got in my way, until I found an all-American word, 'Washington,' which even though I pronounced it as 'Vashington,' everyone knew what I meant. Kitsch and Washington stuck with me for the next 40 years. A psychologist would probably have a field day with the combination Kahnweiler–Kitsch–Washington."*

Spelling my name — notably the "DJ" beginning, which often has resulted in misspellings like D'Jerassi or even O'Jerassi — has been a burden, which I had to tolerate since my arrival in America, but hardly ever had to face in Austria or Bulgaria, for the simple reason that in German Dj automatically generates the same pronouncement as the word Djibouti, for instance, and in Bulgarian it never arises since it is spelled with the appropriate Cyrillic letter Ж leading to the same sound as the beginning of Djibouti. Most of the time, I did tolerate it, at times out of pride, at times out of impatience, but in the early years also out of dread that it would automatically lead to the next question, "where are you from?" or "what kind of a name is it?"

In point of fact, I can't give a reasonable answer. My father hypothesized that it was of Spanish origin (after all, the Sephardic Bulgarian Jews could all trace their origin to the expulsion from Spain during the fifteenth century Inquisition) as a compression of De Jerez — "coming from Jerez". The related Jerassi, Gerassi, Tscherassi, Çerassi, and probably even Çeraci variations are simply different spellings of the same Sephardic surname that can be found nowadays in Israel and earlier also in Turkey. The Sicilian Geraci, even found in the name of the Sicilian city Geraci Siculo, is almost certainly a homophone without Sephardic connection. At least I hope that this is true, given that Antonio Geraci, one of the major Sicilian Mafiosi of the late twentieth century, was the notorious capo in Partinico. Questioners, of course, always expected a one-word answer, which, as I just showed through one lengthy paragraph, I could never produce. An answer such as "it is of Spanish origin, but comes from Bulgaria", would always lead to complicated explanations and further questions, none of which I wanted to pursue. Quite recently, in a deeply personal book of poems (*A Diary of Pique*) I did include a long poem which started as follows:

"How Do You Spell Your Name?"

D *as in David.*
J *as in Joseph.*

"Stop.
All I want is your last name."

"That's what you're getting!
Let's start all over again."

D *as in David.*
J *as in Joseph.*
E *as in Elizabeth.*
R *as in Robert.*
A *as in Alice.*
S *as in saccharin.*
S *as in saccharin.*
I *as in Ida.*

A litany repeated thousands of times
Since arriving, 1939, on these shores.
The text always the same;
But why these names?

David, Joseph, Robert? I hardly know them.
Elizabeth, Ida? Unknown.
Alice? My mother.
Why *saccharin*? Is this the key?
Sweet, but synthetic.
(Why not? I'm a chemist.)

I shall skip the remainder of the long poem, because the entire question about my name is really only an entry to what now follows: a risky path, replete with lengthy quotations from my subsequent novels, because in retrospect, it is now clear to me that in my first three novels, I was just behaving like a moth flittering around the lighted candle. But in the next two novels I headed straight toward the fire of Jewish identity and practices, because in *Menachem's Seed* and especially in *NO*, Judaism plays an indispensable role in the overall story. And why do I use the word *risky* in describing what is about to follow? Because only as I pursued the Jewish thread in my novels — the one place where I indulged in open, though not deliberate, conversation with my subconscious — did I choose topics which truly related to my circuitous examination of my Jewish identity. It would be pointless (because it would be incorrect) not to search in these novels for my inner thoughts, but how can I do so effectively without quoting extensively from them, something that many readers will simply consider ostentatious peddling of my novels? They would be wrong, because what follows is simply a stream of consciousness supported by an avalanche of self-quotations, not unlike the excess of *Schlag* covering so many otherwise delicious Viennese desserts. But just as such excess *Schlag* can readily (though sometimes regretfully) be removed in one swoop with a spoon, so the excerpts cited below can simply be skipped by the reader. But just as such Viennese desserts without *Schlag* aren't anymore the true Viennese culinary creations, so the absence of my self-quotations will not quite correspond to the Jew I wish to describe.

Overt Jewish Themes in Fiction

With this plethora of warnings and guilt feelings expressed openly, I shall now embark on that risky path. Thus, in *Menachem's Seed*, I wanted to give a verifictional account of the Pugwash Movement in which I participated for nearly twenty years — an enterprise that was nearly unknown to the general public or even most scientists until it got a sudden flash of public recognition through the joint award with one of its founders, Joseph Rotblat, of the Peace Nobel Prize in 1995. Pugwash was founded in 1957 during the coldest period of the Cold War and offered a forum where scientists from opposing countries could discuss solutions to world problems, initially nuclear disarmament, but later also many other conflicts that diplomats at that time could or would not touch.

This sounds harmless as well as irrelevant to supposedly autobiographical musings, until it is noted that for my novel, I chose as an example of the areas Pugwash, though not standard diplomacy, could explore one of the first meetings between Israelis and the PLO in Munich in 1977 — an encounter in which I participated as an observer and where the key Israeli representative was Shalheveth Freier to whom I dedicated my novel. His identity was explained in the novel's afterword:

> *Shalheveth Freier, to whom this book is dedicated read early versions of several chapters, but he died unexpectedly on 27 November 1994 before my novel was completed. Among numerous positions in Israel, he served at different times as Director General of the Israeli Atomic Energy Commission, as Director of the Scientific Department of the Ministry of Defense, and at the time of his death as vice president of the Weizmann Institute of Science and advisor to the Government of Israel on Atomic Policy. During World War II, he fought in the Jewish Brigade of the British Army and was responsible for smuggling thousands of Jewish refugees into Palestine. In most of these respects, he bears a strong resemblance to my fictional hero, Menachem Dvir. But unlike Dvir, who was born in the former Belgian Congo, Shalheveth Freier was born in Eschwege, Germany and left his native country during the rise of Hitler.*
>
> *I have known Shalheveth Freier for approximately twenty years. Many of our meetings took place at Pugwash Conferences on Science and World Affairs, which are the inspiration of my fictional Kirchberg Conferences. For over one year, I had numerous interviews with him on the subject of this novel, both in Israel and London. I was also fortunate in being the beneficiary of his legendary abilities as a correspondent. I would like to end this tribute to an extraordinary man by quoting an excerpt from one of his letters.*
>
>> *[12 May 1994] If your novel extends into 1981, let me know. I was the only intermediary between the Israeli and Soviet governments from 1970 to 1985, and the aftermath of the bombing of the reactor was really quite dramatic with certain understandings reached on the extent to which Israel should, and would allow itself, to be punished. My Soviet partner was then Mr. Primakov, at the time Director of the Institute for Asian Research*

of the Soviet Academy and now Director of Russian Foreign Intelligence. [Note: On 9 January 1996, Yevgeni Primakov was appointed Foreign Minister of Russia in the Yeltsin cabinet.]

What I did not mention in this afterword or, in fact, anywhere else is that Freier and I may well have crossed paths unknowingly as infants in, of all places, Sofia, Bulgaria, where his father, an Orthodox rabbi from Berlin, attempted to establish a functioning Orthodox community. He failed and returned shortly thereafter to Berlin, but Freier and I once joked that it was not unlikely that the baby carriages holding the Vienna-born Carl and the Eschwege-born Shalheveth may have been pushed in 1924 through the same park in Sofia.

Weizmann Institute Archive

Shalheveth Freier, 1990

Why did I pick the name *Menachem Dvir* as the pseudonym for Shalheveth Freier? *Shalheveth* is not only a rare name, but also a uniquely Jewish one and I wanted to select a name with similar distinction. *Menachem* was clearly a suitable candidate, but what made me choose it? Its origin in my personal memory bank started at the first Wolf Prize ceremony in Israel in 1978, where I received the prize in chemistry — an event that was boycotted by much of the Israeli scientific establishment. I am about to embark on a shaggy dog story by again quoting from a novel of mine, in this instance from *NO*, which is also sited largely in Israel, and where I made a cameo appearance in typical Alfred Hitchcock fashion in a telephone conversation between Renu Krishnan and her American advisor, Professor Frankenthaler from Brandeis University. Since it was Prime Minister Menachem Begin who presented me with the Wolf Prize instead of the expected President Katzir of Israel, I chose his first name for my novel:

Renu gave a précis of the letter she had mailed the day before, adding a brief description of the Wolf Prize ceremony and of the disapproving response of the Israeli scientific community.

"Here you have the only country with a biochemist as president, yet Ephraim Katzir managed not to appear." Renu knew quite well that this was the kind of scuttlebutt the Prof. throve on. "The Prime Minister, Menachem Begin, had to take his place. I was surprised at the intensity of the objection, but one of my younger colleagues at Hadassah explained it for me. She pointed out that the Wolf Foundation made one mistake. They should have had at least one Israeli among the first group of winners. She predicted that as soon as an Israeli scientist gets a Wolf Prize, and they think they're in line for $100,000, the locals will show up. But you should have seen — "

"You said 100K?" Frankenthaler interrupted. "The Wolf Prize? Never heard of it." Ordinarily, Frankenthaler would not have admitted to such ignorance. But the six-digit dollar figure his favorite postdoc had just dropped got the better of him. "What are those categories again? Who does the nominating? Who won?"

"Two men from the Midwest — I think Wisconsin and Illinois — in agriculture. And two in math — one from Russia, the other from Germany. I'm afraid I don't remember their names. And there were three in medicine who got it for their work on the histocompatibility antigens."

"Snell from Bar Harbor?" Frankenthaler interrupted.

"Yes. For his work in mice. Dausset from Paris and Van Rood from the Netherlands shared it for their work in humans." [N.B. Two years later, Dausset and Snell received the Nobel Prize].

"Who else?"

"Ah," exclaimed Renu. "For me, the most exciting was in physics: it went to Chien-shiung Wu from Columbia. I always heard that she should have shared the Nobel Prize with Lee and Yang for her contributions to the discovery of non-conservation of parity. It was nice to see a woman win it. Alone," she emphasized.

"Is that all? I'm surprised they don't have a prize in chemistry."

"I nearly forgot," exclaimed Renu. "Chemistry had a single winner with a crazy spelling... Djerassi, that's who got it."

"You mean Isaac Djerassi for his work on methotrexate treatment in leukemia? I met him last year in Pennsylvania. The Israelis ought to have been pleased by that choice — he's a Bulgarian Jew, who was trained at the Hebrew University before moving to the States. But I would have expected him to get it in medicine, not chemistry."

> *"Not Isaac Djerassi. Carl. From Stanford. He got it for the chemical synthesis of the Pill. I don't know why I didn't mention him first. I once went to a lecture of his on the future of birth control — dismal according to him — when I was a graduate student in biochemistry at Stanford. A good speaker, but a bit arrogant, I would say."*

This finally brings me to the family name *Dvir* of my fictitious *Menachem* and thus to Liko Djerassi alias Liko Dvir, whom I already mentioned at the beginning of this chapter. At the Wolf Prize ceremony, my father (as well as my stepmother, Sarina, my daughter Pamela, and my future wife, Diane Middlebrook) accompanied me and on one of the following days he convened a gathering of the large ex-Bulgarian Djerassi clan who were then living near Jaffa. Present were also Liko's parents and the story of my cousin, his death, and his assumption of the name *Dvir* arose. Clearly, those family facts were then implanted in my subconscious to eventually surface some eighteen years later as I was searching for a new Hebrew family name for my novel.

A reader can fairly argue that this extended story about Shalheveth Freier and Menachem Dvir does not really contribute anything to my seemingly increasing Jewish awareness other than to show that I had visited Israel and incorporated various incidents and personalities in the novel. In fact, I had traveled there on numerous occasions since the middle 1950s, but always as an American and not a Jewish scientist being invited to lecture at the Weizmann Institute in Rehovoth, the Technion in Haifa, and the Hebrew University in Jerusalem. But in this instance, the questioner would be wrong, because the key female character in *Menachem's Seed* is an American WASP, a youngish widow named Melanie Laidlaw, who, after an adulterous affair with the married Menachem Dvir at a Kirchberg (read *Pugwash*) Conference, will desperately try to convert to Judaism — a literary invention that required from me an in-depth exploration on how Jewishness is defined by the Israeli rabbis. Here is how the story goes. It is Melanie's feelings of guilt about having deliberately gotten pregnant, in spite of the fact that Menachem was infertile due to a nuclear accident, through the newly discovered ICSI technique and had hidden her action that propelled her to consider converting to Judaism prior to delivery, so that her child would at least be born a Jew, since Jewish religion requires that the mother be Jewish.

Until that moment, I had not really known how one converts to Judaism nor the restrictions and obligations that such a step entails. Was I, a born "Jew," subconsciously wondering for the first time in my life how to remove the quotation marks by a "Jew" to Jew conversion? In any event, I embarked not only on extensive reading on rabbinical law, but interviewed four Reform rabbis — two of them female — in San Francisco, London and Jerusalem. A distillation of these interviews is herewith reproduced in an excerpt from *Menachem's Seed* — not as literary overkill, smelling of book promotion, but because it is only here that I first faced seriously some issues of the naked word *Jew* without the burdensome quotation marks. The following dialog from the novel occurs between Melanie, who had finally worked out how to convert to Judaism through a Reform rabbi, who was willing to condone what the Orthodox or Conservative rabbis would not approve, and her friend, Felix Frankenthaler — a secular Jew who in this scene is clearly my fictitious alter ego. (It should be noted that in my own four rabbinical interviews as background for the novel, three of the rabbis disapproved of such a conversion based on Melanie's motivation. Evidently, it pays to shop around, if one needs to convert to Judaism.)

> *"I'll be damned," said Frankenthaler. "Have you converted?" He sank back in the sofa.*
>
> *"Not yet, but I'm well on the way. Your advice about calling JTS [Jewish Theological Seminary] turned out to be just the ticket. They referred me to a Reform rabbi — a woman — who admitted me to her conversion class."*
>
> *"You just told the rabbi you wanted to turn Jewish so that your son could be born a Jew?"*
>
> *Melanie hesitated. "No," she finally said. "The rabbi would not have agreed. I'd already found that out from a Conservative rabbi at JTS."*
>
> *"So?"*
>
> *"So from the same Conservative rabbi I also learned what questions a potential convert to Reform Judaism would be asked. When I heard what those five were, I concluded I could agree to them in good faith — more or less."*
>
> *"Five? No more, no less?"*
>
> *"Five. First — and this one the rabbi will ask at the actual synagogue ceremony — is whether I do all this out of my own free will. In my case, the answer is obviously yes, since no one is forcing me."*

"What else?"

"Whether I have given up my former religious affiliation. Since I never had a formal one, there's nothing to give up. Also whether I pledge my loyalty to the Jewish people and to Judaism; and then, of course, whether I promise to bring up any children of mine in the Jewish faith. I don't see any problems with any of them — particularly not the last."

"That leaves one more."

Melanie stared in surprise. "Were you counting? I must have skipped one. Anyway," she shrugged, "I told the rabbi I wanted to do this as quickly as possible. She assigned me a tutor, to whom I go three times a week, so we can cover the required material as quickly as possible. I want to be finished next month — just in case Adam decides to appear prematurely."

"I hope I'm not being intrusive," said Frankenthaler, "but I — a born Jew — have no idea what the conversion process is all about. Could you tell me what you have to learn?" He leaned forward in anticipation.

"Sure," Melanie said good-naturedly. "I'd love to show off what I've acquired. First, I have to learn about the holidays: Yom Kippur, Rosh Hashanah, Chanukah, Pesach ... even you must know those. And of course the Sabbath. Then the life-cycle ceremonies — birth, Bar Mitzvah, marriage, death. And Jewish literature, like the Bible, Halakah — the Jewish Law, the Midrash, the Talmud and others." She waved her hand. "I'm just starting on the Responsa *literature, which I find the most intriguing. These are rabbinical answers to Jewish questions that they keep right on issuing, even today. Let me show you. I've got the book on my night table."*

Melanie appeared with a paperback, entitled American Reform Responsa: Jewish Questions, Rabbinic Answers. *"It would be difficult to think of a question that hasn't been posed ... or hasn't been rabbinically answered." She flipped open the table of contents. "'Surgical Transplants'; 'Eulogy for a Suicide'; 'Jewish Attitude Toward Sexual Relations Between Consenting Adults'; or here, 'Artificial Insemination.' Clearly, this subject is not to be found in Talmudic or rabbinical sources. So the interesting aspect of the* Responsa *is how modern rabbis deal with modern topics. Or even old ones." Melanie's finger ran down the page. "'Masturbation.' Here is the text of the question."*

Melanie moved next to Frankenthaler on the sofa, so that they could both read the text. "'What does the tradition say about masturbation?" she read. "'Are any distinctions made between males and females, young or old, married or unmarried?' That's not a bad question, is it?" She closed the book.

"So what was the answer?"

She gave a dismissive shrug. "I'll lend you the book when I'm through."

Frankenthaler raised his palms. "No thanks. I already have too much to read. Just abstract the answer for me."

"Masturbation is neither harmful nor sinful. In men or women. At least that's the Reform view."

"Is that all they're teaching you?"

Melanie glanced at him. Was he mocking or was he interested? "I have to learn all the basic prayers — "

"In Hebrew?"

"Not with this rabbi. Maybe with others. All I need to learn are the Hebrew names and key Hebrew words involved in those prayers: Kaddish, Shema, Amidah … And finally, in addition to the liturgy, I am also getting some exposure to Jewish history, from the time of Abraham and David all the way through the establishment of the modern state of Israel."

"Are you taking an exam?"

"I will in the end. But my rabbi insists on meeting with me privately on occasion. Since I'm in such a hurry with my conversion, she wants to be sure that I don't take any shortcuts. Actually, she doesn't use the word 'conversion.' In her congregation, it's 'confirmation' or 'affirmation.' At our last meeting, she gave me a small quiz, which I passed brilliantly." Melanie snorted with mock deprecation. "She asked me which Jewish holiday was the most important. I was about to say 'Yom Kippur,' but then I thought that's too simple."

"So what did you say?"

"The Sabbath! And it turned out to be the right answer, because it's the only holiday already mentioned in the Ten Commandments."

"So you're really turning into a Jew." He shook his head in wonder.

"That depends on your definition. But I like my conversations with the rabbi. We aren't just talking about rules and prohibitions; we discuss values — the importance of family, education, and community service in Judaism — as well as small tidbits. Like the difference between the Gregorian and Hebrew calendars. By the way, you do know, don't you, that all Jewish names mean something."

"I suppose so."

"I asked the rabbi about the meaning of 'Menachem.' It means 'consoler' or 'comforter.' Rather apt, I thought."

"Tell me Melanie," Frankenthaler asked quietly. "Menachem Dvir is the father, isn't he?"

As shown above, finally at age seventy-three, I, the "Jew" Carl Djerassi learned under the pseudonym Felix Frankenthaler, what some of the Jewish definitions of a real Jew were. And why in the preceding excerpt did I have Melanie mention that she had read the *American Reform Responsa: Jewish Questions, Rabbinic Answers*? Because I consulted that text as well as the *Code of Jewish Law* by Rabbi Solomon Ganzfried, and did so not only in preparation for this novel, but also for the next, entitled *NO*. In that last volume of my "science-in-fiction" tetralogy. I used the character of Renu Krishnan, an Indian Hindu, who during a short research sojourn in Israel, had fallen in love with an Israeli scientist, Jephtah (in Hebrew "he will open") Cohn whom she eventually married. But in the process she learned what in my opinion are the truly absurd restrictions imposed by Orthodox theocracy in otherwise modern Israel. The following, rather long excerpt, from that novel shows how I used Renu's words to express my own disapproval.

"Here, take a look at the book while I bring the rest of the dishes to the table."

"I can't believe it," Jephtah burst out. "Code of Jewish Law *by Rabbi Solomon Ganzfried. What made you get this book? Where did you find it?"*

"In addition to working with Jews, I'm now sleeping with one." She ogled him from the stove. "I thought I should learn something about Jewish religion and practices. A woman clerk in a bookstore recommended this one."

"But it's so Orthodox," Jephtah blurted. "It's the last thing I would have recommended. I wish you'd asked me."

"I wanted to surprise you ... like using ten candles." She started to light them. "Also, we've hardly ever talked about religion ..."

"How much have you read of this?" He waved dismissively toward the book.

"Mostly I skimmed it. But there were sections on women, on menses, on sex — mostly when not to have it — that I read in detail. I never realized Jews had to cope with so many prohibitions. It seems that according to Jewish law, what is not prohibited is restricted."

"Orthodox Jewish law." He picked up the book, but then slammed it down. "The haredim."

"Relax, Jephtah, and sit down." She led him to the chair. "What are the haredim?"

"'Those who tremble' — presumably with the fear of God — the ultra-Orthodox," he grumbled. "And you can see that I tremble — with irritation."

"Relax," she repeated. "I'm not turning Orthodox. But I couldn't help reading some of it because I actually found it ..." she hesitated, "... funny or at least entertaining. But let's eat. This is the first Shabbat meal I've ever prepared — even if it is Judeo-Indian cooking."

Jephtah grinned. He never held grudges for long. "I see the Judeo part," he pointed to the kosher wine bottle and the challah, "so where is the Indian?"

"Coming up, my Beersheba nabob." She lifted the cover of the serving dish. "Chicken Korma," she announced, waving the cover to fan a whiff of cardamom in Jephtah's direction.

"Delicious," pronounced Jephtah, who had dipped his fork into the serving dish before Renu had even laid down the cover, "but the author of your book back there," he gestured with his thumb without even glancing at the magenta colored volume, "would not have touched it. Chicken and yogurt." He shuddered in fake horror. "I can hardly think of a more serious violation of Jewish dietary laws."

"All right," Renu covered the plate. "No meat for you."

"You can't scare me," Jephtah laughed and uncovered the chicken. "I've been known to eat pork loin, baked ham, bacon and eggs — you name it. Though not in Beersheba," he added.

With the dishes piled up precariously in the small sink and the candles still flickering, Renu and Jephtah moved to the sofa, which seemed hardly wide enough to accommodate two adults and Rabbi Ganzfried's Code of Jewish Law. *Renu picked up the book.*

"I want to tell you why I found some of the sections amusing. I didn't mean this in a demeaning way — Hindus are broad-minded when it comes to other religions. But what would you call this?" She flipped some pages until she found chapter 150. Assuming a censorious tone, Renu proclaimed: "'A man should accustom himself to be in a mood of supreme holiness and to have pure thoughts, when having intercourse. The intercourse should be in the most modest possible manner. He underneath and she above him is considered an impudent act. Jephtah,"

she giggled mournfully, "we have behaved impudently. Wait," she put her hand over Jephtah's mouth. "Just one more, from chapter 151, which applies to our research. I quote: 'One is forbidden to bring on erection or to think about women. A man should be extremely careful to avoid an erection. Therefore," Renu snickered, "'he should not sleep on his back with his face upward, or on his belly with his face downward, but sleep on his side, in order to avoid it.'"

She let out a final guffaw before turning to Jephtah. To her surprise, he was barely smiling. "What's the matter? Do you think a rabbinical interdiction will be issued against our research on MUSA and nitric oxide?" She grinned conspiratorially. "Maybe it's lucky that the chemical symbol is NO."

Renu took Jephtah's chin into her right hand and turned him around until they were facing each other. "What's the matter?" she repeated. "I thought you'd be roaring with laughter."

"Of course, it's funny on one level, but not necessarily in Israel, and especially not in Jerusalem. If the haredim ran this country..." He shook his head. "They may not control what we do in bed, but they certainly could and would affect our daily lives. Just read the chapters in your book dealing with the Sabbath laws. Someone wise once called them 'mountains suspended from a hair.' The Torah lists thirty-nine major classes of acts forbidden on that day. He reached over to take the book out of Renu's lap. "Let's look how applicable some of those prohibitions are to life in 1980." His finger ran down the table of contents. "Here we are: chapter 80, 'Some Labors Forbidden on the Sabbath.'"

Jephtah turned the pages at a speed that made even skimming difficult. "'It is forbidden to shake off snow or dust from a black garment, but it is permissible to remove feathers from it with the hand,'" he read, and then went on, turning page after page. "There are ninety-three of them! I never knew that." He slammed the book shut. "But forget about erections, pious thoughts at intercourse, and who lies on top of whom — "

"God forbid!" Renu interrupted with a typically Israeli intonation that caused even Jephtah to laugh.

"That's just what the haredim say He did," he grinned. "But let me finish my rant. Let's take two crucial issues, marriage and childbirth, to show how hypocritical and politicized our system has become — almost a theocracy. Suppose you wanted to marry an Israeli Jew. You couldn't.

Not in Israel, because only religious marriages are recognized here and you would also have to be a Jew to partake in one."

"So? I could convert. Hindus can be very open-minded about that."

Jephtah stared at her. "You would?" he stuttered.

"I might — theoretically, that is."

"Bu …." he said.

"But what?" she laughed.

"But nothing, I guess," he said lamely. "You'd have to do it the Orthodox way," he pointed to the magenta volume that had fallen to the floor. "They don't recognize other conversions."

Renu's eyes followed his finger. "No," she shook her head. "Not Orthodox. But there must be some alternative."

"To conversion? Here in Israel, none."

"I meant in getting married. You don't have to marry in Israel."

"Ah yes," he nodded. "You can fly to Cyprus for a civil marriage. That's the closest place."

"And when I return with my husband?"

"Under Jewish law, your marriage would not be recognized, and your children would not be considered Jewish." Seeing her shock, he rushed on. "Let me tell you why I'm so irritated by all this hypocrisy. If your parents were Jewish — say they were Cochin Indians — and you never practiced any form of Jewish religion, never went to synagogue, never followed even minimal Sabbath rules, not even what you and I just did — you could still marry legally in Israel. But if you, a Hindu, underwent a Reform conversion — for instance at Brandeis in the States — and fully practiced Reform Judaism, you and any child you might have would still be goyim.*" He shook his head in disgust.*

"At least, he wouldn't be a bastard," Renu said consolingly.

Jephtah threw his hands over his head. "Bastard? Let me tell you something about Orthodox bastardy, what we call mamzerut. *To be fair, the concept of* mamzer *according to Jewish law is often more liberally understood than elsewhere: a child born out of wedlock to a Jewish mother or to two Jewish parents is not a bastard, provided the union can theoretically be legitimized. But — and this is a monumental 'but' — if the mother is already married — in other words committing adultery — that child is considered a* mamzer.*"*

"Why just the mother? What if the father were married?"

"He can screw around as much as he wants to," Jephtah exclaimed bitterly. "Though," he raised a warning finger, "the Orthodox law can

> *also be more cruel. A* mamzer *can never marry a Jew, meaning that such a person can never be legally married in Israel, and that* mamzer *stigma can never be erased. It is passed on to all future generations: according to the Orthodox, for all eternity a* mamzer *can only marry a* mamzer."
>
> *"Maybe there is a historical reason — "*
>
> *"Maybe? Of course, there is. There always is with us Jews. In this instance, the historical reasons — say two thousand years ago — made sense. I can respect history without having to accept that all bible history is unalterably applicable today. But that is exactly the view of the* haredim."
>
> *"How do you know all this about* mamzers*?" Renu was trying to defuse Jephtah's anger, but she was also curious.*
>
> *"Because I am one."*

I shall not further continue with Renu's and Jephtah's remonstration, because what I have described through quotations from my last two novels are simply factual observations through the words of my two surrogates about the rules and restrictions of Orthodox Judaism for which I have neither understanding nor sympathy. Furthermore, they still do not tell anything significant about Carl Djerassi's thoughts on his own Jewish identity. But they do demonstrate that rather late in life, I started to consult rabbinical legal reference works, which quite frankly staggered me in terms of the eventualities they were intended to cover. Some seemed to be of potential relevance such as "Caesarean on a Dead Mother," "Judaism and Homosexuality" or even "Predetermination of Sex," (written in 1941 — some fifty years before ICSI was invented that now makes such predetermination feasible!) but what about "Marriage with a Mother's Half-sister," "Status of an Uncircumcised Retarded Adult" or "How to Shampoo the Hair"? An uncircumcised retarded adult having his hair shampooed seems hardly a subject to worry about, but subconsciously, at least some of this reading may have been prompted by my desire for acquisition of factual knowledge in preparation for a dialog that I could only hold with myself in my writing but never out loud — hence the numerous quotes from my fiction — and had never pursued further until my watershed year of 2006, a year I would have gladly skipped if the choice had been mine.

I shall briefly summarize that traumatic year, which, as happened occasionally at other times, led to dramatic changes in my life. In this instance, on Christmas Eve 2005 — medically the worst day of the year in the UK, since everything closes down until the beginning of January — I broke my hip and shoulder while trying to catch the last train of the evening from Oxford to London. As the other passengers rushed by me to board the departing train, I was lying dazed and in great pain on the wet concrete platform. If my grandson had not been with me, I might well have lain there until the following morning since not a single person was then in sight. To compound my problem, when the ambulance arrived to take me to the Radcliffe Infirmary at Oxford, the radio suddenly crackled with the news that a fire had broken out in the Infirmary's kitchen and directing the ambulance to take me instead to a provincial hospital some thirty miles away. My grandson had to leave the following morning for the USA and after an emergency operation I was left alone for a week in that hospital. Alone, because my wife, who had earlier been diagnosed with an incurable liposarcoma had to depart shortly thereafter from London for Germany where a surgeon had offered to undertake an eleven-hour long debulking operation, which eventually prolonged her life for nearly two years through the removal of over ten pounds of tumor, an operation that the American surgeons had deemed impossible. Thus, while I was virtually helpless on crutches for some weeks in London, she was in a German hospital that I could not visit. My despondency was compounded by reflecting that some years earlier, my father had died at age ninety-six — not as a result of illness, but through a bone infection after having broken his hip. Those events and my realization that my wife and I would henceforth be living under the Damocles sword of her not-too-distant death, led me as a survival technique to a workaholic phase of writing a book at age eighty-three that was totally different in style and in subject matter from anything I had ever attempted before — a book dedicated to her. (Though not living long enough to see the beautifully produced and heavily illustrated volume, she did read each chapter as it was written and was greatly pleased by it.) The style was almost entirely dialogic and its overarching theme the question of Jewish identity.

Four Jews on Parnassus

The book, bearing the title *Four Jews on Parnassus — A Conversation: Benjamin, Adorno, Scholem, Schönberg*[14] was not fiction but factual biography with only superficially disguised major autobiographical components.

---

[14]See footnote 5 on page 7 for full bibliographic details.

But what had decided me to embark on biographical research on four Central European intellectuals of the last century who were neither scientists nor individuals to whom I had ever paid much attention other than listening on occasion to Schönberg's music? The following excerpts from the preface to *Four Jews on Parnassus* explain my choice:

> *Why did I pick this particular foursome? Because all four belonged to the peculiar subset of German and Austrian bourgeois Jews of the pre-World War II generation who often were more Berlinish or Viennese than their non-Jewish compatriots. None was religious; some of them were entirely secular. This is also the generation and social subset to which I belong, and my personal experience with the indelible effects of growing up as a secular Jew in Vienna in the 1930s made me want to examine the range of the meaning of "Jew" through four individuals who responded so differently to that label. Sometimes even non-Jews such as Paul Klee — an important though silent character in my book — fell under suspicion in that era of vicious anti-Semitism, and were branded: it was enough that their vocation or creative output resembled that of their secular Jewish counterparts.*
>
> *But there is more to my choice of these four European Jews: I recognized themes in their lives that I also wanted to examine in my own as I approach its end. Just as in my own autobiography, I decided to deal with my four subjects through selected sketches, which are not necessarily chronologically connected. In this instance, my choice was based on themes that in my opinion have hitherto either been largely underrepresented or even misrepresented in their otherwise overdocumented biographical records. And even more importantly, I have chosen to characterize my subjects by writing dialogue for them. The five episodes could, therefore, be categorized as scenes in a prose docudrama, because (with two stipulated exceptions) every nugget of biographical information I disclose is based on historical documentation, at times even on direct quotation, derived from the bibliography or the personal interviews that are acknowledged at the end of my book.*

Defining or classifying this book is difficult. Is it pure biography presented in docudramatic form? Rather than quibble about it, let me point out that it covers a number of underrepresented or even ignored aspects of those men's lives — notably their relation to their wives (the subject of a subsequent radio play of mine that was broadcast several times by the

Austrian Radio ORF), the importance that Paul Klee played through one of his most famous works, and some puzzling aspects of the frequently discussed but never proven contents of Benjamin's lost grip which he carried with him until his suicide in 1940. Here I will solely limit myself via selected quotations to the autobiographical component that led me to take up this subject in the first place: illuminating the question of Jewish identity in the absence of a common religious label through these four men, who exemplify the range of such identity reigning in Central Europe prior to World War II.

> *Theodor W. Adorno as the prototypical German Jewish non-Jew; Walter Benjamin, as a complicated vacillating German Jew, who never really came to terms whether he was a German Jew or a Jewish German; Gershom Scholem, as the committed German Zionist Jew; and Arnold Schönberg, as the Austrian Jew, converting to Protestantism for professional reasons, who returned to Judaism when he recognized the futility of such all too common sham conversion. And finally a category of Jew that arose especially around the turn of the last century, the Central-European non-Jewish Jew, of which Paul Klee represents the example par excellence.*
>
> *There is, of course, another overriding presence, namely that of Carl Djerassi, since I belong to the same subset of secular, middle class, Austrian and German Jews of the pre-World War II generation as my four protagonists. After my immigration to the USA at age 16 as a fugitive from Nazi Austria, I was totally focused on assimilation. Flaunting or even hinting at my Jewish origin was far from my mind — not unlike Adorno upon his arrival in the USA. Decades passed before I, the typically non-reflective, workaholic scientist, acquired a taste for introspection. With it arose questions about Jewish identity that I am attempting to answer in this book through the putative words of Adorno, Benjamin, Scholem, and Schönberg. Thus, while the biographical facts attributed to them are rigorously documented, what they say and how they say it had to pass through the very fine mesh of my own psychic filter.*

I decided to cover a number of those issues in the form of posthumous conversations that my four "Jews" conducted on Parnassus, the commonly accepted metaphor for the ultimate recognition of literary, musical or intellectual achievement. Arrival on this exalted peak demonstrates that the process of canonization is complete and these four men unquestionably were and remain canonized. What unites as well as distinguishes them are the quotation marks around the words "Jewish identity." Rather than describing those differences in standard prose, I shall now do so through the much more concise and personalizing form of direct speech selected from *Four Jews on Parnassus* — a book of which I am inordinately proud for deeply personal reasons. Suffice it to say that in terms of Jewish identity, I find my own identification most closely associated with that of Adorno — a conclusion that I shall summarize through a potpourri of selected quotes with a finale in Vienna, the city where I was born a Jew, labeled a "Jew", and expelled as one with the quotation marks intact.

> ***SCHOLEM*** *Let's face it. These days, anti-Semitic really means anti-Jewish. And before even thinking about changing that cultural abomination, we would first have to question our assumptions.*
>
> ***SCHÖNBERG*** *Assumptions about what?*
>
> ***SCHOLEM*** *Is a Jew the same as a Jew in quotation marks? And can that person even choose to remove the quotation marks? I am not referring to the Nazi terror when the quotation marks became first a yellow star on their clothing and then an indelible mark on their skin, but before. Already in 1832, Ludwig Börne wrote: "Some reproach me with being a Jew, some praise me because of it, some pardon me for it, but all think of it." So what does it mean to be a Jew?*
>
> ***ADORNO*** *But to whom? The Jew-hater or the Jew himself?*
>
> ***SCHÖNBERG*** *For years, Kandinsky and I were friends who corresponded warmly until I detected an anti-Semitic undercurrent. Here's what I wrote him in 1923: "I finally got it and I'll never forget it, namely that I'm no German, no European, barely a human being — at least the Europeans prefer the dregs of their compatriots to me — but a Jew."*
>
> ***SCHOLEM*** *Why did you write this? Kandinsky was no anti-Semite!*
>
> ***SCHÖNBERG*** *True. He tried to make up by writing to me, "It's no great luck to be a Jew, Russian, German, or European. It's better to be a man. But we should strive to be a superman ... that's the duty of the few.*
>
> ***SCHOLEM*** *And what was your answer?*
>
> ***SCHÖNBERG*** *There was nothing to answer. Kandinsky could afford to make that statement, because quotation marks around the words*

Russian, German or European do not exist! But I still saw them around the word Jew.

SCHOLEM *In that case, why not start with what the word Jew… but without quotation marks … means to each of us? But mark you … in the non-religious sense! Up here on Parnassus, we ought to have a wider perspective.*

BENJAMIN *This is a dangerous topic… even on Parnassus.*

SCHOLEM *Dangerous, but also intriguing. Let's start with Paul Klee. After all, he's been the silent companion of our earlier conversation.*

SCHÖNBERG *(surprised) You mean Klee was Jewish?*

ADORNO *"Klee" doesn't sound Jewish. What was his name before?*

SCHOLEM *(laughs) There you are! You two have already satisfied one criterion of what it means to be Jewish: not only is a Jew someone who continues to ask himself what it means to be a Jew, but then always wonders whether the other person is Jewish. And immediately suspects another name behind the current one if that one doesn't sound Jewish.*

SCHÖNBERG *So was Klee Jewish or was he not?*

SCHOLEM *As early as 1919, when he applied for a position at the Stuttgarter Kunstakademie, he was spurned as Paul Zion Klee. And when the Nazis dismissed him in 1933 from the Kunstakademie in Düsseldorf, some called him a Galician Jew and others a Swiss Jew.*

SCHÖNBERG *But that doesn't make him one. People also called him "the Bauhaus Buddha" but that didn't make him a Buddhist.*

SCHOLEM *I'm surprised to hear you say that. Our friend Wiesengrund … or should I now call him Adorno? … was baptized at birth a Catholic. For religious Jews, he wasn't even a Jew since his mother was a Corsican Catholic. Yet when Hitler came to power, Herr Wiesengrund-Adorno left Germany for Oxford and then the USA because he couldn't get a teaching job in Frankfurt. For the Nazis he'd become the Jew Wiesengrund. Or what about you, Herr Schönberg? You converted to Protestantism when you were 24.*

SCHÖNBERG *"Baptized"… not converted!*

SCHOLEM *A subtle difference.*

SCHÖNBERG *Not to me. Baptism was simply a union card … for a better job. Not unlike Gustav Mahler becoming a Catholic in order to serve as Opera director in Vienna. Conversion is generally based on conviction.*

SCHOLEM *(with irony) Since you were evidently convinced that baptism would help, didn't that make your baptism a conversion? But more to the point, your children, born in Austria, were Protestants at birth. The anti-Semites called all of you Jews, which made you a Jew.*

SCHÖNBERG *We left … and I never returned. In 1933, before immigrating to California, I reconverted in Paris … in the true sense of the word … with Chagall as my witness. I found it was not too late to recognize that by restoring my Jewish self-confidence, I also restored faith in myself and my creative capacity.*

SCHOLEM *It took you 35 years to learn that once a Jew, always a Jew. To the true Jew-haters, neither baptism nor conversion ever counted …. And especially not if you don't take the precaution of also changing your name … preferably a generation or two earlier.*

BENJAMIN *In Germany? Name changes were a pointless exercise … at least during our time. About as permanent and effective as a fig leaf.*

SCHOLEM *In that case, why did you, my dear Teddy Wiesengrund, change yours?*

BENJAMIN *Don't be so tough, Gerhard. Name changes are an all-too-common stigma among Jews.*

SCHOLEM *Surely it's no stigma when a Jew changes his name from Gerhard to Gershom to emphasize that he is a Jew.*

BENJAMIN *Gershom, with you I stand corrected.*

ADORNO *(bitter laugh) But I forgot to change my accent. (Seriously.) We all know that it wasn't a matter of forgetting… it was the impossibility of not forgetting German…*

BENJAMIN *Kafka put it well when he talked about the three impossibilities confronting us German Jewish writers: "the impossibility of not writing; the impossibility of writing in German; and the impossibility of writing in another language." (Points to Adorno.) That was our problem… yours and mine.*

ADORNO *First, do you know what the rector of the university that refused to even let me teach there in the 1930s wrote less than 20 years later to the Minister of Education? "The expedited award of a professorship is politically and humanly warranted."*

SCHOLEM *(sarcastic) Tell me, Herr "Reparationsprofessor," weren't you embarrassed, if not insulted, by the title of "Wiedergutmachungsprofessor" and the years it took before you became a non-sanitized real Professor who just happened to be Jewish?*

BENJAMIN *For me, it was the German language ... and later also French ... that kept me in Europe until it was too late. (Back to Adorno.) And it must've been the language that brought you, the emigrant Adorno, back to Frankfurt in 1949 when hardly any German Jew dreamed of returning to that ruined country.*

SCHOLEM *I solved my Kafka problem by proving that it is not impossible to write in another language. I learned Hebrew back in Berlin, when it was not fashionable ... and I still a teenager. Most of my writing on the Cabala was conducted in my adopted language.*

BENJAMIN *When it came to my personal Judaism, the fatal problem was my continually vacillating attitude between important and problematic. The problematic always kept me from learning Hebrew.*

ADORNO *(to Scholem) Before we go off in still another direction, I must reply to your implied accusation: that I returned to Frankfurt ... still largely in ruins ... for a Herr Professor title. You all should hear the real reason ... especially my esteemed maestro Schönberg, who never returned to either Germany or Austria!*

SCHÖNBERG *Yes, tell us. Because returning was always on my mind, since professionally, the Americans treated me miserably. Still, I didn't have it in me to return to Europe.*

ADORNO *I did .... But then I am different from you three.*

SCHOLEM *We all are different ... and it's that difference we're discussing.*

ADORNO *I returned as a German Jew.*

SCHOLEM *What does that mean? That you're first a German ... and then a Jew?*

ADORNO *Rather than debating with you the difference between a German Jew and a Jewish German —*

SCHÖNBERG *(irritated) Everybody is talking about differences, without explaining them. What, Herr Scholem, was suddenly different? And what was different about you, Theodore (emphasizes next letter)* **W**. *Adorno?*

ADORNO *I needed to come back to my place of birth, to the site of a magic childhood, to the country where I could again write in my own language, rather than in English … the language I only learned as an adult so I could write at best as well as the others.*

BENJAMIN *I have not interrupted any of you, but now I must. Teddie, I understand the feeling about language … but there is more to it than wanting to return to the dreams of your childhood … however paradisiacal it may have been. You have written about those reasons in words I only learned years later. Do you remember them?*

ADORNO *Of course. I think too sociologically to see Fascism as the expression of the German national character. Rather, I interpret it as the consequence of a socio-economic development, and therefore not as a concept that the entire German nation is to blame.*

SCHOLEM *Good God! This sounds like a lecture … not unlike Arendt's generalization that Eichmann and the holocaust were the results of broader political and economic developments … and to that extent almost unavoidable. Let me remind you how one of the biggest German newspapers, the Frankfurter Allgemeine Zeitung, greeted your return: "He is Theodor Wiesengrund-Adorno, born in Frankfurt, who had received his habilitation at his home town university and later went to America …. And note that contrary to the Americans, <u>Mister</u> Adorno, the Germans never let you forget that you were born a Wiesengrund. (Continues loud and sarcastic). "<u>And later went to America</u>!" It rather sounds as if you went voluntarily ... perhaps even on vacation.*

ADORNO *I can respect your feelings, but you will have to respect mine. I, who immigrated to the USA trying to disguise any Jewish connection … even though intellectually or psychically there was very little to disguise … returned openly flaunting my Jewish connection. I simply wanted to return to the site of my childhood in the belief that in the final analysis everybody wants to recover his childhood in some changed sense.*

I didn't underestimate the danger and complications of that decision, but to this day I have not regretted it. I made no concessions to the prevailing politics ... I fought them and I contributed to the resulting changes. And I am not ashamed to admit that academic recognition ... as a German Jew ... pleased me, as did the response of the students. Anyway ... there you have it. The reasons for my return ... as a Jew to whom religion meant nothing.

SCHÖNBERG *I think it's time to give this "Who is a Jew?" question an Austrian twist that you are unlikely to find in Germany.*

SCHOLEM *(interrupts) I know exactly what you are going to say: "I decide who is a Jew." Those were the words of Karl Lueger, Vienna's most beloved mayor ... not very different from what Jean-Paul Sartre said years later, "A Jew is a person who others say is Jewish. It is the anti-Semite who makes the Jew" —*

SCHÖNBERG *(interrupts) And don't forget that Lueger was also an avid Jew-hater of the pre-Hitler days. No, I have something much more nuanced in mind. Note, I'm talking about Vienna ... where few things are expressed directly!*

ADORNO *A city I once loved and then despised. Better be careful ... you're addressing three German Jews.*

SCHÖNBERG *(to Scholem). Remember when you complained about the absence of a stamp honoring you?*

SCHOLEM *On that front, I had plenty of reason to complain.*

SCHÖNBERG *I understand. But what do the following have in common? Karl Kraus, Hugo von Hofmannsthal, Ludwig Wittgenstein, and a couple of composers, Gustav Mahler and Johann Strauss.*

ADORNO *Which Johann Strauss?*

SCHÖNBERG *Why not both to make it an even half dozen and not argue whether father or son wrote better waltzes? (Turns to Scholem.) Even though there is no stamp of you, let's take stamps as a sign of Jewish identity. Not Israeli stamps ... but Austrian ones. All six have appeared on Austrian stamps.*

SCHOLEM *So what? Bach has been on an Israeli stamp and Mozart on an Austrian one. It doesn't make them Jews.*

SCHÖNBERG *Be patient! Not so long ago the Austrian post office sponsored an exhibition on Jewish personalities on Austrian postage stamps, entitled "Stamped? Branded!". And that all of them were included there as Jews with 35 others!*

SCHOLEM *Johann Strauss Jewish?*

ADORNO *I pride myself at my knowledge of music, but that I did not know.*

SCHÖNBERG *Neither did I ... until I saw the catalog of that stamp exhibition. It turns out that the grandfather of Johann Strauss the elder was a baptized Jew. So what's good enough for the modern, politically correct, Austrian post office ... not some Nazi remnant ... ought to be good enough for us here ... especially since there is no post office on Parnassus. Karl Kraus, born a Jew was baptized a Catholic at the same age as I turned into a Protestant, yet, like me, he dropped that union card some decades later. Mahler never concealed his Jewish origin, even after becoming a Catholic, but always said that being a Jew gave him no pleasure ... that he felt as if he were born with only one leg or one arm. Or Wittgenstein, whose grandparents were mostly Jews, although his parents were Catholic and Protestant, respectively. He was given a Catholic burial, yet confessed to having allowed others to underestimate the extent of his Jewishness.*

BENJAMIN *What are you driving at?*

SCHÖNBERG *That there is no threshold beyond which one is not Jewish. Once contaminated—*

SCHOLEM *What a disgusting word.*

SCHÖNBERG *I used it deliberately.*

BENJAMIN *I can think of a more elegant way of reaching the same conclusion than licking the back of some Austrian stamps. Consider Moritz Goldstein ... whom Gerhard here already mentioned. Remember his famous essay of 1912: "The German-Jewish Parnassus?"*

SCHOLEM *Considering where we are holding this debate, Goldstein's arguments seem to be more relevant than the Austrian post office's. All the stamps demonstrated that you are considered a Jew because of parentage ... a synonym for "race" or "blood" ... and for anti-Semites, parentage goes back ad libitum and ad nauseam.*

SCHÖNBERG *So what did Goldstein have to say?*

SCHOLEM *We, the Jews, administer the intellectual property of a people that denies us the qualification and the capacity for doing so.*

SCHÖNBERG *What else did Goldstein say?*

SCHOLEM *Basically, that those German Jews up on Parnassus got here on their merits and may have accomplished it all in Germany ... or in Austria ... and in the German language, but they are here as Jews and not as German Jews. Jewishness is not necessarily based on religion or history, but on the fact that Jews live among non-Jews who call them Jews. Assimilation is impossible for such pariahs. Even though the German Jews per capita had made greater contributions to Germany than the Germans, had won half the German Nobel Prizes, had led German culture by becoming some of the greatest poets, artists, stage directors —*

BENJAMIN *Rather than argue, let's sharpen the definition. Nobody would disagree with the statement that fundamentally Jews are neither Europeans nor non-Europeans.*

SCHOLEM *More than that. The Jews never lived within a border ... not until recently when the State of Israel was created. And even there, the precise border or even its validity is still up for debate.*

ADORNO *Isn't it curious that in the pre-Nazi days, whenever religion had to be specified ... schools, jobs, passports ... the Germans or Austrians used at times "israelitisch", a label that was not a religious one ... like Catholic ... but a nationalistic one, assigning us to a country that hadn't existed for nearly 2000 years.*

BENJAMIN *You have moved the argument to geographical borders. I want to return to cultural ones. We're really talking about a very special subset of Jews — not just European, but in fact German ... and here the borders are precisely defined ... and what is even more important, accepted by that subset. In other words, the type of German Jews we represent and which Goldstein had addressed.*

SCHÖNBERG *Don't forget Austrian.*

BENJAMIN *Or more precisely, Viennese and Berliner and Frankfurter Jews. Some wise man once said Jewish sameness is riddled with otherness. So let's address this point: take the etymology of Hebrew.*

ADORNO *Now we are getting somewhere: the literary mind with his academic glasses!*

BENJAMIN *Abraham, not Moses, discovered monotheism. He came from across the river and was therefore a crosser, an ivri … ergo a Hebrew.*

SCHOLEM *(ironic) Praise the Lord! Walter Benjamin finally understanding his roots.*

BENJAMIN *Listen … my parents, who were Christmas Jews with a Christmas tree every December, would have sympathized. When they had reconciled themselves to the fact that their son would become a writer, they said "it would be good if not everybody noticed that he was a Jew." Eventually, I humored them by assuming several synonyms aside from writing under my own name. Is that simply a reflection of the sorry fact that in some way we always wanted to be culturally assimilated? Become parvenu Jews as Hannah Arendt called such Jews … even though assimilation during my 20th century life as a German Jew was ultimately impossible? That I was the sort of pariah Jew, who regretted that he could not become a parvenu, while at the same time ashamed of that desire?*

ADORNO *Since the words "pariah" and "parvenu" as applied by Hannah Arendt to German Jews are suddenly bandied about … first by Gerhard and now by you. Walter … let's remember exactly what she said about* <u>*conscious*</u> *pariahs. In her eyes, these were persons, "who were great enough to transcend the bounds of nationality and to weave the strands of their Jewish genius into the general texture of European life … who tried to make of the emancipation of the Jews that which it really should have been: an admission of Jews* <u>*as*</u> *Jews to the rank of humanity, rather than a permit to ape the gentiles or an opportunity to play the parvenu." (Addresses Scholem) Even though you have distanced yourself so firmly from her, you must agree that she was right in pointing out that in the process, these conscious pariahs in Europe were not only marginalized by European society, but in the end by the Jewish community as well*

SCHÖNBERG *What exactly was this all about?*

SCHOLEM *A catharsis.*

SCHÖNBERG *You think we needed it? Up here, on Parnassus?*

SCHOLEM *Precisely here, because here it occurs so rarely.*

ADORNO *A catharsis usually leads to relief. I don't feel relieved.*

BENJAMIN *(to Scholem) Gerhard, when you and I held such conversations by correspondence, there was always the unspoken assumption that somehow it would benefit others. That it would be published ... that it would lend itself to analysis by others, to generalizations —*

SCHOLEM *The catharsis I am thinking of is personal and Parnassian. The only non-debatable conclusion is that we were discussing our respective personal definitions of what it means to be a non-religious Jew —*

BENJAMIN *You think I know it now?*

SCHOLEM *At least more than you did before.*

BENJAMIN *In that case, summarize it for me.*

SCHOLEM *A person is a Jew when he calls himself a Jew or is so identified by others —*

BENJAMIN *(interrupts) How does that differ from a Catholic ... or a Protestant?*

SCHOLEM *You didn't let me finish. <u>And</u> ... please note, I'm saying "<u>and</u>," not "<u>or</u>" ... who has some Jewish ancestors —*

BENJAMIN *(interrupts, impatiently) Catholics have ancestors!*

SCHOLEM *Of course, they have. But how far back? With Catholics or others, the choice is up to them. Not for Jews. Non-Jews and especially the Jew-haters will not offer you that option. They'll go back for generations upon generations until a Jewish stain is encountered ... Because for them it was an indelible stain ... invariably labeling them as perpetual outsiders. That's why Wittgenstein or Johann Strauss — father and son — find themselves in an Austrian stamp collection featuring Jewish personalities and labeled "Stamped? Branded!" It's the post office that made that decision, not Wittgenstein or Strauss.*

SCHÖNBERG *So what's new? It's others that label us as Jews ... be it the Jew haters or Jew lovers ... even if we have never entered a synagogue in our lives.*

Vienna

In the above dialogs, Gershom Scholem called their discussion of Jewish identity a catharsis. I put these words into his mouth because in my mind such a long-overdue personal catharsis was the principal reason why I wrote that book in the first place. I am purposely ending these excerpted dialogs with the Austrian post office exhibition, which I also mentioned in my chapter on *Heimat(losigkeit).* It is still another example of someone else — be it an institution or a person — deciding whom to call a Jew. In this case, of course, it was meant benevolently, rather than in the brutal "I decide who is a Jew." manner of Vienna's former hyper-anti-Semitic mayor, Karl Lueger, whom Vienna has honored in unprecedented overkill fashion through two sites on Vienna's most beautiful and most important street, the Ringstrasse. One by placing his huge monument on a square, the Dr. Karl Lueger Platz, near one end of the Ringstrasse, and then naming a portion of the Ring near the other end — in front of the university, which at one time was also a hotbed of anti-Semitism — "Dr. Karl Lueger Ring."

I am the only living person who is represented in that *Stamped? Branded!* exhibition and thus in a position to make a slight confession. The catalog cover was already printed before "my" stamp was officially released, which was the reason that it is included in the text but not on the cover. Yet through the help of Gabriele Seethaler, the illustrator of all pictures in *Four Jews on Parnassus,* "my" stamp was also smuggled unto the catalog cover near the lower left hand corner as shown in the accompanying picture.

For a Jew — especially one driven out of Vienna as a "Jew" after the Anschluss — the terms "stamped" and "branded" have horrendous double meanings, which the Austrian Post Office used deliberately, a gesture I deeply appreciate. Yet in the final analysis, any true return to the city of my birth must be based on a stripping of the power to assign such quotation marks. Removing the words "Dr. Karl Lueger" from the street name in front of the university with its own anti-Semitic history might be a proper first step.

That last sentence was composed by me in late March of 2012. To my utter surprise, a few weeks later, I learned that the Viennese municipal authority had decided to change the name of the Dr. Karl Lueger Ring to "Universitäts Ring." And two months later, I received an honorary doctorate from that University. It was not my first (that happened sixty years earlier

Gabriele Seethaler

Cover of bilingual Austrian Post Office catalog "Stamped? Branded!" Jewish Museum, Vienna, 2004.

in Mexico) and I have already been informed, barring sudden death, that it won't be my last. But there are special circumstances, which can only apply to the University of Vienna and no other — both of them related to the themes *Heimat* and *Jew.*

When one gets a couple of honorary degrees, they are usually from institutions with which the recipient had an intimate relation: as an alumnus, a long-time professor or a major donor. With a couple of dozen or more doctorates, the reasons are different since one can't possibly have intimate connections with so many different universities. Then it is a form of generic acknowledgment for an important achievement or a position unrelated to the institution. With all other universities which have honored me — in North America, Latin America, or Europe — I fall into both categories, yet the University of Vienna is different. I have never studied at that University, having been driven from the city in my early teens. But my parents both received their doctor's degrees there; in fact they met in Medical School as fellow students, later married, and produced me! Obviously, I would never have been born were it not for the University's existence. More pertinent is the fact that I would have attended the University of Vienna with 100% certainty if the Nazis had not taken over and I would have turned into a practicing physician, not a researcher. I say this because both my parents, after their early divorce, had their treating rooms in their residencies. Hence until age 14, I "lived" in a doctor's office and would have been totally imprinted through that setting. In a certain sense, I now feel that I have finally graduated from my true alma mater, but in a field I would never have studied there. I am not a framer nor displayer of diplomas — a professional habit of physicians I disdain — other than theater posters of performances of my plays which cover the walls of my studies in San Francisco, London and Vienna.

But who knows? I may some day make an exception with the grandiose Latin diploma I was presented in a truly thoughtful ceremony with a refreshing institutional *mea culpa* on the part of every speaker, including the Bundeskanzler who made a surprising appearance; it was he who pointed out that during the Nazi takeover, overnight hundreds of distinguished Jewish professors were ejected, an academic disaster from which the University of Vienna, now having an enrollment of close to 90,000 students, has never recovered. As I sat in the spectacularly ornate *Festsaal*, a Schiller quotation from my teenage *Gymnasium* days crossed my mind: *Spät kommt ihr, doch ihr kommt. Der weite Weg, Graf Isolan, entschuldigt euer Säumen.* [Late you come, yet come you did. The long road, Count Isolan, excuses your delay.] The road has indeed been very long, but at least it now seems to be heading in the right direction.

"Professor of Professional Deformation"

Professor is a weighty title, full of respect, at times of pedantry, and often even of fear. Yet, even at the risk of being accused of excessive infatuation with quotation marks, let me point out that just as in the titles of some other chapters, garlanding them around the word *Professor* adds nuances way beyond sheer weightiness. In wearing the gown of professor for over six decades, it is the nuances that have increasingly occupied me. What follows is an account of how some of the nuances have affected my professional and hence also personal life. But first a comment about the title.

Déformation professionelle is an elegant French expression, generally meant in a pejorative way. Wikipedia defines it as "a tendency to look at things from the point of view of one's own profession rather than from a broader perspective. It is often translated as 'professional deformation'. The implication is that professional training, and its related socialization, often result in a distortion of the way one views the world."

I, however, have chosen to interpret it in a complimentary way as shown in one of the most joyful poems in my recent (2012) rather gray poetry book, *A Diary of Pique.*

Deformation Professionnelle

Somehow
Déformation professionnelle
Sounds much more elegant
Than any English translation.

Leave it to the French
To make it sound
A rare malaise,
A tantalizing vice.

The *déformation professionnelle*
I speak of
Is Leonardo's malaise
That was his virtue.

Or Samuel Johnson — so exquisitely deformed.
Sans déformation professionnelle,
Merely another Dr. Johnson.
Deformed, THE Doctor Johnson.

I am looking for a job
As Professor of Professional Deformation.
If you know of a vacancy,
Call me. Collect.

Polymaths like Leonardo da Vinci or Dr. Samuel Johnson, who were never content to stay within the boundaries of any single art or profession, were for me heroic models worthy of emulating. As I became interested to go beyond the standard practices of a chemistry professor, I got deformed and in the process expanded and raised the level of my professorial and professional life in a way that I need not defend. On the contrary, I consider this deformation tantalizing as described in my poem and the responses of my students have often confirmed this. It is open to question whether all of my chemical colleagues at Stanford University, where I have served as professor longer than any other member of the chemistry department for over half a century, are of the same opinion. I say this because, next to physics, chemistry is the hardest of the hard sciences, the rock on which the biomedical, environmental, and material sciences all build their molecular edifices; it is also the most insular of the hard sciences. Sadly, many of its academic practitioners proudly (though in my opinion inadvisably) erect high if not actually impenetrable walls that prohibit meaningful intellectual interaction with non-scientific departments and not many attempts are made to bridge that gulf. Although chemists feel continually defensive in this time of rampant chemophobia, the majority is unwilling to go proselytizing among the scientifically untrained public, even within academia. The academic chemical community offers few brownie points for such missionary service. With this shadowy comment out in the open, let me proceed, but this time in chronological fashion, in marked contrast to the approach I have used so far in most chapters.

Since my initial professorial life started in chemistry, I need to discuss first what kind of chemist Carl Djerassi is. In my "chemical" autobiography of 1990, *Steroids Made it Possible,* the intended readership was entirely scientific and the text to that extent largely inaccessible to ordinary readers. In my 1992 general autobiography, I spent a long chapter, *What Kind of Chemist Are You*? answering that question in what I then took to be a more digestible fashion, but now two decades later, I find it still too dense, meaning still "too chemical." Consequently, I have chosen a compromise: since that autobiography is out of print and undoubtedly unfamiliar to many present readers, I have constructed a potpourri of excerpts from that chapter to properly describe my professorial life as a pedagogue dreaming of *Déformation professionnelle* and then attempting to actually realize it.

"What kind of chemist are you?" is basically a tourist's question. My reply, therefore, will be a guide's reply: descriptive, anecdotal, at times metaphoric, on occasion historical. But that presupposes two other questions: "Why did you become a scientist?" and "Why did you remain one for so long?" to which the answers are simple and pithy: "Serendipity," to the first, and "Excitement, curiosity and ambition" to the second.

I am an organic chemist, meaning that I deal with molecules containing carbon chemically bonded to hydrogen. This sound simple, until one realizes that it encompasses the chemistry of life through millions of natural and synthetic substances ranging in molecular weight from sixteen for the simple gas methane (one carbon atom and four hydrogens) to proteins and polymers that may exceed molecular weights of one million. To describe my own chemical persona, it is easiest to start with subdividing the field of organic chemistry itself between theoretical and experimental organic chemistry, the latter being my own. Of the latter's various subdivisions, I shall consider only two: synthesis and structure determination. All of my industrial research — first at CIBA in New Jersey the year prior to entering graduate school and four years following my doctorate as well as the subsequent two-year stint at Syntex in Mexico City — was in synthetic organic chemistry; whereas the overwhelming part of my academic research, starting with my first cactus studies in 1952, was in one way or another related to the elucidation of the chemical structures of natural products.

In January 1952, I packed our belongings into our Chevy and drove with my second wife, Norma, and two-year old daughter, Pamela, from Mexico City, through miles of cactus landscapes, to Michigan. I had finally been offered an academic post, a tenured associate professorship at Wayne University, with the promise of a full professorship within a couple of years. Again, I was proceeding against the advice of my friends and colleagues up north. Two and a half years earlier, they had advised me not to go to Mexico for chemical research. Now they thought I was crazy to leave a well-paying and productive research career at Syntex in sunny Mexico to go to cold and slushy Detroit. Wayne was not exactly on the top of the academic totem pole; in fact, as far as graduate programs were concerned, it was then fairly close to the bottom — a city university with a primarily blue-collar constituency. But it was the only academic offer I had received; and, at age twenty-eight, I thought it was time to find out whether the dreamed-for academic career was indeed what I wanted.

"Old Main" Building of Wayne State University, Detroit

When I arrived in Detroit, Wayne had not yet become Wayne State University. Except for one new science building, most of the classrooms, offices and laboratories were located in Old Main, a huge, nineteenth-century building that was formerly a high school, and in an assembly of private houses that had been taken over by the university as the urban slum enveloped the campus. Since there was not room in the new science edifice for all faculty members, I, as the newest arrival, ended up in Old Main. I have not visited all American university chemical laboratories, but I have seen my share of them, and Old Main was clearly one of the dingiest. Moreover, it had a distinction that must be unmatched in American university circles: the stockroom, from which all necessary chemicals, glassware, and other instrumentation had to be secured, was in the new science building; to reach it, my students had to make many daily trips across one of the busiest, four-lane streets of Detroit, irrespective of rain, snow, slush and speeding traffic. My first Chinese postdoctorate fellow,

Liang Liu (subsequently, under the name Huang Liang, a director of the Institute of Materia Medica of the Chinese Academy of Medical Sciences in Beijing), was once followed into the lab by a policeman after she had dashed across in a heavy downpour while the traffic light was still red. She must surely be the only chemist in America to have received a jaywalking ticket while doing research.

Fortunately, the stockroom was generously stocked, the instrumentation more than adequate and the chemistry library superb. Old Main was hot in the summer and overheated in the winter, but it did not leak; the water and electrical supply worked; and while the lab benches were old, they were serviceable. And the graduate students were hard working: the work ethic of Detroit in the 1950s could be felt in the students who went to Wayne to get an education, not to play around. I managed to secure research support from organizations like the National Institutes of Health, the American Heart Association, the National Science Foundation, the American Cancer Society and various pharmaceutical companies, such as Merck and Schering. Within a couple of years, I had over a dozen graduate students and postdoctorate fellows in my research group. Having now described how I first reached the portals to a professorship, let me return once more to the earlier mentioned subdivision of synthetic versus structure elucidation in organic chemistry in order to explain how I moved from one to the other.

The best synthetic chemists are both architects and builders, and in the scientific literature often see their work anointed with terms like *beautiful*, *elegant* or *dazzling*. As architect, the chemist designs a strategy for synthesizing a complex molecule which may involve permutations of dozens upon dozens of separate chemical steps. As builder–engineer, the chemist devises new chemical reactions and discovers new synthetic reagents. Both architect and builder know precisely what the edifice should ultimately look like.

The structural elucidation of a new natural product, however, entails elements of secrecy and suspense that are absent in synthesis. When I first entered the field of structure determination, it was a chemical variant of the game "Twenty Questions." Although the questions do not mean anything to a non-chemist (Does the substance contain only carbon and hydrogen? Does it contain oxygen atoms? How many? Does it contain nitrogen? Any other heteroatoms? Is it saturated or unsaturated?), they make for a winnowing process through which one can eventually create a picture of a substance's chemical constitution. Perhaps a more useful metaphor for the process is that of entering a pitch-dark room with the aim of ultimately determining its contents, the precise location of the furniture, the color and composition of each item. Some people might enter brusquely, bumping into a chair or a table, which they would then touch in order to derive some idea of its dimensions and composition — wood, plastic, upholstered material. Others might be more cautious and systematic: they would start by feeling their way around the wall, perhaps counting the steps so as to determine the overall dimensions of the space, crossing the room at fixed intervals to determine the rough location of certain objects before focusing on them one at a time. Someone may come equipped with a penlight capable of illuminating only small areas; or with a powerful flashlight to get a quick picture of the whole room and its contents. Or somebody with a flash camera and a wide-angle lens may with one picture record the contents of the entire room and in color to boot.

Organic chemistry started in the nineteenth century as an attempt to determine the structures of chemical substances that, isolated from plant and animal sources, had interesting biological properties; only when this was accomplished would the synthetic chemist, the architect–builder, enter the field. The early methods of structure elucidation, the feeling and stumbling process in the dark room, could deal only with relatively simple chemical structures. But, as more refined penlights and then flashlights became available, ever more complicated chemical structure problems were solved in less and less time. The great age of natural products chemistry was the period from 1930 to 1960, when all the important steroid hormones, vitamins, and antibiotics, along with a host of other biologically significant molecules, were isolated and their structures established. I had the great luck to enter natural products chemistry during its glory days when a fiendishly clever chemical game of "Twenty Questions" was still *de rigeur*, although I later observed its eventual decline through the ever-increasing use of sophisticated "flashlights." The main techniques — the ever more powerful flashlights — were ultraviolet and infrared spectroscopy, nuclear magnetic resonance spectroscopy and mass spectrometry. The principles behind these techniques were discovered by physicists, who also developed the first instruments to detect them. But it was the chemist interested in structure elucidation, rather than in synthesis, who first applied these new physical tools to the solution of organic chemical problems. The reason is obvious: the architect–builder already knows what the room should look like. Only a person entering a dark chamber is interested in flashlights and much of my subsequent work in academia actually focused on improving such illuminating devices so as to make detailed chemical examination superfluous.

I had by then attracted a passel of postdoctorate fellows as diverse and exotic in background as the plants we analyzed. Chemists from England, Scotland, Switzerland, Japan, Italy, India, China, Mexico, Israel, New Zealand, Australia, Costa Rica, and Brazil worked together in one huge lab in the Old Main building of Wayne State University and variously assaulted the English language. Shortly before the arrival of my first Italian and Japanese postdoctorate fellows, Riccardo Villotti from Rome and Tatsuhiko Nakano from Kyoto, Jim Gray from Glasgow had entered my laboratory. He had a Scottish burr that, in terms of intensity and purity, bordered on incomprehensibility even to Americans. Villotti and Nakano had hardly ever been exposed to spoken English. The three shared a lab bay, and for many weeks most of their interchange was restricted to sign language or written communication. When Villotti finally gained sufficient confidence in spoken English, he inquired hesitatingly of Gray, "Jim, your native language, it is what?"

My first Brazilian postdoctorate fellow was Walter Mors, who was fluent in English. Before returning to the Instituto de Química Agricola (located in the Rio de Janeiro botanical garden), he asked whether we could possibly start a US–Brazil cooperative venture similar to what I then had under way between my group at Wayne and the Instituto de Química of the National University of Mexico. While directing Syntex's chemical research in Mexico City, I had several Mexican university chemistry students doing thesis work with me (one of them, Luis Miramontes, synthesized the first few milligrams of norethindrone, the progestationally active ingredient of the Pill as I described *The Bitter-sweet Pill*). A number of these Mexican chemists then formed the nucleus of a small research institute, the Instituto de Química, which was established on the university campus and partly funded by Syntex. When I left for Detroit, I maintained a collaborative program with that group; I felt that the best way to establish an academic research center in Mexico was for the chemists to work on local problems. The academic counterpart to the industrial example of Syntex, with its production of steroid hormones from Mexican yams, became the structure elucidation of natural products from a variety of Mexican plants, especially those that had a history of indigenous medicinal use, and from cacti. I asked my first British postdoctorate fellow at Wayne, Alan Lemin from Manchester, to spend a year at the University of Mexico training a group of young Mexican chemists in the techniques and methodologies we had just developed in Detroit. I was able to provide guidance for this program during my frequent consulting trips to Syntex in Mexico City. Just as explorers or astronomers have the privilege of naming newly discovered territory, chemists can do so likewise with newly isolated natural products. Prompted by some form of linguistic masochism developed during those Mexican days, I introduced into the chemical literature some real tongue twisters such as *tlatlancuayin* and *cuauchichicine*, based on the Aztec names of their plant progenitors.

During the 1950s, the Rockefeller Foundation supported our Wayne–Mexico collaboration. As a result, the American chemical journals suddenly found themselves publishing Mexico-originated research not only from Syntex, but also from the Universidad Nacional Autónoma de México. (My very first honorary doctorate in 1953 came from that university. Whenever I wear at some special academic occasion the Mexican black hexagonal hat, crowned with a blue powder puff and tassels hanging down on all sides, I think of the time my Columbia University friend Gilbert Stork tried to photograph the occasion, and the flashbulb exploded in his hand as the rector of the university was placing that silly-looking hat on my head. Owing to recent bombings in Mexico City, everybody responded with panic — as the newspapers later reported — as if this were another terrorist attack. On seeing my friend's look of utter shock, I found myself incapable of saying a word, not even "*Muchas gracias,*" and just sat down, wiping tears of laughter from my cheeks.)

The Rockefeller Foundation generously supported the cost both of the frequent trips I began taking to Brazil, and of the annual residencies of the postdoctorate fellows from Detroit who performed joint research with Walter Mors's group in the Rio botanical garden. The first such scientific ambassador, Ben Gilbert, met his future wife in Mors's lab and remained in Brazil, where he heads an important effort to make Brazil less dependent on imported drugs. We decided to concentrate on alkaloids from the rich Amazonian flora, which were known to display a variety of pharmacological effects on the central nervous system. The productivity of this collaboration, which continued for over a decade, even after I had moved from Wayne via Mexico to Stanford, was impressive in terms of both chemical accomplishments and human intangibles. The latter included a number of humorous incidents of which two are worth retelling. One of my longest postdoctorate collaborators and eventually also very good friend, Ben Tursch (born in the Congo, he was educated in Belgium, where eventually he became professor, expert diver, sophisticated collector of primitive art and world authority on the taxonomy of marine mollusks, among other ventures) spent some of his time in our US–Brazil collaborative project and was the person who managed to persuade the Brazilian Air Force to lend us an old B-27 bomber to transport a large quantity of sea cucumbers, which he had collected in North Eastern Brazil and which were in the process of rapidly decomposing, before reaching our laboratory in Rio. This was not the only assistance that the Brazilian military provided. On several occasions, we needed rapid replacement of some instrumental spare parts, which would have taken months before passing through Brazilian Customs, at that time probably the greatest hindrance to sophisticated research in Brazil. The Brazilian military attaché in Washington helped us "smuggle" some of that equipment on their bi-weekly flights from Washington — an enormously helpful gesture that we did not dare to acknowledge in our eventual scientific publication.

Another amusing event involved the propensity of Brazilians, especially women, to have some of the longest strings of names known to me compared to, for instance, the single names of Indonesians. Knowing the obsession of scientific journal editors with saving space and hence preferring just initials of forenames, Tursch and I insisted in one of our joint publications in the *Journal of Organic Chemistry* to use the full name, Gloria Berenice Chagas Tolentino de Carvalho Brazo da Silva, of one of our collaborators. I still remember cracking up when the proofs arrived and noticing that the editor had inserted commas after each third name, thus converting her nine-word name into conventional triads. As a fierce proof reader, I restored Gloria's name to its nonary glory.

In the late 1960s, when I chaired the Latin America Science Board of the National Academy of Sciences, I proposed the formation of a US–Brazil chemistry program — modeled after our more modest natural products collaboration — in which half a dozen major professors from Stanford, Caltech, the University of Michigan and the University of Indiana participated. Programs in synthetic organic chemistry, inorganic chemistry, physical chemistry, and polymer chemistry were instituted at the universities of São Paulo and Rio, and involved over a dozen young American postdoctorate fellows. Chairing this program over a period of years has remained one of my most joyful memories of how science transcends geographic and political boundaries, an experience that led me subsequently to propose extending such collaboration to Africa, with specific reference to Kenya and Zaire (now known as the Democratic Republic of Congo).

In addition to exploring the "dark rooms" of many natural products from Mexican and Brazilian plant sources in my first academic years in Detroit, my research group also started to develop new "flashlights" rather than just use existing ones. The first one I chose was a logical outcome of my earlier steroid research in Mexico.

Many natural products, and all naturally occurring steroids, are "optically active," or capable of existing in mirror-image forms. If light is passed through a solution of such an optically active molecule, the plane of polarized light will be rotated either to the left ("levo-rotatory") or to the right ("dextro-rotatory"). Only one of these mirror-image forms retains the biological activity: for instance, naturally occurring dextro-rotatory testosterone is responsible for all the androgenic properties of this male sex hormone, whereas the levo-rotatory antipode (available by synthesis) is biologically inactive. At the time I worked in Mexico, individual steroids were characterized by various physical parameters, among them the "optical rotation." The conventional way of accomplishing this latter measurement was to determine the extent to which the angle of polarized yellow sodium light is rotated upon passage through a solution of the substance in question. I wondered about the possibility of defining the optical rotation of a steroid not just at this single (visible) wavelength — the yellow sodium line frequently seen in fog lights — but at many different wavelengths down into the ultraviolet range of the spectrum. The resulting plot of wavelength versus angle of rotation is called an "optical rotatory dispersion" curve.

When I started my new academic position at Wayne, one of the first research proposals I submitted to the National Science Foundation was for funds to construct a "spectropolarimeter," which would allow us to carry out such measurements on steroids. In a series of investigations, which involved many graduate students and many Ph.D. man-years, we were able to convert such measurements — which by then had entered chemical jargon under the acronym ORD — into a powerful flashlight. In the end, ORD proved useful not only for steroids but also for the exploration of many other classes of natural products. One of the most important applications was the "establishment of the absolute configuration" of optically active molecules, the specific determination of the mirror-image formulation of a given substance. (That work was the principal reason I received in 1958 the American Chemical Society's Award in Pure Chemistry.) For the first few years, it took us at least three hours to measure one ORD curve. Toward the end of that decade, we secured an instrument that recorded these measurements automatically, thus reducing the time of measurement to a few minutes.

But if this period at Wayne State University had many peaks of intellectual joy, it also had ever deeper and wider valleys of physical pain. My knee, as a result of an earlier ski accident in Bulgaria, was by then so painful, especially in the Michigan winters, that I consumed daily at least two dozen aspirins. Finally, a biopsy reaffirmed an earlier suspicion that I suffered from a tubercular infection in my knee joint. When I learned that henceforth I would have to use a brace and crutches, I decided to have the joint removed and my left leg fused, and to have that operation in Mexico City during a two-year leave of absence from Wayne while serving as vice president in charge of a greatly expanded research department at Syntex. (The company had just been acquired from its Mexican owners by Allen & Company, a New York investment banking firm, and "taken public." Over the next fifteen years, Syntex's stock, though traveling somewhat on a roller coaster, was one of the great success stories on Wall Street.) I took with me to Mexico a group of postdoctorate fellows from Wayne to start what eventually became the first industrial postdoctoral fellowship program in a pharmaceutical company. Several of them stayed at Syntex — first in Mexico, and later transferring to Palo Alto as the company followed me to California — and became department heads, vice presidents, and in the case of the late Albert Bowers, the eventual chief executive officer of the company.

My initial two-year leave of absence from Wayne State University became a three-year stint at Syntex, which never brought me back to Detroit except for a personally moving honorary degree ceremony in 1974. The reason I did not return to Detroit was not dissatisfaction with Wayne — the university that had launched me on my academic career. Rather, toward the end of my second year in Mexico City, by which time I was pain-free and about as mobile as one can be with a fused knee, a professor from my graduate school days at the University of Wisconsin, William S. Johnson, was offered the chemistry department chairmanship at Stanford. Would I be interested in joining him? he asked one day over a crackling long-distance telephone call. By 1959, I had published a great deal, had won my share of awards and honors, was a full professor at Wayne State University though on leave of absence in Mexico, and hence could negotiate from a position of strength. So I flew to San Francisco, drove down the Peninsula to Palo Alto, and met the legendary Frederick Terman, then Stanford's provost and the man generally recognized as the creator of the Stanford Industrial Park and of Silicon Valley. While many academics over the years had been suspicious of my bigamous professional life, Terman was not. Just two years earlier, Stanford's medical school had moved from San Francisco to the Palo Alto campus, with a dramatic shift in emphasis toward the basic medical sciences and research rather than just focusing on health care delivery. Terman felt that the presence of a first-rate medical school and an upgraded chemistry department would encourage biomedically or chemically oriented industrial enterprises to join the electronic and computer companies in the Stanford Industrial Park. In his eyes, my industrial connection with Syntex made me attractive, not suspect.

Johnson and I decided that we would come either as a pair or not at all. Terman was not fazed by our decision, and was only temporarily taken aback when we refused even to consider moving into the space that Stanford intended to renovate for us in the existing chemistry building. To me, this structure, which had survived the 1906 earthquake, ominously resembled Wayne's Old Main. I felt that I didn't have to demonstrate again that exciting and productive research can be performed in an ancient building, and was ready to show that new, up-to-date facilities wouldn't be a hindrance either. In no more than eight weeks, Terman found a donor in the person of the chemical industrialist John Stauffer, who, together with his niece, agreed to fund the building that would be the bait to bring Johnson and me to Stanford.

Since it would take nearly a year to build the new Stauffer Organic Chemistry building, I decided to sit out the building phase in Mexico City by extending my Syntex stay. During this time, Syntex generated a flow of new drugs that pharmaceutical companies many times our size could not match. (Our greatest coup, and indirectly the most meaningful accolade extended to us, was when Eli Lilly, then one of the two biggest American pharmaceutical companies, committed itself to fund for a five-year period 50% of our research, with the choice of research topics and ownership of patents remaining with Syntex, provided Lilly would get co-marketing rights for any inventions.) During those three years, in addition to laying most of the clinical groundwork for our norethindrone Pill, we also created a second progestational agent (chlormadinone, which Lilly eventually marketed as an estrogen-free contraceptive which is still being marketed to this day in Germany by a local pharmaceutical firm), the powerful anabolic oxymetholone, the best-selling topical corticosteroid Synalar, a systemic corticosteroid related to prednisone, and finally dromastonolone propionate, a steroidal breast cancer palliative, which Lilly brought to the US market in the early 1960s.

Concurrently I directed — largely by semi-weekly long distance telephonic marathons and bimonthly visits to Detroit — my academic research group in Detroit in various structure-elucidation projects in the field of antibiotics, alkaloids, terpenoids, and, most important, optical rotatory dispersion. During that third year in Mexico City, in addition to writing a lot of articles, I even managed to complete my first book, *Optical Rotatory Dispersion: Applications to Organic Chemistry.* This wasn't the *tour de force* it might seem: the bulk of the contents dealt with our own research; the entire literature was at my fingertips. Furthermore, by then I had stopped doing laboratory research. Like virtually all scientists who wish to climb rapidly up the academic ladder or, for that matter, head into the higher echelons of corporate management, I was directing a large group of researchers, in this instance industrial as well as academic, which made it almost impossible to continue working or even dabbling at the bench. The choice is usually clear-cut: to work personally in the lab, one has to do it alone or with a small team. If a large group is needed — because one is in a hurry or wishes to attack various problems concurrently — one should stay in the office or the library. For me, the choice was never in doubt: not only did I always put a great value on time, but invariably I wanted to work simultaneously on a variety of projects and do so, furthermore, side by side in two worlds, the academic and the industrial. Since 1952, my lab coat, figuratively and literally, never got dirty again.

One aspect of my publishing contract with McGraw-Hill was unusual and perhaps even unique. When the publisher invited me to prepare this first monograph dealing with organic chemical applications of optical rotatory dispersion, I insisted on a penalty clause, whereby my royalties would escalate by 1% for each week the book's appearance might be delayed beyond my six-month deadline, but as a *quid pro quo* I offered a similar royalty reduction for each week's advance appearance prior to my requested publication date. To everyone's surprise, the McGraw-Hill lawyers accepted my proposal, provided I agreed to return the corrected page proofs from Mexico City within twenty-four hours of their receipt. This was supposed to prevent a horror scenario, whereby my royalties might escalate to unprecedented heights were I simply to sit on the page proofs. As a final compromise, the publisher set each chapter in print as it was received, rather than waiting for the entire manuscript. I managed to finish the book in time by sticking to a rigid Monday–Wednesday–Friday writing schedule, and McGraw-Hill was equally diligent. The royalties from the book eventually paid for a swimming pool at my new house in California, whose steps were set in Mexican tiles reading "built by optical rotatory dispersion."

My aspiration for an academic career, I am now certain as I cast my mind back to my twenties, was predicated largely on my yen to conduct research on my personal intellectual turf without apparent outside interference or control. At least that was my initial plan, right after receiving my doctorate not yet twenty-two years old. To accomplish that, I intended to establish a sufficient reputation through publishing significant research first in industry so as to start at an advanced level rather than at the beginning of the academic ladder as instructor or assistant professor — a plan that turned out to be rather naïve, because until the 1960s or perhaps even somewhat later, moving from industry to academia was a one-way street proceeding in the wrong direction, even though a few chemists had shown that it was not entirely impossible. My original dream about the supposed freedom of life in academe, especially nowadays, was also naïve, because the search for monetary support for one's research is so tough, time-consuming, and even demeaning that it constitutes a form of control frequently more oppressive than that always assumed to exist in industry. While this presumed freedom for research, coupled with academe's nebulous aura of prestige, was the chief attraction, the prospect of teaching also appealed to me and that is the subject I wish to explore for the remainder of this chapter.

Teaching Chemistry

The reasons why I was invited to Wayne University were solely my credentials as a researcher, since I had never before taught anywhere, not even as a teaching assistant in graduate school. I was expected to focus on graduate education to upgrade the caliber of the chemical doctoral program, because it is especially there that the results of cutting-edge research are most commonly melded with pedagogy. But like a baby seal entering water, I felt at home the day I faced my first class. Some of that self-assurance came from the fact that in the deeper sense of the word, I had already served as a highly successful teacher: I had taught my laboratory colleagues in Mexico City, primarily by precept and example, how to do chemical research. Thus, classroom instruction to graduate students in a subject with which I felt highly comfortable did not seem that different from laboratory mentoring. But as a result, strange as it may seem, I have never in a teaching career spanning over five decades had to prove my mettle on the most difficult terrain, that of large undergraduate classes. While I do not consider this a plus, I do not find it necessary to offer excuses, because my debt — and I do recognize it as such — towards the large undergraduate segment of university students was paid later as a chemist but outside of chemistry in my experiments in pedagogic *Déformation professionelle.*

Thus, during the first twenty odd years of my professorial life, I stuck exclusively to advanced chemistry courses dealing with my personal research forte: the chemistry of natural products and the use of advanced physical methods as aids to structure elucidation with a particular emphasis on those methods — optical rotatory dispersion, optical circular dichroism, magnetic circular dichroism, mass spectrometry, and ultimately also the application of computer artificial intelligence techniques to organic chemical problems — to which my research group had made fundamental contributions.

These may sound like dry topics — and, indeed, they can be. Students have been known to enroll in a poetry or history class out of curiosity and then get hooked on the subject by a brilliant teacher. Even chemists have been known to walk into poetry classes on impulse. But students don't open the doors of chemistry or physics lecture halls unless they have to take the course or have already decided to get some exposure to the subject. This is even truer of graduate students in chemistry: they have already made their career choice; at that stage, a professor might influence them, by good teaching, to focus on a certain chemical specialty or, by dull or obtuse lectures, push them into another one. I always took my classroom teaching seriously, especially since the formal teaching load at Stanford was never onerous. But I didn't want only to stimulate the students; I also wanted to involve them in a variety of pedagogic experiences — to experiment with variations on the usual one-way flow from lecturing professor to note-taking student. As I reflect on that early period in my teaching, I realize that it is in chemistry that I became infected by *Déformation professionelle*, although its real burst occurred outside my own discipline.

Pedagogic Experiments in Chemistry

My first pedagogic experiment was also the most ambitious. In the fall of 1962, I offered a graduate course on recent advances in organic chemical synthesis. In order to reduce the subject to manageable proportions, I decided to use steroids as the instructional template because during the preceding decade the synthesis of natural steroid hormones and their analogs represented the ultimate in complexity and sophistication. Few newly discovered synthetic organic reactions were not immediately applied to the steroid field. It was as if, for purposes of illustrating the state of the art to a budding chef, I limited myself to French *nouvelle cuisine.* I would not touch on Chinese, German, Greek, or Indian cooking; yet the new chef would learn how to prepare all courses in a meal, from soup and salad, to entrée, to dessert, through exposure to a newly introduced methodology. I decided to go a step farther, however; I asked the apprentice chefs to write a cookbook, each student being responsible for one chapter.

My course at Stanford was obligatory for Ph.D. candidates in organic chemistry, so I knew in the spring who the students would be in my autumn class. Before the summer recess, I convened all sixteen students and offered them a choice of sixteen topics in the field of steroid synthesis for detailed scrutiny. I provided each student with a quick survey of their selected subject and with leading references. I then asked them — in lieu of any examinations at the end of the course — to spend the summer going through every issue of sixteen international chemical journals (American, British, German, French, Swiss, Japanese, Canadian, and Czechoslovakian) covering the last ten years and to look for every article relevant to their topic. In culinary terms, one student would search all recipes for soups, another for sauces, a third for fish, and so on. At the beginning of the fall quarter, the students presented me with the first drafts of sixteen chapters — mostly a series of staccato phrases, with much of the information presented in terms of chemical structures — which we then studied together over the course of the quarter.

About a hundred copies of this text were distributed to steroid chemists throughout the world, soliciting their comments and criticism. The response was so enthusiastic that the following quarter, the sixteen authors polished their chapters, and two of the students redrew the chemical structures in a form suitable for commercial publication. The book appeared in 1963 under the title *Steroid Reactions: An Outline for Organic Chemists,* "prepared by sixteen graduate students at Stanford University under the editorship of Carl Djerassi." The name of each chapter's author appeared at the end of that chapter; for most of the students, it was their very first professional publication. Over half of these authors eventually became full professors in various universities and I like to think that this experience contributed to their career choice. The book was an instant success, and we all agreed to assign the significant royalties to the university to use in the construction and furnishing of a small seminar building, which on all Stanford University maps is now identified as the "Chemistry Gazebo" (see page 95 for a photograph). It is sad that by now, almost exactly fifty years later, virtually nobody in the present chemistry department knows nor cares about the uniquely cooperative origin of a small building in which everyone at one time or another had absorbed or transmitted the gems of chemistry.

Not all my courses were devoid of examinations, though I favored open-book or take-home exams over the standard true-or-false or multiple-choice questions; I was not interested in promoting rote learning or rewarding a capacity for memorization. I wanted students to face real-world facts: that time is the most expensive commodity, and that to solve difficult problems as quickly as possible, one needs to know *where* to look for the answers. This thinking eventually brought me to the most time-consuming examination of them all — at least for myself. I even gave it a name: the maximum-leverage test.

I used this exam in the early 1960s in a course dealing with contemporary methods of structure elucidation of complex natural products. At the time, this activity required a fair amount of chemical experimentation and intuition; the use of physical methods was not yet well developed or all-pervasive. In terms of the "dark room" analogy I introduced earlier, our flashlights were not yet powerful enough, and the *modus operandi* was still "Twenty Questions." The aim of my course was to teach students chemical *Fingerspitzengefühl* — the intellectual fingertip sensitivity that, for instance, distinguished the real medical diagnosticians from the hacks during the days when diagnostic medicine was still dependent more on "Twenty Questions" than on laboratory analyses and sophisticated scanning techniques. For examinations, I gave students chemical puzzles to solve at home or in the library. Creating such puzzles is not easy, and one day, while starting to think about the midterm exam, it occurred to me to give myself a break. Why not teach the students how to ask questions rather than simply how to answer them? When the midterm day of judgment arrived, I appeared empty-handed. As the nervous group looked anxiously around the room for the examination questions, I informed them that for our next scheduled meeting I wanted from each student a set of questions that could be used for an open-book, take-home exam and would cover the contents of the first half of the course. I would grade them on the appropriateness and the pedagogic value of their questions. The students were at once jubilant and disturbed. It seemed all so easy, but was there a catch? Indeed, there were two.

First, as the students soon found out at home, it isn't easy to pose test questions to someone who has access to every type of reference book; it is harder even than generating a crossword puzzle for people with a good dictionary and a thesaurus. As I expected, many of the questions the students asked were much more difficult than the ones I would have raised; many students confused trickiness, even deviousness, with subtlety or true insight. Second, when the students handed in the questions, they learned a new definition of *leverage.* I proceeded to hand the questions out to different students, being careful that no pair of students got each other's questions. I now asked them to *answer* these colleague-produced questions. When they had, I asked the original questioner to grade the answers, and the respondent to grade the quality of the questions. Finally, I graded each of the answers as well as the fairness of the grading the students applied to each other. Anyone patient enough to penetrate this thicket of grades and countergrades will appreciate how time-consuming the process was for me — each student had a different examination — but also how many different grades I generated out of one such examination. The vast majority of students agreed that in the process they had learned not just chemistry but also pedagogy.

When the time of the next examination approached, the end-of-term test, I told the students that I wanted them again to present me with a set of questions, which were this time to cover the material of the entire course and to be answered within two hours in class. When the day of the final examination arrived, I collected the questions and then handed them out to the various students. I had presented the first set to one student, the second to the next, and was about to offer a set to the third student, when the first called out. "Professor Djerassi, these are my own questions!" I ignored the interruption and proceeded with the distribution. "Professor Djerassi," the second student then complained, "you made the same mistake with me!" "And with me!" chimed in the third. It gradually dawned on the class that this had been my intention all along. Most of the students were delighted, but a significant few were horrified. They had never worried about the answers — after all, this was not an open-book examination; they thought they would demonstrate their virtuosity by the complexity of the questions they had asked. Hoist with their own chemical petards!

By the 1970s, structure elucidation had largely been transformed into an exercise in the judicious use of physical methods, to which I have referred to metaphorically as "flashlights"; and most of my chemical lectures focused then on that exercise. This was also the time when our research on computer-aided structure elucidation was at its most intense, conducted jointly with the research groups of Joshua Lederberg, a Nobel laureate and chair of the genetics department, and Edward Feigenbaum, the chair of computer sciences and one of the early pioneers in the field of artificial intelligence. I thought that it was time to expose a diverse group of chemists, rather than just my own research co-workers, to the power and limitations of knowledge engineering, otherwise known as "artificial intelligence." Instead of lecturing about the various spectroscopic methods and their applications, I told the students that I would take for granted their working knowledge of the various physical methods. If their knowledge was deficient, they could repair it on their own, using the standard texts. What I wanted to do was to demonstrate how diverse information — the various portions of the dark room illuminated by different flashlights — is put together to make a whole picture, and how one ensures that only *that* combination corresponds to the correct spatial arrangement of the room's contents. The computer programs our artificial intelligence group had developed were designed to accomplish a task for which the computer is best qualified, and that is, at the same time, most difficult to perform manually: the *exhaustive* generation of all possible structural candidates consistent with the *isolated* bits of information collected with the different flashlights. Once all of these structural straw men have been assembled through intuition and knowledge, the chemist can usually design a few key experiments or measurements that will demolish experimentally all but one of these alternatives, leaving only the correct structure behind.

Our software programs were written in user-friendly English and thus immediately accessible to the students. I asked each student to search the chemical literature for a publication in which the structure elucidation of some natural product was based on conclusions derived from a variety of physical methods (flashlights), but not confirmed by the unambiguous method of X-ray crystallography (equivalent to a color photograph in my dark-room analogy). After each student had selected such a paper, I asked that they subject the literature data to scrutiny by our computer program.

Did the computer agree with the chemist's conclusion that no other structural alternative was consistent with the published data? Or was some straw man lurking in the background which had not yet been eliminated by the evidence at hand? Had a straw man possibly been overlooked by the chemist? This hands-on approach to real-life problems would, I hoped, effectively illustrate the power of computer-aided checks and at the same time offer the class a view of the wide variety of structure-elucidation problems that were being studied all over the world.

The results of this pedagogic experiment were even more dramatic than I had anticipated. Without exception, each student discovered that the evidence cited in the published literature was consistent with at least one structural alternative that the authors had not considered: in one instance, the computer generated over two dozen structural straw men that had not been eliminated by the experimental evidence in the literature! I wrote to each author — in Japan, Italy, Spain, England and North America — describing the student's (really the computer's) conclusions and asking whether the author had any comments about the ambiguity of his or her published results. Most authors gave the expected, "Ah well, but..." response and then cited some additional spectroscopic or other experimental data that were not contained in the publication, but that the outraged author had dug up in order to demolish one of the computer-generated alternatives. My answer was, of course, that these data should have been in the paper in the first place. Some authors, however, never replied — possibly out of shock or chagrin; and I used those examples for a further educational experience for the students. I asked them to come up with the most time- and material-saving experiments that would differentiate among these remaining structural alternatives. A third group of respondents pleased me most: they wanted to know how they could get a copy of the program.

Undergraduate Teaching Away from Chemistry

It was also in the early 1970s — a watershed in chemistry's approach to structure elucidation and in my own attitude about classroom teaching — that I finally started to lecture to undergraduates. Why is it that as a young teacher I concentrated solely on graduate students, often older than I, whereas in the impending twilight of my professorial career, my principal educational constituency has turned into undergraduates? As I have begun to realize in the course of writing about the teaching component of my life, my gradual conversion from "hard" scientist to one with softer overtones occurred largely in the classroom.

In 1969, I had published my first "public policy" articles forecasting the decline in contraception research and the associated costs to society. I soon realized that the only way to reverse this decline would be to create a better-informed citizenry, and that the media, notably television, would never accomplish this through their present mode of spending a minute or two of precious broadcast time on grossly oversimplified sound bites dealing with complicated issues requiring extended discussion and critical thought. Precisely at that period, as it happened, an innovative new undergraduate program was established at Stanford with financial support from the Ford Foundation. The Program in Human Biology was designed to combat the increasing scientific illiteracy of our population during a time when most public policy issues had acquired technological or scientific aspects. Much of that illiteracy can be traced to the poor quality of our high school education in mathematics and the hard sciences, which still manifests itself as a fear of such subjects among undergraduates at even the most prestigious universities.

One way to offset this tendency is to emphasize the less physically oriented sectors of science, notably the biological ones, and to do so on the most anthropocentric and thus most persuasive front: the study of man. Not surprisingly, considering the composition of university faculties at that time, all the founders of Stanford's Program in Human Biology were men: the geneticist Joshua Lederberg, the pediatrician Norman Kretchmer, the population biologist Paul Ehrlich, the neurobiologist Donald Kennedy (later to become commissioner of the FDA in Washington, and then president of Stanford University), the sociologist Sanford Dornbush, the psychologist (and later Stanford's provost) Albert Hastorf and the psychiatrist David Hamburg. They devised a two-year core curriculum, that would both enable students with minimal exposure to the physical sciences to become proficient in biological and social science areas, and be followed by two years of advanced courses in more specialized fields — all this to be superimposed on the regular liberal arts requirements of the university. These senior professors also served as the principal lecturers in the courses.

The response of the students was remarkable: introductory classes, which were scheduled in lecture halls accommodating some fifty students, had to be transferred to auditoriums for four hundred. Within a few years, The Program in Human Biology became one of the most popular undergraduate majors at Stanford, selected by students whose goals were medicine, public health, law, environmental sciences and politics — precisely the constituency I wanted to address in terms of contraception and population issues. Since no chemistry professors were participating in the program (nor have any done so since), which by then had attracted a highly interdisciplinary faculty, I volunteered to offer a course for advanced undergraduates under the rubric "Biosocial Aspects of Birth Control" — a course that eventually led to a total change in my life as a classroom teacher. I chose that topic because I felt that birth control affects nearly everybody: people have used it, will use it, or are, at the very least, against it.

Of my several aims, the most important was to encourage students to think seriously about public policy and with real problems in mind. At a time when Stanford offered no formal undergraduate public policy courses, I felt that at least my professional background, bridging academia and an industry highly concerned with risk–benefit considerations, would qualify me for such teaching. I did not want to limit myself to prospective scientists; the future legislators and formulators of public policy are unlikely to come from that guild. By using the modifier *biosocial,* I hoped to make it plain that I was emphasizing the "softer" and broader aspects of birth control, and thus to attract students from a wider circle. I omitted all course prerequisites other than the requirement that students be in their last year and thus competent in at least one relevant discipline: religion, psychology, sociology, anthropology, economics, political science were some of the departments to which I proselytized. I knew that in biology and chemistry I would find the premedical candidates. Never in my career as a chemistry professor had I looked for customers; now I found myself becoming a promoter. I composed a one-page broadsheet in which I outlined the purpose of my course and the manner in which it would be taught. Attached to it was a questionnaire, which every interested student was asked to complete. I was curious not just about their academic qualifications but also about their social and geographic backgrounds, especially their exposure to travel and life abroad. I had a special educational experiment in mind, for which I needed a special group: equally distributed by gender, and with adequate representation from various ethnic, social and religious backgrounds. I limited enrollment to forty. Since over eighty students completed the questionnaire, I was able to start with a highly select and motivated group.

I like to think that it wasn't only the subject matter that attracted the students, although 1972 was the height of the sexual revolution, and contraception a topic that either interested or antagonized almost everyone. I like to believe it was the unusual structure of the course I described in my blurb. There would be no examinations, I announced, and my formal lectures would end after two weeks. During that time, the students could pick among a series of population groups, whose birth control options they would then study in depth in groups of six or seven. The emphasis would be on projected improvements in birth control, with each student examining the chosen population group from a particular disciplinary standpoint. A typical task force might include majors in pre-medicine, pre-law, economics, religion, anthropology, chemistry and psychology. The students would organize their research together, but each task force member would write a separate chapter of the group's report from her or his professional perspective. The main purpose of my course was to demonstrate that the concept of an ideal, universal birth control agent or approach was a chimera — in retrospect, an obvious conclusion, but one I had paid little attention to during my days as a "hard" scientist in the 1950s and early 1960s. Because of the tremendous divergence of different populations, what is appropriate for one group or even one individual may not suit the next. I wanted the students both to see that what the world needs is a kind of contraceptive supermarket, and to propose, through their own research, what some of the components of that supermarket might be. In the event, my first class in Biosocial Aspects of Birth Control picked the following seven population subgroups: white, American college students, typified by the majority of Stanford's affluent student population; Chicanos in San Jose, a politically and economically disenfranchised group of Catholics; Puerto Ricans in Manhattan, a similar group on the East Coast; people in the lower-income strata of Mexico City, a group related to the preceding two in economic status and religion, but living within their own political setting; Egyptian peasants in the Nile Delta and Indian slum dwellers in Calcutta — two third-world constituencies existing in quite different religious and political settings — and finally, a group representing the "women's liberation" position.

This first class in 1972 turned out to be an important educational experience for me as well as for the students and in retrospect was my true entry into the *Déformation professionelle* that I had redefined in the beginning of this chapter in complimentary terms. Both the students and I worked extremely hard. After the second week, once I had completed my lectures — illustrated with many slides, and each lasting for nearly three hours — I met twice a week separately with each of those seven task forces. During those sessions, I questioned each student about her or his research progress; I provided students with key contacts and encouraged them, using a modest financial kitty provided by the Human Biology office, to use long-distance phone calls as the most rapid way of extracting information from government bureaucrats here and abroad (remember that this occurred in the pre-fax, pre-computer days and a time of expensive long-distance calls!). Most important, I insisted that the students collaborate. Although all important social and technical advances in real life are the result of interdisciplinary team efforts, we tend not to incorporate that concept formally in our undergraduate curriculum. Our entire grading and evaluation system emphasizes individual performance and competition; collaboration among students is explicitly or implicitly considered cheating. In the student evaluations of my course, this task-force approach was voted the most original and worthwhile learning experience. (Even nineteen years later, a student from that first class, now equipped with both M.D. and Ph.D. degrees, wrote what every teacher adores to read: "'Biosocial Aspects of Birth Control' was the most important course I have ever taken in my career as a student.... You taught me how to fish instead of simply giving me a fish to eat when I was hungry.") I had no difficulty in evaluating the individual accomplishments, since everyone wrote a separate chapter, but these contributions had to be integrated within the entire group's report; each student had to know what every other member of the group was writing.

The climax of the course was the presentation of each task force's conclusion to the rest of the class and some invited guests. Each group had available three hours — half for the formal presentation, the other half for questions and answers. This was another time when I could evaluate each student's performance: by the manner of the presentation, by the incisiveness of the questions and by the perspicacity of the response. It was during these presentations that the students really surprised me. I had given them *carte blanche* in presenting their conclusions, provided every member of the group participated in some fashion and thus had the opportunity also to be questioned. The first task force used a *laterna magica* format, similar to what Czech filmmakers employed with notable success at one of the world's fairs — a device that pushed some thespian button in the other groups. From then on, students used everything from skits to full-fledged dramas. Even though I taught this course only every two years during the 1970s, word spread among the students about these presentations, and subsequent classes tried to outperform each other. Was this the time when the seeds to my subsequent playwriting career were sown, recognizing the extreme pedagogic value of live "case histories"?

Two of the most memorable presentations in the mid-1970s were given by task forces dealing with birth control problems among black Americans. On each occasion, all but one of the members of these task forces were black. One of these events was organized by Brenda Jo Young, subsequently a practicing psychiatrist, who, in those days of unisex student garb marked by blue jeans and hiking boots, stood out in her high-heeled shoes and fashionable dresses. Her group took over the main chemistry lecture room to stage an imitation mini-rock concert with strobe lights, raucous music and clever skits illuminating the attitudinal differences between blacks and whites. As usual, I sat in a corner seat in front, so that I could also observe the audience by making just a half-turn. Just as one of the students was wildly dancing on top of the lecture table, I noticed the partially open rear door and the horrified face of our departmental vice-chairman. Somebody having reported to him that something close to bedlam was going on in the auditorium, he had rushed down. Only my cheerful wave from the front row made him withdraw.

The second task force on black Americans wrote and performed a tragi-comic drama, which effectively demonstrated several basic factors that they felt determined birth control alternatives chosen by an American black, urban population: the high teenage pregnancy rate; the nonjudgmental and generally supportive attitude of black parents and grandparents in respect to teenage pregnancy; young black males' general lack of interest in effective birth control; and white social workers' relative ignorance of black family interactions. The role of the white social worker was played by a light-skinned black student, who took it for granted that the young teenage woman would have an abortion — only to find upon coming to visit the family to arrange for the procedure, the boyfriend, the girl's parents and a grandparent all sitting in a modest living room and planning the birth of the baby. The woman who had assumed the role of the pregnant teenager subsequently became pregnant herself shortly before entering medical school. I was both pleased and proud when I learned later that as a single mother she had successfully graduated as an M.D.

The most ambitious projects were conducted by my third class in the fall of 1975 and the spring of 1976. By then I had received feedback from two classes who, having taken "Biosocial Aspects of Birth Control" as a one-quarter course, had complained about its extreme time pressure and workload. Most of these students claimed that it demanded more work than any other class in their undergraduate experience, and I certainly found the same to be true of my professorial life. Since each student worked on a different project, and the bulk of the quarter involved one-on-one meetings, I had to be prepared to cover an extraordinarily wide range of subjects. At the end I had to read, criticize and grade the final reports from each group, which were usually at least one hundred pages long and frequently included several hundred references. This was a process that I could not delegate to others, and I found myself spending solid days when I did nothing but read and jot down marginal comments on these volumes — an experience that could not help but sensitize me further to the numerous sociopolitical and cultural issues associated with birth control. Additionally, these class papers — many of them quite sophisticated — also contributed to my knowledge of diverse population groups, as illustrated by a comparative study of the birth control problems of three Chinese populations: in San Francisco's Chinatown, in Taiwan and in the People's Republic of China. I learned most from this third class, when the course was spread over two quarters, with some of the students using the Christmas break for a type of field research not often available to undergraduates.

I had contacted the Rockefeller Foundation to ask whether it would fund, as a one-time experiment, the travel expenses of my human biology class for some exploratory research in more distant locations. Until now, my students' research was limited not only by time pressure but also by the financial resources at their disposal. These consisted of reimbursement for telephone calls and local travel within a hundred-mile radius of San Francisco. Students who had chosen Egyptian, Indian or other distant population groups had to depend on library resources or past foreign travel experiences. The Rockefeller Foundation, being especially devoted to supporting research in developing countries, agreed to fund this educational experiment because of an interest in birth control in general and its application to poorer populations in particular. As a result, I was able to organize the largest class of all — with ten task forces — and to offer each group the opportunity to send at least two, and sometimes all, members to sites irrespective of distance. Geographically, the three most ambitious projects involved populations in Kenya, Java and rural Mexico, but the American-oriented projects were also interesting. For instance, the Chicano task force, consisting of four students named Martinez, Ramos, Renteria and Rios, decided to conduct a comparative study of three Chicano communities in Denver (second and even third-generation Mexican-Americans), El Paso (a floating population on either side of the border) and Los Angeles (a site of numerous illegal immigrants who spoke no English). Several members of another group, which chose Native Americans, spent a couple of weeks in New Mexico with a tribe with which one of the anthropology students had established contact. A third group picked a rural setting in the South — Cherokee County, North Carolina, the home of one student — and provided a generally unfamiliar view of the educational and public health restrictions operating there.

Two particularly interesting choices had no geographical, but rather a functional definition: one dealt with the birth control problems of carriers of genetic diseases; the other, with those of developmentally disabled persons. Reading all of these reports was a monstrous task, and to address it, I secluded myself one drizzly weekend during California's rainy season at my ranch home. Soon wearying, on the spur of the moment I picked up my umbrella and took to my outdoor hot tub, where I floated naked without dropping a single page into the water. Of course, the steam was not without effect, but I never confessed to the students why the pages of their reports came back slightly curled.

The oral presentations of these groups were on the whole impressive — luckily, for so was the composition of the invited audience. The medical director of the Rockefeller Foundation flew out from New York for several presentations, and the chairperson of the California State

Assembly's subcommittee on health came to those on genetic disorders and developmentally disabled persons, in view of then-pending hearings on these topics in Sacramento. The report of the Indonesia task force, which dealt with market and social incentives for contraceptives in Java, got one of the students a job with the Agency for International Development in Washington; the World Health Organization hired a member of the Kenya group, on the basis of his field report, for a summer internship in Geneva before he entered medical school. Sharon Rockefeller, the wife of the present senator from West Virginia, attended the presentation of the Kenya task force, since she was also a trustee of Stanford University and acquainted with one of the student members of that group who had brought back some pink condoms with the logo of a spear-carrying Masai. It so happened that a reporter from *People* magazine was present for an article on the unusual nature of my course. "Not exactly an apple for the teacher," the reporter wrote in *People*, "it was a box of pink condoms. Djerassi was delighted." The group graphically demonstrated to the audience the gulf in attitudinal differences between *us* and *them* by passing around fried termites, a Kenyan delicacy, and challenging everyone to taste them. I seem to recall that Mrs. Rockefeller was one of the few to take up the offer.

When I started teaching the birth control class, the distribution of the students by gender was approximately equal. By the late 1970s, fewer and fewer men were choosing to enroll; and by the early 1980s, at most 20% of the students were males. This lack of male interest in the subject was even more pronounced in an offshoot course I first sponsored in 1983. I had observed that the one task-force topic always selected by each class dealt with what I called at that time "women's liberation position." Aside from its obvious timeliness, there was another reason it was such a popular choice. I always drew attention to the quotation by the anthropologist Margaret Mead that I have already cited in the chapter on the Pill:

> *[The pill] is entirely the invention of men. And why did they do it? ... Because they are extraordinarily unwilling to experiment with their own bodies ... and they're extremely willing to experiment with women's bodies ... it would be much safer to monkey with men than monkey with women ... Now the ideal contraceptive undoubtedly would be a pill that a man and a woman would have to take simultaneously.*

I invariably asked the students to start out with Mead's position and then to present evidence either for or against it. The students who considered Mead's thesis justified were asked to develop a realistic proposal for ameliorating the perceived male dominance in birth control research. As the years passed, and especially when Stanford introduced a special Feminist Studies Program, I decided to create a specific course under the title "Feminist Perspectives on Birth Control." I have since given it five times, once at Bard College in New York, because I was interested in contrasting the perspective of some students in a small eastern college with that of their western counterparts in a large university. With a solitary exception, all the students in those classes were women.

Have men suddenly stopped believing in birth control? Has "yuppification," the present student generation's preoccupation with material goods and professional advancement, made the men relegate birth control to a low priority? I believe that the real reason is that the students of the 1980s are all children of the Pill generation of mothers. The Pill has made many important social contributions, not the least of them that birth control became an accepted topic of dinner conversation. But, concomitantly, it has also created a social atmosphere in which one more responsibility — this time that of contraception — has fallen on the shoulders of women. Many women, of course, accepted it eagerly as an important sign of emancipation and freedom from male dominance, but one of the consequences of that achievement has been a collective shrug of male shoulders, an outcome I regret deeply.

I find it both disheartening and amusing that it was the women in my class who played the biggest role in making condoms available at Stanford University. In view of the fact that approximately 40% of all condom purchasers are now women, I thought it only appropriate to encourage some of the students in "Feminist Perspectives" to focus on that form of contraception. At a time when AIDS was still considered a misspelling for the dietary pill AYDS, I wrote in an article dealing with teenage pregnancy in the US that if high schools and colleges were to make condoms readily available to students, they would not only be contributing to public health but also helping to teach young men at the most impressionable stage of life to accept some responsibility for contraception. Among the papers women in my class wrote were critical feminist evaluations of condom advertising and promotion. For instance, instead of a phallocentric terminology like *Sheik* or *Ramses* or *Trojan* — or, for that matter, the spear-throwing Masai on the box of Kenyan condoms the *People* writer reported — why not call a brand of condoms *Cleopatra*? And instead of blue and green and orange, one of my feminist warriors in class asked, why not color the Cleopatras gold? (I wish I could claim that the subsequent commercially available gold-wrapped condoms were stimulated by that report but that would be wishful fantasy.)

In 1980, two women task force members examined what it would take to introduce condom dispensers at Stanford University — and received a first-class lesson in academic bureaucratism. The dean of student affairs sent them to the acting deputy vice president for administrative services and facilities, who suggested they see first one of the university's legal counsels, and then the athletic director since they were thinking of gyms as a possible location. Even the university's ombudsperson was of no help when the students reported the horrified response of one librarian after they had suggested the library's toilets as a suitable site. "Just imagine all the high school students who would come to the library to get condoms!" he exclaimed. I could hardly think of a better use for condoms: reducing teenage pregnancy while increasing literacy, even at the price of finding a used condom or two on the floor among the stacks. Not until 1987 did a feminist Gang of Five, dealing with over-the-counter contraceptives and led by my students Shirley Wang and Jennifer Yu, succeed in wearing down Stanford's administration and getting the first condom dispensers into some of the toilets. Were it not for the fact that I practice what I preach by having had a vasectomy many years ago, I would have been one of the first customers. Thus, the box from Kenya still rests unused in my huge collection of condoms into which I dip for occasional demonstration purposes. I am probably one of the few persons who could claim the cost of condoms as an income tax deduction for my professional activities as a teacher.

Ethical Discourse through Science-in-Fiction

My just-described transition of a chemist teaching pure chemistry to that of a chemist applying his pedagogic skills to aspects of socio-biology and feminist studies was not only a jump in subject matter. It meant that I was now dealing with undergraduates at a time when they were not yet necessarily fixated on a given major and hence more receptive to see their outlook broadened and thus even persuadable to consider another field. The biggest attraction for me, however, was that in contrast to chemistry classes, I now had students from a wide variety of disciplines, notably from the social sciences and even humanities, students who if they had any contact with a chemistry professor would have had it solely in a huge, impersonal freshman lecture auditorium. Now we met for months in my office in small groups or even one-on-one.

The next jump was even bigger though clearly logical if not inevitable. If one intends to deal with societal issues, a focus on professional ethics sooner or later must enter the picture. It so happened that around that time, I served on a specially convened interdisciplinary panel, mostly from biomedical disciplines, that had been assembled in Washington by the National Institutes of Health (NIH) to consider guidelines for the obligatory implementation of courses dealing with ethical practices in research and in the medical sciences. Soon thereafter, this became a requirement for all individuals supported financially through training grants of the NIH, although this was not extended to ordinary research grants from the NIH or the National Science Foundation (NSF). I not only supported these moves but felt that sooner or later, such a course requirement should also be instituted in chemistry and the other basic sciences. But when I approached the chairman of our department with an offer to organize such a course, the reception was unexpectedly negative: "Why open a Pandora's box?"

I probably could have forced the issue, but instead I turned to the Medical School as a potentially more receptive site. It so happened that years earlier I had served on the Medical School's selection committee for a new dean, a committee that also included a student representative, Thomas A. Raffin. In the intervening years, Raffin had graduated, entered the Medical School faculty and by the mid-1990s had become a professor as well as co-director of the Stanford University Center for Biomedical Ethics. He was not only receptive to my offering a course within that program, but was also intrigued by the approach I had planned to take as already indicated by the title I had proposed, "Ethical Discourse Through Science-in-Fiction" which was left unchanged in the Medical School course catalog under

"Medicine 256". Because of its unusual character — how many courses in the hard sciences involve fiction-writing and are taught by fiction-writers? — I describe it below in some detail through excerpts from one of my by-now out of print works.

The impetus for the course was the unanticipated success that my first science-in-fiction novel, *Cantor's Dilemma*, enjoyed as an academic textbook. The paperback is still being reprinted once or twice annually, primarily as a result of adoptions by American colleges and universities in courses dealing with ethics in research, the primary focus of the novel's plot. Graduate schools of business administration have long learned the advantage of employing "case histories" in their curricula, which makes the adoption of my novel easily understandable. But ethical or behavioral deviance in research generally involves individuals rather than impersonal corporate entities, which means that scientific ethical case histories quickly run into concerns about violation of privacy as well as into manifestations of the whistleblower syndrome. I wondered whether student-generated "science-in-fiction," in which all aspects of scientific behavior and of scientific facts are described accurately and plausibly yet disguised in the cloak of fiction, could be used to illustrate ethical dilemmas that frequently are not raised for reasons of discretion, embarrassment or fear of retribution.

The course was restricted to graduate or postdoctoral students, because I wanted stories based on experience, or on events that the writer knew enough about to fictionalize with authority. The first time I offered that course, fourteen graduate students and postdoctoral professionals from twelve different departments (ranging from chemistry, biophysics and computer sciences to genetics, infectious diseases and psychiatry) enrolled to compose short stories dealing with ethical issues in science or medicine. The stories had to be handed in on the first day of class (meaning that they were written during the preceding quarter break) and were then distributed to all class participants without authorial identification — thus permitting unrestricted debate. The balance of the course dealt with in-depth three-hour long examinations of the ethical or behavioral problems raised by these stories — discussions that frequently bordered on fireworks. In all my decades-long years of teaching at Stanford University, this was the most exciting classroom I had ever experienced. Aside from offering the students a veil of anonymity that removed most obvious restraints, the course had allowed them to pick the topics that concerned them most rather than being restricted to ethical problems chosen by the instructor.

Several doctoral thesis supervisors (especially from chemistry) were not too sympathetic to their students taking what they considered Djerassi's "Mickey Mouse" course: instead of spending time on literary fantasies, why not spend those precious hours in the laboratory? Such attitude is by no means unique to professors. It is shared by many students in the hard sciences such as chemistry, where "ethics" courses — if taught at all — are not taken very seriously, the tacit assumption being that "hoaxing, forging, trimming, and cooking" (Charles Babbage's words in his famous admonition of 1830 to the Royal Society) in research is a pathology encountered mostly among biologists or clinicians. I have now taught "Medicine 256" four times, with students from well over a dozen departments, but during those years only two chemists, both female, dared to desert a small portion of their seventy-hour work week for such a "soft" exercise. Yet one of the two, Shirley Lin, now a professor at the Naval Academy, received a $500 payment from *Chemical & Engineering News*, the largest chemical news magazine in the world for her story, *Meeting*, which the editor introduced with the following words: "it is the first time in C&EN's 75-year history that we have published a short story." Another story — among the best from that first class — subsequently appeared in a medical publication, anonymously at the request of the author. In this one, a young physician finds herself in the painful role of having to advise a semi-literate, pregnant Albanian refugee to undergo an abortion, while she herself is facing the heavy decision whether to abort her own first pregnancy. The story was autobiographical. Several students had tears in their eyes when it was discussed in class.

In addition to creating a forum for open disclosure and debate, my course also addressed the question of how scientists might communicate better with their colleagues and the general public. This discussion led to an extraordinary literary experiment of which I am still very proud. I had the class attempt a group composition modeled after the Japanese Renga (linked verse in which stanzas are composed by two or more poets in alternating sequence, often as a form of competition) for a short story dealing with some scientific ethical dilemma. Each paragraph was to be written by a different student without knowing the identity of any predecessor author, each being allowed two days to compose her or his paragraph. Once the fourteen-paragraph "A Science Renga" was completed, each student was asked to add a fifteenth paragraph, thus generating fourteen new endings. After distributing all variants to the class, the "winner" was selected by secret ballot. Though bearing the names of all authors — a feature common among scientific papers but virtually unheard of in literary publication — none knew who had contributed what segment.

The Renga format bears an interesting resemblance to the process of scientific co-authorship, which also has its collegial and competitive aspects (indeed I used the idea in my second novel, *The Bourbaki Gambit*). But the Renga experiment of the class was a "purer" collaboration, since each author was associated with the entire enterprise though with no identifiable individual component. I decided to see whether a scientific journal would have the guts to publish this story by starting on the very top, with *Nature* — the science equivalent of submitting a first short story to *The New Yorker*. But the *Nature* editor bit within a week (itself virtually unprecedented) and "A Science Renga" under the names of fifteen authors appeared in the June 11, 1998 issue of the journal. It was the first piece of fiction that *Nature* had ever published — at least knowingly so — since its founding in 1869. So unusual was this event that a major French newspaper, *Libération*, featured it on a full page. It may also be the first short story in literary history that bears the name of fifteen authors. But why fifteen, when there were only fourteen students?

Though dangerously tainted by overtones of a shaggy dog story, the answer is relevant to the subject matter of "Medicine 256", since "senior" authorship is so often the source of ethical conflicts in science. Whose name should come first? Some students suggested the device of "honorary" authorship, so common in science yet still not sufficiently condemned; in other words to compromise with "Djerassi et al." "It was your idea," some of them said. "You organized it and even got it accepted in *Nature*. Put your name in and put it first." Of course I rejected that alternative since adding one's name to papers where one did none of the actual work was one topic the entire course had meant to address critically. Alphabetical order was the next obvious alternative, a common enough approach that is also fraught with complications. Here is one such case, taken from *Cantor's Dilemma*.

> *"Most people in the field — including Celestine — would consider me the senior author. It's not necessarily the first name in a list of authors, although some senior researchers feel very strongly that their names must always appear first. Others always use an alphabetical order — "*
>
> *"Well, it's not true in our lab," Stafford mumbled, "it's always alphabetical." This was the only serious bone of contention in Cantor's group. Lab gossip had it that no Allens or Browns had ever worked with Cantor. There had been an exchange fellow from Prague, named Czerny, but that was the closest alphabetical proximity to "Cantor" that anyone remembered until Doug Catfield had arrived last year."*

Alphabetical precedence among the fourteen authors of "A Science Renga" would have meant that Dina L. G. Borzekowski, a postdoctoral fellow from the Stanford Center for Research in Disease Prevention, would appear as senior author in the *Nature* index. But was that fair, when all deserved equal credit? Ordinary mortals outside the scientific community are often astounded by our preoccupation with names on papers and the complicated solutions we sometimes devise, especially since many professors — especially "honorary" authors — now place their names last. But what about the front of the queue? Here is another relevant excerpt from one of my novels, this time the last volume, entitled *NO*, where Celestine Price, a fictitious chemist, discusses journal authorship with her former boss, Michael Marletta (in point of fact, a real person, and now President of the Scripps Research Institute), when Paula, a non-scientist, intrudes.

> *"What we were talking about, before you came," Celestine turned to Paula, "was the subtlety of how to apportion credit among all the authors. These days, four or more authors is par for the course in any competitive field of chemistry or biology. Having settled who is last, the question now is who comes first."*
>
> *"Really?" said Paula. "I'd think you'd pick the person who has done most of the work."*
>
> *"You think that's easy? That's exactly what we've been discussing. Recently, John Scott from Portland published a real first in* Science. *He had five co-workers, all women — a real harem — but what made it a first was that the first two names listed in the article were marked with an asterisk. Can you guess what the footnote said? 'These authors contributed equally to this manuscript.'"*

"Brilliant," exclaimed Paula.

"You see?" laughed Marletta.

"Brilliant?" Celestine snorted. "Suppose the first asterisked name had been Smith and the second Price. I would have gone to Scott to point out that in any citation, that article would be referred to either as 'Smith et al.*' or 'Scott* et al.*' To me,* 'et al.*' does not mean 'equal.'"*

"So what would you have had Scott do?"

"Ah," grinned Celestine. "As a first try, I'd have separated the names Smith and Price by an equal sign rather than a comma. But since no editor would allow that, I'd have told him to do it alphabetically."

"You mean Cantor's system? Why should Smith agree to that when your name starts with a P?"

"Fair enough. That was also Michael's point. So I asked why not toss a coin? And you know what the fair-minded Professor Michael Marletta said?" Celestine poked him lightly with her index finger. "Why don't you tell Paula."

"In my lab, I decide such issues, not the drop of a coin."

I felt that none of these conventions would work with our Science Renga. When the paper appeared in print, my unilaterally chosen first author was one "Alfred N. Alston Jr." whose name was followed by an asterisk that did not indicate a department address but rather the fact that he was deceased.

Nature never caught the discrepancy — fourteen students but fifteen authors in alphabetical order, headed by Alston — but an interviewer on an NPR radio show did, and asked me to explain who the fifteenth author was. "It's an anagram," I admitted, and then challenged the listeners of the program to come up with the answer by e-mail. The first correct respondent was promised a signed copy of *NO*. I had barely returned home to find a correct answer waiting for me: "Leland Stanford Jr.," the person after whom our university is named, and now the author under whom "A Science Renga" will be found in perpetuity in the annual index to volume 393 of *Nature*. I considered Leland Stanford Jr. a fair compromise, since our university would not have been founded — and our Science Renga never composed — had he not died prematurely whereupon his bereaved parents founded the university named after him. The student authors were so jubilant to find their names in *Nature*, an addendum to their professional biographies that many of their professors could not boast of, that none objected to my unilateral decision about senior authorship.

I am almost finished with my drawn-out record of this pedagogic experiment, but not quite. One of the participants in my course, a talented poet, E. Weber Hoen, decided to compose an abstract of the original "A Science Renga" (since abstracts are *de rigeur* in every published scientific article). But he did so in the form of a Shakespearean sonnet, where each of the fourteen lines corresponded to one of the paragraphs of the Renga! The "old goat" of the title is the professor in the short story, who was terrified of being scooped by his younger disciple. I quote the last six lines of Hoen's sonnet, which interested readers can look up in its entirety by simply Googling for it on the web:

Old Goat
It is height you desire, and with that, truth,
to shake your beard on an eternal view,
as if from there you might behold your youth.
The rain, though, has you blind. Below, like you,
the young conspire in fear against their king.
Goat, you are old. You have not learned a thing.

When I first read the sonnet, I realized that my age clearly qualifies me as an "old goat." But I am luckier than Hoen's old goat. I have learned a thing or two by turning into a Professor of Professional Deformation, a claim I could not have made if I had remained a pure, unsullied chemist.

Science-in-Theater as Pedagogy

By now, it will be clear to any reader that my ever-extending exercises in professorial professional deformation were related to the changes in my own intellectual interests. Thus, it was the shifting interest in the 1960s from my earlier focus on contraceptive "hardware" like the Pill to contraceptive "software" — the cultural, political, religious, economic and legal aspects of human birth control — that made me pay attention to broad policy issues. Once having entered that thicket, it was only a small step to incorporate that interest in creating one of the first public policy courses in the newly established Human Biology curriculum at Stanford. My subsequent ambling through "science-in-fiction" caused me to first embark on the above-described "Ethical Discourse through Science-in-Fiction" experiment and then, in 2001, to organize a sophomore seminar entitled "Science-in-Fiction is not Science Fiction" in which students had to pick from a list of science fiction novels, which did not only include mine, but some early classics like Sinclair Lewis's *Arrowsmith*, C. P. Snow's *The Search*, and William Cooper's *The Struggles of Albert Woods*, as well as the more recent John Updike's *Roger's Version*, Simon Mawer's *Mendel's Dwarf* and Jennifer Ball's *Catalyst*. During the seminar, less attention was dedicated to the actual outline of the novel than to the known or hidden scientific background behind them that served to distinguish them from science fiction. During the three years I taught that seminar, ending with my eightieth birthday, I had some superb students including Tonyanna Borkovi and Joshua Bushinsky, who, as I mention in the *"Writer"* chapter, served for nearly a year as valuable research assistants in my archival research on Isaac Newton and some Restoration playwrights of that period.

After an interregnum of a few years, during which I did not teach at Stanford but primarily dedicated most of my creative time to playwriting, I started to wonder how to convert this latest infatuation of mine into an innovative educational experience for undergraduates. As I will show, it was not easy to accomplish this, but by now I was eighty-five years old and obviously impatient. With the actuarial clock in my mind clicking louder and louder, I simply got cracking on a typical *if the mountain will not come to Mohammed, Mohammed will go to the mountain* approach.

What I proposed to offer, and eventually managed to accomplish, was an undergraduate seminar with the title "Science-in-Theater: A New Genre?" My saying "managed to accomplish" clearly implies some difficulties and there were some. First of all, while Stanford is a superb university with generally wonderful facilities, this unfortunately does not apply to the drama department, whose theatrical space and equipment are, quite frankly, deplorable. It is not surprising, therefore that its educational emphasis is placed primarily on playwriting and theater history rather than performance. One need only compare the respective theater departments of Yale and Stanford, or for that matter of Stanford with a number of other large state universities, to see the validity of my judgment. And since Stanford is one of the richest and best-endowed American universities, financial resources are not the key reason, but rather the allocation priorities which I consider grossly skewed in favor of the sciences and professional schools compared to the Humanities and Arts. The final problem is that, at least in science departments such as chemistry, there was no interest to include any sort of "drama" into some of the undergraduate courses as demonstrated by the fact that when I suggested using one of my pedagogic plays (*NO* dealing with the chemical Nitric Oxide) in a freshman chemistry class, the proposal was turned down for lack of time in an already overcrowded curriculum of standard obligatory lectures. (Fortunately, my alma mater, the University of Wisconsin, saw this very differently and has used *NO* precisely for such educational purposes.)

Second, the drama department was extraordinarily territorial and not exactly welcoming to outsiders, notably from one seemingly overburdened with scientific credentials. The following concise description of the projected course raised no open opposition, but little enthusiasm:

> *Scientists operate within a type of tribal culture where rules, mores and idiosyncrasies are not taught through specific lectures or books but, rather, are acquired through a form of intellectual osmosis in a mentor-disciple relationship. Is that also the reason why, until recently, scientists were hardly ever "normal" characters in plays other than being represented as Frankensteins or nerds? But during the past dozen years, more and more intellectually challenging plays have appeared on the Anglo-American theater scene in which scientific behavior and even science are presented accurately. Has this happened because of didactic motivation on the part of some playwrights or because the intrinsic theatricality of science and its metaphoric significance has been recognized? These issues will be discussed and partially viewed (via videos in the instructor's San Francisco home) through an examination of a number of plays, some of which were written by the instructor. A short play-writing experiment will also be conducted.*

The statement in the penultimate sentence "plays, some of which were written by the instructor" did not change my outsider status, since none of these plays had ever been read, let alone performed by the drama department in spite of some earlier efforts of mine. Yet I wanted to have my course cross-listed in the catalog between chemistry and drama to initiate a hitherto never attempted collaboration between two such departments, where the respective chairs had never even met, so as to encourage undergraduates from diverse disciplines to enroll. Fortunately, this time I had the great luck that the then serving chair of our department, Richard Zare, was not only a superb chemist (true also of most of our past chairmen) but, *mirabile dictu*, also a real theater aficionado. He immediately agreed to my proposal and by offering to have all of my professorial salary paid by chemistry, the drama department agreed to the cross-listing as a free hitchhiker.

Nancy L. Tolin

Two views of San Francisco living room where the Stanford undergraduate seminar "Chemistry-in-Theater" was taught

There was another unusual statement in my proposed catalog entry for my course description, namely that the (three-hour)-seminar would be held in my home in San Francisco, nearly an hour's drive from the Stanford campus. I did not make this suggestion for the sake of my personal convenience, especially since I still have an office at Stanford, but for other reasons. The vast majority of Stanford undergraduates spend preciously little time in San Francisco, a type of cultural parochialism also displayed by a fair number of professors. But this should not be a reason not to try to combat it. Second, as I already indicated in *Heimat(losigkeit)*, I live in a large beautiful apartment with one of the greatest views of the San Francisco Bay and city. I also have some attractive and even important art in my abode and exposing students with frequently very little familiarity in this area to such an environment seemed commendable. However, the most important factor was the excellent projection facilities that I had in my large living room, which meant that I could show audiovisuals of a type that virtually no drama department possesses to students reclining comfortably on a couch or easy chair.

And to what audiovisuals am I referring? I have had the good fortune to have had all of my plays published in book form and translated as well as performed in several languages; in addition, I have always tried to acquire archival videos of such performances. This has enabled me to expose students or audiences at some of my academic lectures on "science-in-theater" to an aspect of theater that is virtually never taught, but to which I refer in the *"Writer"* chapter: while a film, however important or magnificent, can only be shown world wide in its original form except for dubbing or different subtitles, a play can be adjusted in terms of space, form of presentation and most importantly textually so as to accommodate different cultural and even political facets that are encountered in different countries.

As an example, I use *Oxygen*, the play that Roald Hoffmann and I wrote jointly and which is so far available in twenty languages, as an example to demonstrate this unique aspect of the theater's flexibility. I show students a short three- or four-minute excerpt from a beautifully filmed production of the play at the University of Wisconsin Theater, which the Wisconsin Science Initiative under Professor Bassam Shakhashiri had made available for commercial distribution, and then show the students exactly the same scene from some highly professional productions in Korea, Bulgaria and Germany. I repeat that experiment with my play, *An Immaculate Misconception*, where I compare a fragment from the original American staging with the French, Japanese and Austrian ones. Any observer, even the most experienced theater professional will be startled and excited to see how spectacularly different each identical segment is performed through the interpretative abilities of different theaters. In fact, in most instances, they would not even have recognized it as coming from the same play.

To make this unique seminar possible, I convinced a sympathetic Dean of Undergraduate Education at Stanford to cover the travel cost, so that students could come to San Francisco together in a couple of cars. By scheduling the seminar from six to nine-thirty in the early evening and serving them interesting ethnic food of which San Francisco is famous, I offered the students (limited to no more than fourteen) the ability to interact during their one-hour drive to my home and to then recapitulate among themselves what had just transpired upon their return trip — a bonus in terms of collegiality and togetherness that most courses lack totally. The remainder of the structure of the course was also pedagogically successful as shown by the generally enthusiastic student course evaluations that every Stanford class has to submit anonymously to the Dean's office.

I shall end this description with one amusing incident. While all students had read the plays used in my seminar, each given student had to select one specific one and then, just as I had done earlier in my "Science-in-Fiction is not Science Fiction" class, delve in great detail into the historical and scientific antecedents. But in addition, each student also had to rewrite the last few pages of "their" play by providing a new ending, which the student and some volunteers then presented to the rest of us in the form of a mini-dramatic reading. One of these plays was Michael Frayn's *Copenhagen,* which a very sophisticated freshman from Vienna, Carolyn Schwanzer, had picked. When, after the students' departure, I returned to my computer for another couple of hours of work, it suddenly occurred to me to send Carolyn Schwanzer's new ending to Michael Frayn in London, whose e-mail address I had from earlier correspondence. It was then close to midnight in San Francisco on May 8, 2009, which meant that it was nearly eight o'clock the following morning in London. Apparently, Frayn is an early riser since by one o'clock in the morning my time, while I was still working, the following e-mail message arrived from him: *Excellent. Thank you. It would obviously be in line with the spirit of quantum mechanics to sum over many alternative endings to the play. Anyway, give Ms Schwanzer my congratulations.*

Before going to bed, I forwarded the message to Schwanzer, who was greeted in the morning by a message from the playwright whose work she had deconstructed and slightly rewritten just twelve hours earlier. It is bonuses like this together with engaged smart students that make teaching such a pleasure.

When I wrote the lines of this chapter, I was approaching my eighty-ninth birthday. What other *Déformation professionelle* is still in store for me in my last decade, assuming, of course, that there is still one more to come? I wonder.

"Writer"

The Austrian stamp reproduced in the chapter *Heimat(losigkeit)* bearing my face describes me as a *Romancier* as well as a chemist. *Romancier* is a lovely word, which I am more than happy to embrace, but in this chapter I would like to record my transformation from chemist — who, like all scientists who publish, is *ipso facto* a writer — into a person who decided late in life to don the mantel of novelist and subsequently also that of playwright. I will not be unhappy if it ultimately turns into my burial shroud. But whereas a scientist's writing represents mostly transmission of information and is accepted and evaluated as such, including its didactic component, a *Romancier* would reject outright such didactic baggage, since didacticism, unless well hidden, is often considered the kiss of death by literary colleagues and reviewers. In addition, for a scientific author in information transmittal, content is king, with style counting only as embellishment. Nobody would dare say that of a *Romancier*. I make this point to explain my use of the protective function of quotation marks around the title of this chapter to describe my own literary output during the past quarter century where at least a whiff of didacticism has been deliberately allowed to escape. If Quintus Horatius Flaccus's words, *Lectorem delectando pariterque monendo* [delighting the reader at the same time as instructing him] in *Ars Poetica* is still quoted approvingly two thousand years later as an accurate description of *didactic*, what's wrong with my at least dabbling with what Horace preached?

But rather than starting with the intended subject, I shall digress — as I have done so often in these autobiographical reminiscences — by first presenting a totally factual account of some sexual fantasizing of mine that had already started in my teenage years.

> *Through years of sexual reality, I dreamed ever so often of a woman lover who'd sing while coupling with her man. Once a husky-voiced woman, who'd brought her guitar, started to sing in a stunning contralto. Lying on my bed, exhausted and content, gazing at the naked woman*

strumming her instrument, I was about to ask her whether she could.... But then I chickened out; I was afraid she'd just laugh.

Years later, I happened to go to a performance of Monteverdi's L'incoronazione di Poppea, set in the time of Emperor Nero before he went mad, with Tatiana Troyanos singing the lead role. About halfway through the love-scene on the couch between young Nero and Poppea, the performance assumed such erotic overtones that I began to squirm in my seat. I don't go to operas for sexual titillation; except for an occasional Salome or Lulu, it's the music that excites me. But this was different. Suddenly I realized that Troyanos was the woman of my fantasies who'd walked into Spartacus's tent over two thousand years ago.

I was a relatively late bloomer, a virgin until nearly twenty. But as an otherwise precocious teenager, I made up for it soaking in the delicious warmth of a full bath. Not in one of those modern tubs, where even I — only five foot five — can't stretch out, so shallow that the water barely reaches one's navel, where either shoulders or feet project into the cold. No, my passions throve in a real tub — one of those huge pre-war jobs — where I'd float, water up to the chin, with soapy hands between my thighs, incandescently copulating with Veronica Thwale.

When I met that cool, severely dressed, sexual androgyne — perfumed in civet and flowers, walking down a church aisle, the Decameron camouflaged inside the covers of her prayer book to prove the pleasures of blasphemy — I was smitten for weeks, then months. Veronica, in her twenties, was the deftly cunning courtesan I'd been waiting for and had finally found: in Aldous Huxley's Time Must Have a Stop. God, she was something! Once our passion slipped me forward in that six-foot tub, so that I choked on soapy water.

But the spur to my longest and most lucid dream was The Gladiators — Arthur Koestler's version of the slave uprising led by Spartacus. Don't get me wrong: as time passed, as I became a man, there were months, even whole years, when Spartacus did not exist. But the vision never departed totally from my memory. When I had a lover whose climax always ended in such a long-drawn cry that we could never meet in a hotel, I wondered more than once how Spartacus had handled this in his tent on the plains of Campania. And when I saw Yury Grigorovich's choreography of Spartacus at the Bolshoi in Moscow thirty-five years after I'd read Koestler's novel, I felt a pleasure reawakening.

I still remember where the scene appeared: on a left-hand page, fairly high up, three or four lines from the top. Koestler had sketched

Spartacus's portrait with just a few deft strokes: the tall, slightly hunched body draped with fur-skin; his wandering eyes and cleverness; his freckles; his words that seared your ears as he spoke. And the women he had come to his tent to satisfy his sexual urges — the camp-followers, the impedimenta. But one evening, on that left-hand page, a woman of a different breed came to him. I, the teenage virgin, saw it clearly: the tent flap swaying as she slid in barefoot; a whiff of musk and ointment and female sweat entering with her; the chocolate skin gleaming as she passed the flickering candle; her firm breasts bearing nipples like diamonds on a ring. She kneeled beside Spartacus's reclining figure, peeled off his fur-skin and wordlessly started to caress him. For once, Spartacus took no initiative; he let the woman pleasure him. When she saw him aroused, she proceeded to sing in a low voice, mounting him — the first time Spartacus had ever been mounted by a woman — and with his phallus deep in her, she thrust faster and faster, singing more loudly, in full voice, to her climax. I know what you're thinking. But remember, I'd read Koestler's novel over fifty years ago as an innocent youth who wanted to have his fur-skin removed.

Since Poppea, I saw Troyanos in many roles, from a very male Julius Caesar in Handel's opera, or a flighty Dorabella in Cosi Fan Tutte, to a concert performance in Berlioz's Les Nuits d'Eté. At the end of the Berlioz, I met one of the musicians, whom I knew quite well, and told him how touched I'd been by Troyanos's singing. On the spur of the moment I spilled out the Koestler story — he's the only person who's heard it — and confessed that in Troyanos's Poppea I had finally seen the woman of my bathtub days. I even alluded to my most recent Troyanos fantasy: listening to her while we're both soaking in a hot tub. That's when he caught me by surprise. Did I want to meet Troyanos? he asked; he could arrange an introduction. "Absolutely not," I replied. He seemed taken aback, so I tried to explain that a life-long illusion might be destroyed. What if, screwing up my courage, I blurted out my Spartacus tale to Troyanos and asked, "Do you sing when you make love?" She'd probably fix me with those huge dark eyes, this time full of impishness and irony, and murmur, "Don't all women?" What then? Would I've dared to mention my choice of the song I wanted to hear as she…?

Instead, I went home and decided to re-read The Gladiators, which I hadn't looked at for decades. But I couldn't even find the book; I must have lost my copy during one of my many moves. The local library didn't have it. Finally, I located a worn copy at the university, a 1950 third printing by Macmillan. I took the volume home and skimmed the upper portions of the left-hand pages. That probably took half an hour, because every once in a while I stopped to read a choice paragraph or two. But I was so impatient to find my scene that I kept turning the pages. I found nothing. Well, I figured, maybe this edition is printed in a different format. Back I went to page one and flipped through the right-hand pages. Again nothing. It was preposterous. I'd seen Troyanos in Poppea and everything in me — memories of untold and untellable fantasies, embellished by the wishful inventions of late middle age — told me that she'd been in Spartacus's tent. I was prepared to concede minor imperfections in my memory; but the fact that she had mounted Spartacus and sung, that simply had to be there.

So I took the book to bed and read it, slowly, starting with the prologue: It is night still. Still no cock had crowed. It's a great beginning and I savored every page. That is, every page until I got to page 84 — a left-hand page. There in lines twelve to fourteen Koestler has Spartacus say these words to his fellow gladiator Crixus: 'I have never been to Alexandria. It must be a very beautiful place. Once I lay with a girl, and she sang. That is what Alexandria must be like.'

I've never been to Alexandria and I've never lain with an Egyptian woman. But no temptation on earth would cause me to do so now. Instead, I tore out page eighty-four. "Damn you, Arthur Koestler," I muttered as I shredded it into illegibility. "Damn you."

The above tale about Spartacus and my fantasies about making love to a woman who would sing while we were copulating is absolutely true. Not a word is invented. It is pure autobiography, yet I have only published it as fiction: as the second (and also shortest) short story I ever wrote and as the very first for which I got paid when it appeared in the December 1988 issue of the UK edition of *Cosmopolitan*. While it can be categorized as pure information transmittal, I do not hesitate to identify it as the product of my first stirrings as a *Romancier*. It was published under the title "What's Tatiana Troyanos Doing in Spartacus's Tent?" and in retrospect is about as good a description of my entry into literature as I can conceive. Furthermore, I have never before specifically admitted that as a life-long opera aficionado — who claims to still remember snippets of his very first opera attendance of *The Barber of Seville* at age four at the Sofia Opera — Tatiana Troyanos was without question my all-time favorite among the many divas I have heard and ogled through opera glasses. A couple of years before her premature death in 1993, she gave a concert at Stanford University where I attended a reception in her honor following her stunning recital. I brought with me a copy of my first "fiction" book — fiction in quotes since so much was pure or partial autobiography — entitled *The Futurist and Other Stories* with the intention of presenting it to her. The meeting actually materialized and while her smile upon receiving my proffered gift momentarily made me fantasize that she would eventually contact me to ask what song I would like her to sing in my hot tub, the fantasy remained just that: a delusion by a besotted opera fan.

Tatiana Troyanos signing my short story, "What's Tatiana Troyanos Doing in Spartacus' Tent?"

There is one other reason why I start this chapter with a published story of mine. It is only through my fiction and playwriting that people will learn the truth about me, assuming they are even interested. As I already stated in *Caveat Lector*, for various reasons — some not even understandable to me — for much of my life I have refused to face certain issues in public or possibly even to myself, but have then started to address them indirectly through my fiction. I would urge the reader to view these passages as pages from a special diary that I have kept, initially unconsciously, for the last two decades and which I am now willing to share openly through excerpted highlights in the hope that they constitute a discreet as well as unfiltered account of my persona. Some impatient readers will undoubtedly be inclined to skip these excerpted quotes, which for convenience are printed in italics. But if they do, they will miss most of what I wish to transmit in this chapter.

Short Stories

How did short stories even strike my fancy? As I already recounted in the introduction to my most recent (2012) bilingual poetry collection, *A Diary of Pique: 1983–1984*,[15] my initial motivation was powerful and ugly, namely revenge. In 1983, the great love of my life, Diane Middlebrook, who was sixteen years younger than I, announced that she had fallen in love with someone else — a much younger pseudo-literati rather than scientist — and that we were through. Although her message was gentler and more diplomatic than here stated, its effect on me was a typically solipsistic male reaction: how could she fall in love with somebody else if she had me? Given that until then, I had never written a single line of poetry or any belletristic prose nor had I ever felt the urge to do so, my response was totally unexpected. I decided to demonstrate to her, a sophisticated Professor of English Literature as well as a refined poet, that I could also graze on her turf. The first few months after our rupture resulted in an explosion of poems, which was then followed by a *roman à clef*, *Middles*, whose simple-minded purport it was to demonstrate the supposed lack of romantic judgment on the part of an otherwise highly sophisticated woman.

Fortunately, fate intervened, because exactly one year later, on May 8, 1984 Diane suggested in a letter that we meet again to examine what had gone wrong between us. Since I have already told this tale of loss and reconciliation in my poetry book *A Diary of Pique*, I shall simply conclude with the fact that after a few months we decided to marry and married we remained for twenty-two years until her premature death from cancer in 2007. But before firmly tying the knot, Diane asked me to promise that I would never publish *Middles* and I have stuck to that bargain. In fact, I have never again opened that manuscript for the simple reason that certain plot elements had been imprinted in me to such an extent that they surfaced in another literary life through cannibalization. With a few exceptions, the poems also remained unpublished and unread until two years after Diane's death, when I decided to reread them through the lens of wiser retrospection and concluded that they represented an honest as well as self-critical diary of my *annus horribilis*. I am not ashamed of what I wrote in that free-verse diary and after stylistic as well as emotional revisions, I decided at age eighty-eight, before death could intervene, to publish it under the appropriate title, *A Diary of Pique 1983–1984*. *Middles*, however, will never be published, because it would mean breaking a promise; yet even the brutal motivation of revenge leading to the birth of *Middles* did have some positive consequences.

[15] See footnote 13 on page 83 for full bibliographic details.

Within two months of our marriage in 1985, I was diagnosed with a serious case of colon cancer. The lengthy post-operative phase at age sixty-two caused me for the first time in my life to reflect realistically on the likelihood of death and to pose the question whether I planned to die in my lab coat. My answer was no. I had stayed with chemistry for most of my adult life because satisfying my curiosity gave me a great deal of pleasure: each question answered raised another. And I could live simultaneously in a world of research, with no ostensible utility, and of practical projects potentially benefiting millions of people. So why take off my lab coat? Because lying in that hospital bed, I was confronting death, a topic that I never addressed before. Of course, I knew I would die eventually, but until then, I had been very healthy and still expected to live for some decades. But now, as I asked myself whether I would have done something differently if I had known some years earlier that I was about to embark on the tight-rope walk of cancer survival, I conceded that shedding my lab coat would make sense. I felt that for the remaining years — never suspecting that there would still remain so many — I would attempt to lead a new intellectual life as a writer, very different from what I had done for the preceding forty-three years as a scientist: to explore another creative world, beyond science, beyond research and its applications, with which I wanted to deal directly and to do so in the seemingly most unscientific manner of them all, namely in fiction. Not only does this genre offer the opportunity of moving from the exclusively monologist written discourse of the scientist to the at least partial dialogic style of the novelist, but, for once, it might also allow me full rein to let my imagination roam over autobiographical as well as imaginary terrain without the slightest hindrance imposed by embarrassment or shame. For a scientist, being able to say, "It's just fiction" seemed to me a startlingly refreshing luxury. Without realizing this then, I am now fully convinced that writing (in contrast to, say, playing the cello expertly, which would otherwise have been high on my agenda) is the one occupation which one can acquire autodidactically late in life, because one has accumulated wisdom (or so one thinks) and life-time inter-personal experiences which even the most brilliant young author cannot possess early in life. Still years away from widowerhood, I had not realized that such writing is also the most effective antidote to loneliness.

A somewhat corny metaphor will serve to explain my post-operative decision: my first bite of the tempting apple of literary writing was defiled by the poison of revenge, but having spat that out, the apple now looked unspoiled and still enticing enough for a new bite. Diane was the only person to whom I volunteered that desire; and recognizing that her new husband was a disciplined autodidact, though totally untested on the literary front, she suggested that I first focus on short stories to see whether desire as well as consummation might coincide — a piece of advice that is frequently also offered in academic courses of creative writing. Short stories are short enough that they can be revised repeatedly and, in the process, one learns conciseness and even more importantly not just to give birth to literary darlings but also to acquire the cold-bloodedness of killing some of them. Within one year of that time of introspection and deep depression, but then having fully regained my health, I suddenly found myself heading towards my *annus mirabilis* — in retrospect, a heady experience after the disastrous interval of 1983 to 1984. As I already mentioned in *Heimat(losigkeit)*, in the summer of 1986 I became a faculty spouse offering me ample time for full-time autodidactic literary experiments on the fiction front, while as director of the Stanford Program at Oxford my wife was "working" (the term used by some male chemical colleagues, thus expressing their opinion of what I seemingly was not doing).

What caused me, a scientist from the very hard science of chemistry, to be tempted to cross over into fiction? The gulf between the sciences and the other cultural worlds of the humanities and social sciences is increasingly widening, yet scientists themselves spend precious little time in attempting to communicate with these other cultures. To a large extent this is due to the scientist's obsession with peer approval and the recognition that his tribe offers few incentives to communicating with a broader public that will do nothing for the scientist's professional reputation. I decided to do something about illuminating the scientist's culture for a broader audience, and to do it through the genre I shortly thereafter called "science-in-fiction" — not to be confused with science fiction. For me, a literary text can only be anointed as science-in-fiction if all the science or idiosyncratic behavior of scientists described in it is plausible. None of these restrictions apply to science fiction. By no means am I suggesting that the scientific flights of fantasy in science fiction are inappropriate. But if one actually wants to use fiction to smuggle scientific facts into the consciousness of a scientifically illiterate public — and I do think that such smuggling is intellectually and societally beneficial — then it is crucial that the facts behind that science be described accurately. Otherwise, how will the scientifically uninformed reader know what is presented for entertainment and what for the sake of factual knowledge?

But of all literary forms, why use fiction? The majority of scientifically untrained persons are afraid of science, often murmuring, "I don't understand science," while dropping a mental curtain the moment they learn that some scientific facts are about to be sprung on them. It is that portion of the public — the ascientific or even antiscientific reader — that I want to touch. Instead of starting with the aggressive preamble, "let me tell you about my science," I prefer to start with the more innocent "let me tell you a story" and then incorporate realistic science and true-to-life scientists into the plot.

As I now look back at my body of work among short stories, novels, and plays, I find that I largely managed to stay away from just describing in an entertaining but obsessively accurate way *what* scientific research we scientists perform. Good science journalists can also play that role and frequently perform it very well. But to bridge C. P. Snow's two-culture gap, which has widened to a multicultural gulf since he first raised that contentious issue, it is also necessary to illustrate *how* scientists behave. And it is here that a scientist-turned-author can play a particularly important role. It is here that "crossing over" can really pay off. It is only later that I realized that scientists and especially one of them, Carl Djerassi, would also benefit from such literature since it raised behavioral questions that are posed all too rarely to members of that clan. Exploring their behavior — and thus my own — eventually turned into an openly admitted collective critique or even *mea culpa* with respect to some idiosyncratic aspects of our scientific conduct and values.

As I specifically pointed out in the afterword to my very first science-in-fiction novel, *Cantor's Dilemma*, scientists operate within a tribal culture whose rules, mores and quirks are generally not communicated through specific lectures or books, but rather are acquired through a form of intellectual osmosis in a mentor–disciple relationship. Scientific "street smarts" — in some respects the soul and baggage of contemporary scientific behavior — are absorbed by observing the mentor's self-interested concerns with publication practices and priorities, the order of the authors, the choice of the journal, the striving for academic tenure, grantsmanship, *Schadenfreude* — even Nobel lust. On their own, budding scientists discover the glass ceiling for women in a male-dominated enterprise, the inherent collegiality of scientific research, and the concurrent brutal competition. Most of these issues are related to the desire for personal recognition and even financial rewards, and each is colored by ethical nuances.

To me — as a scientific tribesman for over four decades — it seemed important for the public not to look at scientists primarily as nerds, Frankensteins or Strangeloves. And because science-in-fiction deals not only with real science but more importantly with real scientists, I feel that a clansman can best describe a scientist's tribal culture and even unique behavior. This is the terrain I have now occupied for over two decades and the interested reader can obviously explore it by browsing through some books of mine which have since come out as a result of my ambition to stake a claim in that field. Instead, let me address in some detail the very early short story experiments about which I have written very little — experiments that started with my wife's advice but were first realized through the process of cannibalizing *Middles*. I use the word "cannibalizing" deliberately, because even though I had promised my wife as part of our unwritten marriage contract never to publish *Middles*, I made no promise not to use portions of it in textual or at least contextual form for the simple reasons that initially I wanted to learn the skills of a fiction writer rather than to focus on plots or characters. Plots, events, characters, selected esoterica — those, I felt, I could take without restraint from my life, from the indisputable advantage of having lived for over half a century in turbulent and exciting times, and having traveled all over the world before testing how to present them in fictional, or in my case, quasi-fictional fashion. *Middles*, as a virtually transparent *roman à clef*, contained an overabundance of such autobiographical events.

A Rhino Swallowed My Rolex

One example of such autobiographical plot trawling is the above "What's Tatiana Troyanos Doing in Spartacus's Tent?" It is a description of an autobiographical event, but I actually recalled it only when I worked on *Middles* and that is where I "stole" it from. Having now admitted to this perfectly legitimate form of self-plagiarism (after all, *Middles*, had never been published), I shall proceed to a second example of a short story that really only qualifies as such because it is short. Otherwise, like my fantasies about Tatiana Troyanos, "A Rhino Swallowed My Rolex" is again pure autobiography with the exception of my shooting the rhino. (I have never gone hunting in my life and never killed an animal other than accidentally driving over a skunk, who thoroughly revenged itself in its death throes by stinking up my car for days.) But since it never made it into print in English and is long out of print in German translation, self-plagiarism, though perhaps not self-promotion, does not enter here either. Rather I am reproducing it below in abbreviated form, because it shows again a stubborn bulldog-type quality of mine — at times amusing, at times highly irritating to the person living with me — in that I follow leads beyond what normal curiosity would justify.

> *I was sitting on the balcony, feeling smug. We were taking a break in Panama City during the long trip in 1954 from Peru back to the States. Instead of exploring the city and the canal locks, however, I sat on the hotel balcony waiting for Montezuma (or his Inca counterpart, Atahualpa) to finish taking his revenge on my wife, who lay groaning on her bed in the shaded room. Remembering from my intestinal wars just a few years ago in Mexico that no amount of commiseration could alleviate such cramps, I kept quiet but stayed close by. I had exhausted all reading material; even the last magazine had been squeezed dry. In desperation, I started on the advertisements, which ran the usual gamut from tractors and Piper Cubs to perfumes and furs. Then my glance fell on a Rolex watch ad entitled "Under Greek Waters." A scuba diver — two tanks on his back and air bubbles escaping from his mouthpiece, sea anemones, jellyfish, and kelp visible around him — was reaching for a wristwatch among the corals. A single fish hovered in the background. Accompanying the drawing was the text:*
>
> > *In 1938, an underwater fisherman lost his Rolex Oyster in deep water off the coast of Greece. He could see it clearly in a crevice between two rocks, but could not reach it.*
> >
> > *In 1946, he returned to Greece and took up underwater fishing again. Being lent a self-contained breathing apparatus, he immediately*

> *thought of his long-lost watch. With the aid of the new equipment, he was able to swim down to the sea floor. A short search of the weeds covering the rocks disclosed the watch in the same crevice where he had last seen it 8 years before. After a little attention by a local watchmaker, it kept as perfect time as it had always done.*
>
> *What a tribute this story is to the superb accuracy of the Rolex movement! And how well it demonstrates the perfect protection given to this movement by the waterproof Oyster case.*

"You've got to read this," I called to my wife, brandishing the magazine. With a groan, she rose from the daybed, accepted the magazine, and headed for the toilet. Returning, she stopped at the balcony door to toss me the magazine and a wifely "What now?" look. "Can you imagine anyone buying a Rolex because of this cock-and-bull story?"

"Why not?" she replied. "It seems to be true. Just look at the footnote."

> *(This is a true story taken from a letter written to Rolex by Mr. D. F. Pawson. The original letter may be inspected at the offices of the Rolex Watch Company, 18 rue de Marché, Geneva, Switzerland.)*

But did that make the story true? An offer made in Panama to inspect a letter in Geneva? Rolex, I thought, has met its match. I shall call their bluff. I tore out the ad and, before putting it away, folded it so carefully that even now, many decades after the event, there are no extra crinkles in the yellowed page. "We're going to stop in Geneva next August," I announced to my wife. With a final suffering look, she headed back to the toilet.

Eight months later, on a sunny August morning as we crossed the Rhône over the Pont du Mont Blanc at the point where the river enters the lake of Geneva — the mountains clearly visible in the distance in Lake Titicaca fashion — my cocksureness had started to dissipate, though not yet totally. We had arrived in Geneva in the open back seat of a touring car driven by an English friend, a Cambridge don, to whom I had been bragging for days that I would turn the Rolex corporate offices topsy-turvy. But when he deposited me — hair wind-blown, tieless, my jacket wrinkled like an accordion — on the Rue de Marché in front of the plush corporate headquarters of the Rolex watch company, I turned to my wife. "You may as well stay out here in the car. I'll be back in a jiffy."

"Not on your life," my spouse proclaimed, her decisive voice reflecting a calm stomach. "I wouldn't miss this for anything."

"All right," I grumbled. "Let's get it over with."

We marched into the cool, marbled reception hall. My moist right hand was in my coat pocket fingering the folded TIME ad. "I've come to see the original of this advertisement," I barked at the receptionist, slamming the folded page on the gleaming desk as if I were delivering a papal bull, while hiding my embarrassment behind a mask of aggression. Without a flick of her eyelashes, she unfolded my instrument of challenge. "Please wait." She pointed to a sofa in the alcove, then picked up the telephone. A few minutes later, facing a young man in white coat, I thought for a moment that an asylum employee had arrived. But the laboratory coat was standard garb among the lower classes of Rolex employees, such as this representative of the advertisement department. "Voilà, monsieur," he said, handing me a thick black three-ring binder. In it, I found a series of plastic inserts. The left one held a page from the magazine carrying a Rolex ad; the opposite insert held the letter on which the advertisement was based.

The man left us to our own devices; and for twenty or more minutes, we alternated between giggles and guffaws. There must have been a dozen different ads, although now I recall the subject matter of only two of them. No wonder that I, a chemist, remembered the first. It showed a boiling cauldron, partly tipped over, with liquid goo slowly pouring out onto a tray. "Accidentally, I dropped my Rolex into a vat of boiling lye. Hours later, when the cooled contents were drained carefully, I located my Rolex on the bottom. Once I had rinsed it in water…" The other one was more exotic, involving a canoe trip down the Amazon and a piranha snapping at a Rolex. After a few weeks or months, the ex-owner himself was fishing in the Amazon when he caught a vicious piranha that, mirabile dictu, had the ticking Rolex in its intestines. It was true that every illustrated ad corresponded to an apparently original letter. What the ads had not divulged was that every epistle started with virtually the same sentence: "I understand that if I submit an original story about my Rolex watch, I will receive from you a new gold Rolex Oyster Perpetual."

"Ça va, monsieur?" Our white-coated attendant smiled as he reached for the binder, but I held onto it. "Amusing," I conceded, "but not what I came here for. Where is this ad?" I asked, pointing to my page from Panama. He leaned down and quickly turned the pages. "Excusez moi," he exhaled and fled. Some minutes passed before a second white-coated

employee appeared, this time a woman. "I'm sorry," she said breathlessly, "your copy had been removed temporarily. Here it is. Please examine it." And there it was: the hand-written letter from Mr. D. F. Pawson, on the stationery of a London club, with the usual starting sentence. I made a show of comparing his letter with the text of the printed legend in the advertisement, then ostentatiously copied Mr. Pawson's address on a sheet of Rolex Corporation stationery I had lifted from the receptionist's desk.

As the woman departed with the binder under her arm, I stared at the white page with Pawson's address on the top. Words started to appear in my mind:

> *While camping among the Masai at Ngorongoro Crater, I put my trusted Rolex next to my bedroll. During the night, a rhino crashed through my tent. In the excitement, as I shone my torch onto the scene, I watched in horror as the rhino swallowed my watch.*
>
> *Three years later, after returning to Tanzania with the Duke of Bulloughshire, while taking tea and crumpets in front of our tent, the Duke exclaimed: "I say! This rhino is charging us." Down I dropped to one knee, raising my gun and…*

I had felled the beast with one shot; the jubilant Masai had skinned and quartered the carcass, while hyenas and vultures had already started to circle the remains; two days later we had encountered a completely clean skeleton with only a single shiny object reflected in the sunshine. I was searching for a plausible explanation why my Rolex had remained in the rhino's stomach without having been eliminated with the rest of his daily wastes, when I remembered: hunting is strictly forbidden in Ngorongoro Crater. The Swiss are much too scrupulous. They would catch such a gaffe.

"Let's go," I said to my wife and got up.

"Not to that club in London, I hope?" she asked, a touch of panic in her voice.

This Rolex watch episode was the last autobiographical sketch with which I masqueraded as a budding short story writer — amusing events that had happened to me, but in truth were writing exercises. The rest of them had autobiographical kernels or episodes based on events that happened to me, but otherwise were short stories in the classical sense and eventually also accepted for publication under that guise.

First-Class Nun

The following is a case of borderline-legitimate triple self-plagiarism: my meeting a Carmelite nun for the first time which, though based on a real event, first surfaced in unpublished form in *Middles*, but then in my short story, "First-Class Nun", as well as in my novel, *Marx, Deceased*. It is another example of my innate curiosity which has continued to follow me into my new career.

> *It had been Stephen's first stay at an artists' colony. On several occasions Stephen had been courted to spend time at McDowell or Yaddo, but he'd always turned them down. He was a city person, he had his own study with all the privacy he required — so why go to an artists' colony in the sticks? One of his writer-friends had finally convinced him to give it a try by pointing out that there are other people at artist's colonies — often interesting ones.*
>
> *Stephen, the ever-curious, people-collecting novelist went, and promptly bumped into his first nun — not a hundred-percent nun, he explained to Miriam that weekend, but close to it. Sister Sharon (maybe she wasn't called that yet, but Stephen Marx, the storyteller, had so anointed her) taught creative writing at Maryville, a Catholic women's college; she had published several volumes of poetry and one collection of short stories; and at age 31, she had started to think seriously of entering a Carmelite monastery.*
>
> *One evening at dinner — the best time at artists' colonies, when self-imposed privacy is broken by professional one-upmanship, social grace, and amazingly high-level raconteurship — Stephen Marx had found himself next to Sharon O'Grady. Etiquette never allows artists in the same discipline to admit that they haven't heard of a fellow-colonist. Consequently, the initial probing is diplomatic; its aim to learn as much as possible about one's partner without giving away one's ignorance, and at the same time providing useful autobiographical details oneself.*
>
> *After learning that Sharon O'Grady had been teaching for several years at a small women's college, Stephen had asked, "Tell me, are you ready to move, or are you content in your place?"*
>
> *Sharon O'Grady gave him a long, thoughtful look, before replying. "Yes, I've thought about it for the last couple of years. I'll be leaving Maryville at the end of this academic year."*

> *"Where to?" asked Marx, wondering whether she was moving up or sideways on the academic ladder.*
>
> *This time, her glance was short and amused. "I'm joining the Carmelite order — I'm going to a monastery in Brooklyn."*
>
> *Within seconds — barely sufficient for recovery from total astonishment — Stephen Marx, who had never met a nun, had turned into the novelist-inquisitor. After three and one-half hours of nonstop questioning, the outline of Listen Sister had been drawn in his head. By the end of the week, the first draft of Chapter 1 had been typed on his Olivetti Lettera 22 — his faithful pre-computer writing companion. In his usual manner, he found it necessary to amuse as well as instruct the reader. After all, how many people know about Carmelite nuns and their lifestyles?*

The above excerpt is lifted verbatim from Chapter 18 of *Marx, Deceased* (subsequently republished in German translation under the more telling title *Ego*), and if the name of my imaginary character, the writer Stephen Marx, is replaced by Carl Djerassi and the artist's colony identified as the Djerassi Resident Artists Colony near San Francisco, this would be a verbatim account of what happened to me in the 1980s when I attended one of my weekly dinners with the artists then in residence at the artist's colony I had founded earlier. Just like Stephen Marx, I was so taken aback by meeting my first nun, albeit a nun-in-spe, who seemed so worldly and so unlike my preconceived notions of what a nun should look like that within weeks, she became a character in *Middles* and subsequently the personal and thematic focus of my short story, "First-Class Nun". But unlike my fictional surrogate Stephen Marx, Carl Djerassi decided to do some research on the ground.

In my questioning the prospective Carmelite, I learned that the closest Carmelite monastery (so referred to by Carmelites rather than the expected term "convent") to San Francisco was in Reno, Nevada. Again, I was taken aback, since Reno — other than being the site of one of the University of Nevada's campuses — was in my mind synonymous with gambling and sinning. Two days later, without telling anyone, I took an afternoon flight to Reno and upon deplaning asked a taxi driver to take me to the local Carmelite monastery. He just guffawed, thinking that I was joking. When I repeated my request and gave him the exact address which I had obtained from "my" Carmelite, he became curious. "I've been driving cabs in Reno for twenty years, but this is the first time..." he confessed as we took off in search of the clearly least likely location for a religious retreat. The monastery proved to be in a typical suburban residential area, a sprawling low one-story building shielded by dense shrubbery. I asked the cab to wait, not knowing whether I would even be admitted. When the door was opened slightly by the only nun allowed to speak to outsiders, I quickly mentioned that I came upon the recommendation of "my" nun — a white lie that seemed acceptable for a prospective author turned temporarily into investigative reporter — whereupon I was permitted to enter. Though polite, "that" nun hardly offered me any information that I did not already possess and after some minutes of courteous but essentially useless conversation, I departed and asked the driver to take me to Harrah's Casino, the most un-monastery place I could think of where I planned to spend the night. The notes I jotted down that evening were a peculiar mixture of tame monastery descriptions and loud ostentatious gambling impressions.

Eventually, I wove my meeting with a prospective Carmelite nun into a fictional short story in which a self-assured agnostic suddenly wonders whether he should consult her as a therapist. I herewith present this through an assembly of brief excerpts from the conversation on an airplane, which a man, Michael Brewis, had with a woman he encountered next to him; whom the flight attendant had addressed as "sister." It offers me the opportunity to address in my present autobiographical musings the subject of therapy.

"I concluded that you must be a nun, but... well... you don't look like one."

"You mean I don't wear a nun's habit and I'm not holding a prayer book?"

"I didn't look at your book."

"Well, look at it."

Brewis took one glance at the proffered book cover. The Church and The Second Sex?..."

"You don't know much about nuns, do you? Some of us live in the present." The green eyes sparkled as she continued, "Now that you've met your first one, you're burning to ask the next question, aren't you?"

....

"What made you decide to go from graduate school to a...?"

"You're hesitating," observed the nun. "Is it because you're thinking of a nunnery but don't want to say it? You could always call it a convent..."

"Sister, don't embarrass me. It's just that 'convent' has a medieval ring that doesn't fit you. You're dressed conservatively — but you're not in a nun's habit; you could be an everyday woman, and a very handsome one at that."

She laughed openly. "You really do have all the stereotypical responses. You mean a nun can't or shouldn't be attractive? Besides, you've really only seen my face."

Brewis was not ready to let go. "What about the rest — your dress, the wine, the book you read?"

"That only shows how little you know about nuns."

"I know nothing about them. Actually, I know very little about religious life. You're talking to an agnostic from way back."

"Let me guess," retorted the nun. "You must be a scientist. Am I right?"

"No, but what makes you say that?"

"The almost aggressive tone in which you said 'agnostic from way back.' If you don't mind my asking, what's your occupation?"

"I'm a headhunter," he said matter-of-factly...

...

"Is this how you interview people — the way you're questioning me?"

"No," he said slowly, "I'm not interviewing you. But I'd like to know something about a nun's life — I mean your present life. What brought you here?"

It was her turn to study him. "All right," she said finally. "It won't hurt an agnostic headhunter to learn something about nuns. You probably think only of the dictionary definition of a nun — a woman belonging to a religious order, living in a convent under vows of obedience, poverty, and chastity. Have you heard of the Order of the Carmelites?"

"I've heard of them. But what's special about them?"

"I'll try to make it brief. The origin of our order is somewhat uncertain, but can probably be traced back to a twelfth-century retreat on Mount Carmel. Incidentally, the Carmelite houses are called monasteries, not convents. Because of our hermit origin and emphasis on solitude,

no more than twenty-one nuns live in any one monastery, each occupying her own cell. Nobody else enters such a cell, only the prioress, and she must be invited. This is very different from other orders, say the Benedictine, in which several hundred nuns may live in one convent and sleep in dormitories." Suddenly, she stopped. "I must sound like a tourist guide. Tell me, are you really interested in such details?"

"Fascinated! But I'm wondering when you're going to get around to explaining what you're doing in that dress."

The nun nodded eagerly, as if her tourist guide license depended on her reply. "Most of the time we wear brown, but as you can see by looking at me we don't necessarily wear a nun's habit. Of course, we're not fashion plates. We buy sensible skirts, blouses, sandals — what I'm wearing now was ordered largely out of an L. L. Bean catalogue...."

...

"Our house in Brooklyn is one of the more modern ones. There are others, not far from ours that do not differ much from the Carmelite monasteries of several centuries ago."

"What do the nuns do in your monastery?"

"It depends on the house," replied Sister Olivia. "Some lead a very cloistered life, not unlike what the general public thinks nuns do. Others are very different. In our Brooklyn house there are nineteen nuns, and you'd be surprised how diverse their backgrounds and interests are. I am supposed to be a spiritual counselor... "

"You, a counselor?" The term unsettled Brewis. Only a few days ago, his wife Claire had announced that he should see a therapist. "Damned shrinks," he'd sputtered to her. "What are they going to tell me that I don't already know?"

"I am a counselor — largely for outsiders who wish guidance or advice."

...

"And the vow of chastity?'

"What about it? Remember, women of all ages enter the order. Some are virgins, others have led an active sexual life." There was no hesitancy, no demure look at the floor: she kept her eyes straight on Brewis as she continued. "They know what they're doing. Abstinence from sex is not always so enormous a sacrifice for women as many people, notably men, think."

My God, thought Brewis, I can't believe it. Have she and Claire been talking to each other? He felt embarrassed, but he asked the question anyway. "Do you often talk about this subject?"

"I told you earlier that my primary function in our house is to provide counsel and that I do not just see nuns. There are no boundaries to what a counselor discusses."

"Do men also come to you for guidance?"

"Indeed. Why do you ask?"

Brewis didn't know how to introduce the subject of Claire: the wife who was about to walk out on him. In his professional life, he was essentially a headhunter for men, or at least male types; the few women he'd encountered mostly fell into the same category. In his private life, Brewis hardly remembered meeting a woman where sex didn't actually or potentially enter the equation. But this was totally different.

"You look pensive. Are you bothered by something I've said?"

"No, not at all," he said quickly, almost apologetically. "I was just thinking about your counseling work. I've never had any professional contact with shrinks. I'm sorry, I should've said 'therapists.' A verbal habit — probably some stupid reflex on the part of a supposedly strong man who thinks that he should be able to solve all of his problems by himself and that a therapist is a crutch. Or to put it another way, if you need to unburden yourself, do it with a friend."

"There are times when a friend is the last person. Do you open up often to a friend?"

"I guess not. Perhaps it's an occupational hazard. I spend most of my time analyzing others. I seem to have lost the ability or even inclination for self-analysis.

...

Brewis looked straight ahead. This way he could pretend that he was speaking to himself. "As I listened to you, I was almost wondering what would happen if I, the agnostic headhunter, the impersonal matchmaker, were to talk to a spiritual counselor."

A sudden and remarkable change occurred in the nun's demeanor. It was as if she knew what would follow, and wanted to stop it before the question was actually raised. "Everybody can benefit from therapy and especially from spiritual counseling," she said. And then the curtain dropped. "I hope you won't think me rude, but I've suddenly gotten quite tired." With these words she reclined her seat and closed her eyes.

Brewis began to read, but he couldn't focus on the pages. He kept reflecting on what had transpired. How he'd been on the verge of making an extraordinary request to this nun: to be his shrink, to show him how to talk to his wife. According to Claire, every person has only a certain quota of understanding and he, the headhunter par excellence, was draining his cup of empathy daily at the office. And just when he, who usually questioned others without disclosing anything himself, was ready to relinquish his standard role, the nun entered her cell and closed the door. Brewis simply couldn't figure it out. Eventually, as the cabin darkened for the movie, he fell asleep.

I cite these brief excerpts because at that time I had thought about my own attitude toward therapy. During my long second marriage, I had always held therapy in disdain — an opinion that, strangely enough, my wife Norma also held. In my case, it clearly was a case of misplaced psychic machismo, thinking that I could solve all problems by myself. This attitude may well have rubbed off on my children. Yet years after my daughter's suicide, I wondered whether a neutral professional might not have helped her in a way we all — parents, husband, friends — evidently failed her. My third wife, Diane, was different. Aside from extensive studies of Freudian psychoanalysis, she herself had been in therapy (not Freudian) for helping her cope with chronic alcoholism. Therapy helped, although her complete cure should probably be credited to me — a virtual non-alcoholic all my life — because she realized that if we were to live together, it would have to involve a total break with that addiction. Since our marriage, Diane never touched a drop of alcohol, knowing that one drink would throw her back into the abyss of a dependence on alcohol with which I would never have been able to cope.

I have never consulted a therapist, but since meeting Diane, my attitude toward therapy has certainly changed. During the traumatic year of 1983 when Diane left me for a year, she challenged me on at least one occasion to do so — an event that prompted me to write the following poem, which rested in a locked file for twenty-nine years:

There are No Shrinks in Burma

Dedicated to Aung San Suu Kyi (Oxford, 1983)

"Therapists are needed only
Where friends do not listen,"

Aung San Suu Kyi next to my son Dale at his wedding reception, Oxford, 1984

Said the Burmese Aphrodite at the Oxford party.
Startled, he asked her to explain.
"Burmese lovers ask questions and listen.
In Burma we say, "Tell me all.
Middlemen are not needed by lovers."

He saw the letter in his mind:
"I don't want to see you
until you've been to a shrink."
From the woman he had loved for six years;
The woman who had taken a new lover.

If both had been able to talk,
If both had known how to listen,
Burma would not be so distant.

I did not follow the advice "to see a shrink," not because I found it unreasonable, but because I wanted to pose the questions and work out the answers by myself. At that time, this was probably still due to residual male pride, but during the last two decades, it is another interpretation — quite possibly naïve or even misleading — that has led me to a somewhat different path of therapy. As a teacher and scientist for over half a century, I have acquired the capability to observe with objectivity and to accurately record data. But as a confessional poet, as a *Romancier*, and as a playwright, I believe that I have also acquired an additional perspective. The sort of novels and plays that I have written during the past two decades has allowed me to accomplish the impossible in standard autobiography, namely to bypass the problem of single vision restriction. The unlikely marriage of chemistry (with its highly honed analytical prowess) and the literary arts (with their unsurpassed ability to hide or disguise) has overwhelmed any pre-existing conventional, single-minded emotional filter. With these seemingly strange bedfellows I have turned into a serial autobiographer who, as I already stated in the *Caveat Lector* introduction, is invariably both analyst (as author) and analysand (in terms of the surrogates in my stories). They display *my* problems, *my* idiosyncrasies, *my* faults as well as *my* merits in fictional disguises. But as I have demonstrated through frequent quotes from my writings in earlier chapters and will continue to do so in the remainder of this one, pretending that it is all fiction has given me the luxury of a freedom that no autobiographer can enjoy.

Since I had never before described, nor perhaps even realized, how the various short stories have pushed me into the direction of hidden disclosures of my own persona, let me note that one common thread can be found in all of them, namely one-upmanship — trumping competitors and colleagues alike — an endemic infection among research scientists. Winning an argument is crucial — as is the accompanying admiration and envy it engenders. But even brief reflection will show that it is in fact a human foible that can be found wherever one searches for it. And since I have practiced one-upmanship for most of my adult life — often to my detriment — I thought that the most painless way of confessing that character fault of mine is through browsing in domains that have always appealed to me: food (e.g. "Noblesse Oblige"), sex (e.g. "Research"), art (e.g. "The Futurist"), opera (e.g. "The Glyndebourne Heist"), human companionship (e.g. "Maskenfreiheit"), word plays (e.g. "The Toyota Cantos") and, of course, science (e.g. "How I Beat Coca-Cola"). Even the choice of some of the titles (e.g. "The Psomophile") are a manifestation of one-upmanship, although the ultimate confirmation of its success will only arrive when the *Oxford English Dictionary*'s next edition enters this invented word of mine to define a bread lover or for that matter "The Dacriologist" (a person studying tears). Regretfully, when I Googled for these two words, all I found so far were hits referring to me as the cited source. Much of what I tell in this short story collection, which was published in the USA only in 2013,[16] but already twenty years earlier in translation in German, Italian, Spanish and Mandarin, is autobiographical, though very little of it is biographical: it didn't happen to me as told, but it could have. To that extent, it is much more honest than conventional autobiography, which itself is a form of automythological fiction in which I have also indulged in the past as I have already admitted on several occasions.

[16]C. Djerassi, *How I Beat Coca-Cola and other tales of One-upmanship*. University of Wisconsin Press, Madison (2013).

The Toyota Cantos

Of the short stories of which I only listed titles, one of them, "The Toyota Cantos," is my unequivocal favorite. It deals with a Professor of Italian Literature named Lionel Trippett, a scholar specializing in Dante, who, without realizing, had for years stopped communicating with his wife, Beatrice. While she was on a holiday in Paris, he had a car accident in which the car was wrecked. Here is what happened when he telephoned her from New York:

> *From my hospital bed I said, "Bay-ah-tree-chay, I'll buy you a new car. What kind do you want?" When she replied, "I have one that still runs fine," I probably shouldn't have tried to be funny. I shouldn't have said "Bea, you had one, but I wrecked it." But that was no reason for her to rub it in by asking, "Why on earth did you take my car? You haven't driven a car for years." Of course I haven't, but who needs a car in Manhattan? Besides, when we do need to go by car, she always drives. And who took off for Paris in the middle of the semester anyway? She or I?... I went on to say that the only reason I was still in the hospital was that I'd been knocked unconscious during the accident. They wanted to be sure I had nothing more serious than a broken left arm. This is when she came up with an extraordinary question. My wife of thirty-something years asked me — three thousand miles away in the hospital — whether the front or the rear of the car got demolished. When I told her that it was mostly the front and the sides, she sounded relieved. Only then did she ask whether I had retrieved the bumper stickers. At first, I thought I didn't understand correctly, but when she spelled it out: "B U M P E R S T I C K E R S," I didn't know whether to laugh or get angry. Thank God I didn't laugh, because she really meant it. Instead I said, "Sure, I'll get your bumper stickers," but if I'd known what I'd let myself in for, I would have kept my mouth shut.*

Eventually, he retrieved the car's rear bumper from the dump where the wrecked car had been taken, to discover that three bumper stickers were always arranged in a row, with another set underneath them, and then a third layer. I quote again from the short story to recount what happened next.

> *Suddenly I got goose pimples all over. The word intellection on one of the bumper stickers and the 3 x 3 theme had registered almost simultaneously. For Dante, 3 was the ultimate number: 3 parts to his Divine Comedy; 3 lines to each stanza — the classic terza rima; 33 cantos for each part;*

3 times 3, the magic number 9, to describe his Beatrice. But I bet that few people have ever heard of intellection, even though it's in the dictionary. However, it does occur in a couple of English versions of Dante — John Ciardi's and Mark Musa's translations of the Divine Comedy. Not in any of the others: Cary's, Binyon's, Singleton's, Sisson's, Carleye-Wicksteed's, Mandelbaum's or Anderson's. I know all of them — that's my business. As a student I was brought up on the real classic — Francis Cary's — but Ciardi's in the most colloquial, in fact the only even faintly bumper-stickerish one. My Bea! My fantastic Bay-ah-tree-chay! I was so excited I spread the nine stickers out on the rug and numbered them: 1, 2 and 3 were the original three stickers that I uncovered last from the hair dryer treatment. Numbers 7, 8 and 9, of course, were chronologically the most recent. The one that I saw at the car dump turned out to be Bea's bumper canto number 8.

THE TOYOTA CANTOS

1. *Burning logs, when poked, let fly a fountain of innumerable sparks.* ***Par***. *XVII*
   I do not hide my heart from you. ***Inf***. *X*
2. *The record may say much in little space.* ***Par***. *XIX*
   Mortal imperfection... makes fit for intellection. ***Par***. *IV*
3. *You seem to lack all knowledge of the present.* ***Inf***. *X*
   Why, when every lower sphere sounds the sweet symphony of paradise ... there is no music here. ***Par***. *XXI*
4. *Why does your understanding stray so far from its own habit?* ***Inf***. *XI*
5. *Portion out more wisely the time allotted us.* ***Pur***. *XXII*
6. *Lift your head. This is no time to be shut up in your own thoughts.* ***Pur***. *XII*
7. *Take care. Do not let go of me. Take care.* ***Pur***. *XVI*
   The taste of love grown wrathful is a bitterness. ***Pur***. *XXX*
8. *Richest soil the soonest will grow wild with bad seed and neglect.* ***Pur***. *XXX*
9. *Look at me well, I am she. I am Beatrice.* ***Pur***. *XXX*
   Losing me, all may perhaps be lost. ***Par***. *II*

The time has come to make a confession: at that stage, all I had stumbled upon was the Dante numerology — admittedly a powerful key — and one word intellection, which turned out to be my Rosetta stone. I pounced again on Bea's nine bumper cantos — I rather liked the term and promptly made it part of my private vocabulary — to check for other Dante attributions from Inferno, Purgatorio, and Paradiso. Portions of bumper cantos 1, 3, 8 and 9 turned out to be quite easy — especially number 9 ending with I am Beatrice. By now I was convinced that all had their Divine Comedy provenance.

At least three hours passed, in what proved to be a devilishly delicious game, before I woke up to the fact that I was doing precisely what Bea had been criticizing through her bumper Tragica Commedia. Clearly the point was to understand what she was saying, not where in Dante the quote could be found. I stopped looking for hints in the bumper cantos to Dante's cantos. Instead, I re-read them just for content and the message started to penetrate.

I wrote a good part of this story during a lecture trip to Italy by taking a break for inspiration in Lerici, overlooking the Golfo dei Poeti, where Shelley drowned. Sitting on the balcony of my Pensione, I spent two days searching the entire *Divina Commedia* for bumper stickerish messages. I needed them not only for the above bumper cantos, but also to compose Lionel's Dantean response:

"Oh my Beatrice, sweet and loving guide — "

*I need not tell you where this opening comes from. At one time, it probably would have been typical of me to say that I am not "persuaded to give ear to arguments, whose force is not made clear." But in spite of this quote's origin (**Par**. XVII) it is unlikely to lead us to Paradise.*

*I could have answered: "The only fit reply to a fit request is silence and the fact" (**Inf**. XXIV) and added: "I have no recollection of ever having been estranged from you" (**Pur** XXXIII).*

*I could have complained: "How sharp the sting of a small fault is to your sense" (**Pur**. III).*

*I could have equivocated: "Or have I missed your true intent and read some other there?" (**Pur**. VI).*

*But I am sure that you — who seem to have studied Dante in a way so different from mine — would have replied "the sense of what I wrote is plain, if you bring all your wits to bear upon it" (**Pur** VI). Therefore, I have taken your advice: "Look deep. Look well" and Dante's (**Pur**. VIII): "If you seek truth, sharpen your eyes." Now that I understand "often, indeed, appearances give rise to groundless doubts in us, and false conclusions, the true cause being hidden from our eyes" (**Pur**. XXII), I promise: "I shall from now on no longer speak like one but half awake" (**Pur**. XXXIII).*

*If I "look closely now into that part of me . . . the part with which I see" (**Par**. XX), I realize that I took many things for granted. I was about* *to list these, but you should really hear them from my own mouth. You should question me; perhaps the way Dante was questioned before he was found to be deserving of paradise. I would like to do something that, strangely enough, we have never done in all our years of marriage: reading Dante together. Your "bumper cantos" — this is what I named your messages that, literally and figuratively, I never saw — showed me that there is at least one more way of reading the Divine Comedy. For once, I want to see it your way, because "the most blest condition is based on the act of seeing, not of love, love being the act that follows recognition." As you can imagine, this comes from Paradiso and I would like our joint reading to start with Canto XXVIII.*

I cannot resist ending this letter with a line from Dante's very last Canto: "I feel my joy swell and my spirits warm."

I hope yours will too when you open the accompanying package.

Lionel

And why do I find it appropriate to include this story and especially the cited paragraphs? Because this was written after the termination of a twenty-six-year-long marriage to my second wife, Norma, where lack of spousal communication was surely a contributing factor to a sad rupture. That I learned something from it and hoped not to repeat it again is obvious from the text of Lionel Trippett's message to his wife and from the fact that the entire short story collection was dedicated to my new wife, Diane Middlebrook. At the time I wrote "The Toyota Cantos", Norma was not on speaking terms with me; hence I do not know whether she ever read it, although I now wish she had — or at least Lionel's response.

The present volume is not the place to revisit my second marriage, which in any event I had done as far as discretion allowed in my original autobiography of some two decades ago. Yet I am now reprinting the following excerpt, because it shows in a fair and honest way how deeply such lack of open discourse between spouses can hurt. It starts with an event during one of the shadowiest stages of my marriage to Norma.

When on 10 October 1973 — the very day Vice President Spiro Agnew resigned — President Richard Nixon presented the National Medal of Science to eleven men in the East Room of the White House, I was one of the recipients. My citation read, "In recognition of his major contributions to . . . steroid hormones and to [their] application to medicinal chemistry and population control by means of oral contraceptives." At this festive occasion, attended by the First Lady and cabinet officials as well as by the medalists, their wives, and other members of their families, I had a special distinction about which I learned only two months later when the San Francisco Examiner *published an article with the headline "Nixon game medal to scentist on White House 'enemies' list."*

I was not at all dismayed to find that I had made the White House "enemies" list (of Watergate fame) as a result of both of my involvement in Senator George McGovern's presidential campaign, including serving as a delegate for him at the Miami Democratic convention in 1972, and of my open opposition to our Vietnam policy. I had another distinction, however, among the medalists: I was the only one to attend the award ceremony alone. To the repeated "Where's your wife?" I offered innocuous excuses. How could I tell them that Norma had never *been present at any occasion when I received an award for my scientific work? At such ceremonies I always thought back to other events she'd missed.*

After returning from Kenyon College, where I got an honorary doctorate in 1958 — at that time only my second one and hence still a novelty — together with the poet Robert Lowell and the Episcopalian bishop of Southern Ohio, I told her that the student marshal of the procession had solemnly said to me, "Bishop Blanchard, would you please follow me?" She laughed. But didn't say, "I wish I'd been there." Even during the twilight of our marriage, our overt relationship was civil enough that she could well have attended the Columbia University graduation in May 1975. After Arthur Rubinstein received an honorary doctorate, the entire audience rose in a standing ovation as if he'd just hit the last note at a concert in Carnegie Hall. I was next in line. When the president of Columbia University said that my oral contraceptive research had made its most significant impact on the emancipation of women, the graduating seniors of Barnard, the women's college, jumped up and cheered, followed by a second cresting human wave. "Yeh!" the seniors of all-male Columbia College thundered, right fists thrust in the air. (The interruption was duly noted in The New York Times *and eventually also in Rubinstein's autobiography.) I don't think I even bothered to tell my wife about that student response, which had pleased me so much. However delicious a meal is, it doesn't taste the same reheated.*

Perhaps she would not have enjoyed the early, purely scientific events, like the American Chemical Society's Award in Pure Chemistry given annually to a person under thirty-five, where only chemists and their ilk paraded. But there were commencements Norma might have enjoyed. For years, we went to the theater and opera together; yet when in 1974 I shared a podium at Wayne State University for an honorary doctorate with the actress Julie Harris, or in 1978 at Coe College in Iowa with Sherill Milnes, the operatic baritone, I traveled alone.

The week after I was served my divorce papers in 1976, Norma and I had a marathon duel of words — perhaps the longest in our life together, and surely the most frank. Tight-shut doors were unlocked, a quarter-century of wounds ripped open. Finally, we fell silent, exhausted from the brutal catharsis. But as I checked once more my list of grievances — shorter than my wife's, but still substantial — I found one more, perhaps the most deeply buried.

"Not once did you see fit to come to an event that honored me," I began, "not once in all those years."

"That's not true," she said quietly. And then reminded me of a single occasion, twenty years ago to the day, that I'd forgotten and she'd remembered. The tone of her reply, tinged with sadness, stopped me from pursuing the topic. I said, "I guess it doesn't matter any more.

But, clearly, Norma had known all along that it did matter; yet neither one of us had broached the subject over the years, just as there were many other topics we hadn't raised. Playing the role of faculty wife was bad enough, in the 1950s and 1960s, for an intelligent and highly educated woman, accustomed to independence in her premarital life. Living with a scientist, whose everyday life was conducted in an incomprehensible language, whose workdays lasted sixteen hours, and who brought his mistress home every night, must have been hardly bearable. That the mistress wasn't a real woman, but an intellectual obsession with chemistry, didn't make it any easier to tolerate. Was it really reasonable to expect her to participate in events that publicly honored her husband's way of life? Although it had barely dawned on me in our last exchange, Norma's silent boycott now seems almost a civilized response to protracted grievance. Yet in those years few people would have accepted that state of affairs as legitimate cause for complaint. "What's she bitching about?" they'd have asked. "He comes home every evening. He doesn't drink. He's a good provider. He doesn't run around with other women." ("Are they so sure about that?" she probably would have thought, but didn't say.)

As I already pointed out, writing short stories was the stepping stone to becoming a novelist, a career I followed for ten years during which

I wrote and published five novels. I have already indulged in the "Jew" chapter in virtual overkill fashion telling about them and their autobiographical components; instead, I would like to explain my last jump — to that of a playwright — which I have now pursued for the past fifteen years.

Plays

Through all of my three marriages I have been an inveterate theatergoer. The seeds were probably already sown in my very early teens in Vienna, where I saw, not yet twelve, Lessing's *Nathan der Weise* in the Burgtheater. During my Ph.D. studies at the University of Wisconsin, I frequently attended theatrical performances with my first wife, Virginia, and still recall what was one of the most memorable dramatic experiences of my life, a performance of *Othello* with three of the greatest American actors of that period, Paul Robeson, José Ferrer and Uta Hagen. Virgina was not only a talented amateur actress but was also addicted to theater as much as I was; both of us had the good fortune to then live for four years close to New York City — next to London the unquestionable Mecca of English-speaking theater devotees — where we attended numerous plays. After a theatrical drought of a few years in Mexico City and then Detroit, my second wife, Norma, and I tried to see every play in the San Francisco Bay Area during the heydays of the legendary William Ball's reign at the American Conservatory Theater in the 1960s, but the true immersion occurred during my life with Diane Middlebrook, because for virtually our entire married life, we commuted between residencies near Stanford and our flat in London. In the States, we went to the Oregon Shakespeare Festival every summer where we always managed to see six plays in three days. But such thespian mini-orgies faded in the light of our subsequent annual August trips to the Edinburgh Fringe where I still look back at our record: six plays in one day, starting around eleven in the morning and going beyond midnight. But the rest of the time, it was mostly London with its continually changing overabundance of plays and superb actors. For the last two decades, the theater (mostly in London and Vienna) has been my most captivating avocation and also my most frequent mode of relaxation, often attending up to thirty plays a year.

Of course, seeing plays does not mean writing plays. Until 1996, the idea of extending my authorial ambling beyond fiction had not even occurred to me until one evening in September 1996 when I turned to my wife as we walked out of the National Theatre's Cottesloe Theatre to announce that I was going to write a play myself. It was a decision prompted by the feelings the play we had just seen — *Blinded by the Sun* — had engendered in me as a scientist, since it dealt with the "cold fusion" fiasco which had dominated the scientific as well as lay press just a few years earlier. Steven Poliakoff's play was one of the first recent theatrical works — two years before the appearance and sensational success of Michael Frayn's *Copenhagen* — that could really be called "science-in-theater", an area which I promptly adopted for my next literary bailiwick.

As I have said on more than one occasion, the conviction of many scientifically untrained persons that they are unable to comprehend scientific concepts stops them from even trying. For such an audience, rather than an unadorned lecture, "case histories" can be a more alluring as well as more persuasive way of overcoming such hurdles. If such a story-telling "case history" approach is employed on the stage rather than from the lectern or on the printed page, then we are starting to deal with "science-in-theater."

To write in this genre does not require that the author be a scientist. Since that Cottesloe theatrical event in 1996 or even the somewhat earlier science-related *Breaking the Code* by Hugh Whitemore or Tom Stoppard's *Arcadia,* all subsequent theatrical hits dealing in one way or another with science were written by internationally recognized playwrights, who had gained their scientific knowledge second-hand and used science mostly for metaphoric purposes. Still, why is it that no "hard" scientists, as far as I am aware, have become recognized playwrights, whereas several physicians have made major contributions? Consider Anton Chekov or Arthur Schnitzler, for instance. Is the absence of chemist-playwrights due to the fact that they find it difficult to even communicate with their peers without recourse to blackboard or slides or some other kind of pictographic aid? Or is it because chemists deal primarily with abstractions at the molecular level, whereas physicians spend their days listening to the stories of other human beings? Even the most scientifically invested plays succeed, if they do, because they work at the human level. Or is it that all formal written discourse of scientists is always monologist, whereas the theater is the realm of dialogue?

Perhaps none of those generalizations is the reason, yet it is that last one that tempted me to try my hand at playwriting. I find dialogue so stimulating, both as reader/listener as well as author, that an exceptionally high proportion of my fiction takes place in dialogic rather than in narrative form. I first became intrigued by the idea of doing something with "science-in-theater" when I was working on the third installment of my science-in-fiction tetralogy, *Menachem's Seed*. The science of that novel involved reproductive biology as seen by a childless woman scientist preoccupied with childbearing. Much of the factual information of that novel concerns the current high-priority area of assisted reproduction and the reasons why research on contraception is moving off the main stage.

An Immaculate Misconception

The inherent problems of converting any novel into a play are obvious, yet it seemed to me that building a science drama based on my novel's scientific themes and human conflicts was not unreasonable. In addition, biological reproduction is inherently more dramatic than any abstract chemical concept. At least, those were my thoughts after I left the Cottesloe Theatre at the conclusion of Poliakoff's *Blinded by the Sun*; as we headed home, I found myself starting to think of a title for my first play. When I write, I start toying with the title the moment I have settled on the theme — before I have even worked out the plot. The title *Menachem's Seed* was too suggestive of a dramatic adaptation of my novel. *ICSI*, though describing factually the reproductive technology I wanted to feature in my play and have already described in this book in detail in the earlier chapter on the Pill, offered no subliminal association for a prospective theater audience. *Condom Capers* implied a comedy and *The Purloined Sperm* a mystery. *An Immaculate Misconception*, suggested by a friend, Norma Miller, finally won out, even though the double meaning of "misconception" in English makes this title untranslatable into any other language. I emphasize this point because I did not wish to imply any blasphemous allusion to the religious concept of "Immaculate Conception". The choice of alternative titles in the twelve languages into which the play has by now been translated shows in how many, culturally telling ways the rather contentious subject matter of the play was interpreted. Some of the more extremes are *Rape Under the Microscope* (Serbian) or *This Sperm is Mine!* (Portuguese), while in German and Swedish the title was reduced to the single word *Unbefleckt* or *Obefläckad*, both standing simply for "Immaculate," and thus at least raising an attractive element of ambiguity.

Paulo Matos

Laboratory scene from the Portuguese production (Teatro da Trindade, Lisbon) of *Esse espermatozóide é meu!*

Unlike the professional playwrights I have mentioned so far, who mostly used science for theatrical aims, I started from the opposite side, using the stage for a scientific (and hence at least partially pedagogic) purpose, realizing, however, that for any play to succeed on the stage, it must appeal to a live audience that did not come to be educated, but rather entertained. The sub-title of this first play, *Sex in an Age of Mechanical Reproduction*, is an allusion to Walter Benjamin's famous essay of 1936, "Art in an Age of Mechanical Reproduction." I chose it because I consider the *de facto* separation of sex (in bed) and fertilization (under the microscope) in Europe or Japan one of the fundamental issues facing humanity during the coming century. I picked Benjamin's phrase for a second reason as well: in our preoccupation to conceive, we often forget the product of all the technological skills we utilize, namely the resulting child. Benjamin argues, "The technique of reproduction detaches the reproduced object from the domain of tradition." All the reader has to do is to substitute "child" for "reproduced object" in order to land right in the middle of the ethical thicket that reproductive technologists invariably face: they support heroic efforts by many couples to overcome certain biological hurdles that may very well harm rather than benefit the "reproduced object."

As I have already shown through some dramatic excerpts in the chapter dealing with the Pill, I chose for the didactic component of my play the most ethically charged reproductive technology of them all, ICSI — the direct injection of a single sperm into the egg. I suspect that few will argue with my assumption that everyone has opinions about reproduction and sex, and that most people of theater-going age are convinced that they know the facts of reproductive life. But do they really? I would offer odds that few in such an audience could even answer correctly the following question: while it takes only a single sperm to fertilize an egg, how many sperm must a man ejaculate in order to be fertile? Answer: a fertile man ejaculates in the order of 50–100 million sperm during intercourse; a man ejaculating 1–3 million sperm — seemingly still a huge number — is functionally infertile. Twenty years ago, there was no hope for such men. But now many can become fathers because of ICSI. Yet how many members of the theater audience I wish to attract have heard of ICSI?

My timing with *An Immaculate Misconception* was fortunate. Its premiere occurred in 1998, the year in which Michael Frayn's *Copenhagen* opened and which became such a smash hit as to confer instant respectability on a play dealing with science. My drama, though somewhat didactic — potentially the source of infanticide in the theater — was successful because its pedagogic aims were directed at exciting aspects of reproductive medicine which appeared continuously in newspaper headlines. Thus at the Edinburgh Fringe Festival, which in 1998 celebrated its fiftieth anniversary and boasted over one thousand different performances and spectacles, the BBC's *Edinburgh Nights* — the daily national TV coverage of the most noteworthy events — included my play on the first night, which in turn led to a meeting with the theater director and producer, Andy Jordan. It was he who directed a subsequent radio broadcast as "play of the week" over the World Service of the BBC with a stellar cast that included Henry Goodman, who had just won the annual Olivier Award as the best actor of the year for his stunning performance of Shylock in *The Merchant of Venice.* Just a few weeks earlier, I had watched Goodman through my binoculars in the National Theatre, but now he sat just a few feet across from me projecting my own words into the microphone. It was a heady experience for a first-time playwright.

Writing *An Immaculate Misconception* was hard work. The current version on my web site is the twenty-fourth one, in part because as the play was performed in different countries, I chose to make culturally related modifications for each of them. In the process, I, the self-taught virgin playwright, learned a great deal about crafting plays, and decided to go one step further to explore the pure pedagogic value of such a reproductive case history in the classroom rather than solely on the commercial stage.

For that purpose, I generated a script of a simulated TV interview — a type of "verbal combat" — between a TV moderator and a scientist, which was not intended for the conventional theater, but rather for the classroom to stimulate active debate regarding the ethical issues associated with the birth of children conceived without sexual intercourse. Published in Germany in bilingual (English and German) book form under the title *ICSI*, I labeled it a "pedagogic wordplay" to be used as a classroom text, with students playing the interview roles; indeed, *ICSI* has already been performed in just this way in many classrooms in English, German, Italian, Spanish and Chinese. But it has also been presented at various professional meetings, medical congresses and graduate programs, as well as in a performance on BBC Radio 3, proving that this type of dialogic format can also be employed effectively for instructional purposes outside the conventional classroom.

I had never written anything twenty-four times, but doing so with *An Immaculate Misconception* was an invaluable crash course in playwriting. I do not regret it for a moment, since it gave me the opportunity to meet, talk and work with so many different theater professionals. Through the Edinburgh premiere I met Andy Jordan, who became the producing director of the London premieres of eight of my plays. A second bonus proved to be the German language premiere the following year in Vienna through which I made the acquaintance of Isabella Gregor, who subsequently played such an important role in many of my later plays. In her, I had the good luck to encounter a person who was not only

Andy Jordan

Andy Jordan, the director of seven of my plays in London

"Arguing" as well as in rehearsal with Isabella Gregor, the director of my German language plays

a widely recognized actress in major Austrian and German theaters, but who by then had started on a second career as director. Not only was Vienna the city where I had seen my first plays as a teenager, but in 1999 the premiere was held at the one-hundred-year-old Jugendstiltheater, which, coincidentally, was the site of the most erotic scene in my novel *Menachem's Seed.* No wonder that in my opinion, Isabella Gregor's production of *Unbefleckt* was the most imaginative of all the ones I had seen so far. Subsequently, she became a close personal friend and the director of numerous dramatic readings and of some actual productions of several of my plays.

Oxygen

The relatively rapid acceptance of my first play can in large part be ascribed to the timeliness of the topic and the inherently dramatic aspects of human reproduction that in *An Immaculate Misconception* were presented so graphically — a feature commented upon by many reviewers. But as a chemist turning into a playwright, it behooved me to see whether chemistry can be presented as effectively on the stage as, say, biology, i.e. sex. To test that proposition, I had the good fortune to find a partner, Roald Hoffmann, interested in joining me in such a theatrical experiment (even though he is a theoretician by profession, rather than experimentalist). In 1981, while professor of chemistry at Cornell University, he was awarded the Nobel Prize in Chemistry for his theoretical chemical insights. But unlike most chemists, he has been interested for years in communicating with a broader public and has done so through his own poetry and non-fiction writing.

Our first e-mail exchange was brief and at first hesitant:

Date: 10: 12 PM 11/2/97-0400,
To: Roald Hoffmann
From: "Prof. Carl Djerassi"

Roald—What do you say about our collaborating sometime about a play featuring the wives of Lavoisier, Scheele and Priestley–in other words presenting the collegial competition etc. among the three men through a fictionalized account of their wives? Think about it.
Carl

Date: Tue, 15 Sep 1998 17:41:44-0400
To: "Prof. Carl Djerassi"
From: Roald Hoffmann

Dear Carl
Let me pursue our conversation in Boston. I voiced some hesitations, but I would like to try to write this play together. You have so much more experience in both the writing and production aspects, and I trust your instincts on these matters. So let's give it a shot, with a couple of intensive days of discussion.
Roald

But once we agreed to collaborate, given that Hoffmann was mostly in Ithaca, New York and I in either London or San Francisco, it probably required a couple of thousand e-mails before we completed *Oxygen*, a play dealing with the Nobel Prize that, at least to scientists, is potentially as sexy if not more so than ICSI. In our *Oxygen*, the Nobel Committee decides to celebrate its centenary in 2001 by establishing a new Nobel Prize, to be termed a "retro-Nobel," to honor inventions or discoveries made before 1901, the year when the first Nobel Prizes were awarded. For a change, what's wrong with paying attention to the dead?

Collaborating with Roald Hoffmann on Oxygen at Cornell University, 2000

Our play attempted to deal with two fundamental questions: what is discovery in science and why is it so important for a scientist to be first? To put it even more crudely, why do the scientific Olympics only award gold medals and no silvers or bronzes? Why is it that, in science, being second might as well be last? And yet, why in the end is it even more important to be recognized last? In *Oxygen*, we approach these questions as our imaginary retro-Nobel Committee meets to select, first, the discovery that should be so honored, and then — as it turns out, not a straightforward question — which scientist to credit for it.

Whatever one may think of *Oxygen*, it is without question an unabashed "science-in-theater" play, dealing at it does with the discovery of oxygen and the centenary of the Nobel Prize in 2001. Although we were well known in scientific circles, our names meant nothing in the theater world when we finished the first draft of *Oxygen* in late 1999. After half a dozen staged readings in 2000, the play opened in 2001 with the American premiere at the San Diego Repertory Theater in California; the German one, directed by Isabella Gregor, in Würzburg; and the London opening, directed by Andy Jordan, at the Riverside Studios. Since 2001, the play has been translated into twenty other languages and has had over one hundred independent productions, staged readings and radio broadcasts, with the two most important ones occurring on December 10, 2001 — the exact centenary of the Nobel Prize — by the BBC World Service and the WDR (Westdeutscher Rundfunk). Thus in terms of worldwide theatrical distribution, the "science-in-theater" play *Oxygen* fared far better than the vast majority of science plays written during the past twenty years, though it clearly did not reach the commercial success of some of the major plays described earlier. We convinced Wiley — an important science publisher (rather than a conventional theater publisher) — that *Oxygen* could be considered a valuable book in the field of science history that just happened to be written by two well-known chemists in all-dialogic format. That being true, why wait for theatrical premieres to determine whether *Oxygen* was also a theatrical success before printing the play text in book form — usually a *sine qua non* of theater publishers? The English and German versions of *Oxygen* were released in print prior to its world premiere and reached a sales level of about four thousand copies by the end of that first year. While such a figure might seem piddling if judged by

Stephen King or Danielle Steele standards, it is more than respectable if compared with the annual sales of most plays. In the subsequent two years, *Oxygen* also appeared in book form in ten other languages, with a commercial DVD being distributed by Educational Innovations, Inc.

All this sounds either defensive or boastful; in fact it is both. As pointed out, our scientific credentials meant nothing to theatrical professionals and especially agents, who in any event were and still are terrified by anything smelling of science. It was largely through our own contacts and efforts and especially through the appearance of ten of the twenty translations as books published by literary or academic publishers that it received the theatrical exposure I just listed. The experience with *Oxygen* was for me a bonus and a challenge. A bonus because, without an agent, I learned an enormous amount about play distribution, about the nitty gritty of theatrical contracts and negotiations and, perhaps most importantly, to become accustomed to the spectacular rudeness of so many theater literary managers, who generally do not even acknowledge receipt of a play submission The challenge was to see whether *Oxygen*'s success could be repeated another time.

Euclides Hernandez Penaranda

Spanish-language production of *Oxygen* at Teatro Nacional, San Jose, Costa Rica, 2010

Collaboration with Roald Hoffmann was fun and productive, but also stressful for the simple reason that writing a play or novel together — unless the functions are clearly separated as in a musical between librettist and composer — is inefficient and requires numerous compromises. No wonder that virtually no novels and only very few plays have been written by more than one author. The striking exceptions of the long-lasting Beaumont and Fletcher collaboration in the pre-Shakespeare days or the triple authorship of *Three Hours After Marriage* by John Arbuthnot, John Gay and Alexander Pope in the early eighteenth century are the exceptions that confirm my generalization. Hoffmann and I were realistic enough to construct prior to the start of our collaboration a form of "prenuptial agreement" (without legal advice or legalese language) in which we anticipated certain conflicts of which the most obvious one was irreconcilable disagreement on a given textual wording. Our solution was to depend on an independent referee (from the BBC theater department) to whom we would send our respective versions without disclosing who wrote which one, and to then ask our Solomonic dramaturg to decide. We only had to do this two or three times; eventually, at Hoffmann's suggestion we turned to my wife, Diane Middlebrook, for adjudication about a couple of women's issues, knowing that as the ultimate professional, she would exercise critical rather than connubial judgment. It all worked out, which is one reason why we are still on such friendly terms, but both of us have subsequently stuck to solo play writing.

With two science plays under my belt, I decided to complete a science-in-theater trilogy, which, if one is prepared to slightly stretch the meaning of "science," has by now expanded to a six-part series, interspersed by two plays that have nothing to do with science to prove at least to me that I need not be fixated by my own intellectual heritage. My reason for initially focusing on plays dealing with science is exemplified by my response to a well-meaning, but in my opinion naïve, project initiated in 2000 at CERN, the European Laboratory for Particle Physics, whereby distinguished artists were brought to Geneva to "learn about high-energy physics and [to] respond by creating an original piece of art during this year." One of the artist participants, the Turner Prize winner Richard Deacon, was quoted as saying, "We need to listen to each other, but not necessarily to understand. The misunderstanding in both directions can be creative."

Creative it may well be, but is mutual misunderstanding really fruitful in terms of narrowing the gulf between scientists and artists or does it accentuate that gulf? I am a firm believer in the ultimate virtue of understanding — and if "science-in-theater" contributes to that aim, I will regard my current dedication to playwriting as very well spent. Given that my earlier decision in my "science-in-fiction" novels to focus on behavior rather than just scientific facts has proved to be productive in forcing me to reflect upon my own behavioral practices, I shall now illustrate how I have used the same approach for my playwriting.

One of the main themes in *Oxygen* is the preoccupation by many scientists with priority — one of the most common but also ugliest behavioral features of the scientific community. The following brief excerpt of a conversation between one of the discoverers, Joseph Priestley, and his wife describes succinctly, as really only a play can do, not only the issue at hand but also the accompanying emotional dynamics:

MRS. PRIESTLEY *Why face Lavoisier?*

PRIESTLEY *I must.*

MRS. PRIESTLEY *To prove you told him?*

PRIESTLEY *To show I was first.*

MRS. PRIESTLEY *And Scheele?*

PRIESTLEY *I trust him.*

MRS. PRIESTLEY *He claims priority.*

PRIESTLEY *He did not publish.*

MRS. PRIESTLEY *Yet wasn't he first?*

PRIESTLEY *Perhaps.*

MRS. PRIESTLEY *But that would make you second.*

PRIESTLEY *It would make Lavoisier third.*

MRS. PRIESTLEY *Is that the point? That he was last.*

PRIESTLEY *Indeed.*

MRS. PRIESTLEY *Why?*

PRIESTLEY *Have the world bow to him? When I preceded him?*

MRS. PRIESTLEY *If you were King Gustav —*

PRIESTLEY *God forbid!*

MRS. PRIESTLEY *Still… if you were King… whom would you pick?*

PRIESTLEY *I'd ask… whom would the world choose?*

MRS. PRIESTLEY *Joseph! Answer me… as my husband… not as a clever minister.*

PRIESTLEY *You've always wanted black and white answers.*

MRS. PRIESTLEY *This question deserves it.*

PRIESTLEY *Deserving something does not always lead to getting it.*

MRS. PRIESTLEY *You're not in a pulpit.*

PRIESTLEY *(tired) I published first… which makes me first in the world's eyes.*

MRS. PRIESTLEY *I meant the heart… not the eyes.*

PRIESTLEY *The world has no heart.*

MRS. PRIESTLEY *But you do… you've often opened it to me.*

PRIESTLEY *You're a clever woman, Mary.*

MRS. PRIESTLEY *No… this is your loving wife asking.*

PRIESTLEY*: Before we came to Stockholm, I was convinced… in my heart and mind… that I was first. But now?*

MRS. PRIESTLEY *I understand, Joseph.*

Calculus

Personally, I have also been guilty of fierce competitiveness to establish priority, as is demonstrated so tellingly in the chapter on the Pill. But to believe that this is solely a feature of contemporary science rather than a scientist's imprinted blemish is nonsense, and to make this point beyond *Oxygen,* I decided to exemplify this through a carefully researched and partly unknown aspect of one of the bitterest priority struggles in science history: the thirty-year-old brawl between Isaac Newton and Gottfried Leibniz — arguably two of the greatest polymaths of the seventeenth and early eighteenth century — to prove who had first invented the calculus. I picked that example, because the calculus priority clash — with Newton ultimately charging the other with piracy — has, in the words of William Broad, "been fought for the most part by the throng of little squires that surrounded the two great knights." It is through some of Newton's "little squires" that my play tries to examine one of Newton's greatest ethical lapses.

Newton in his capacity of President of the Royal Society created a commission of eleven Fellows of the Royal Society to adjudicate the conflict. The composition of the committee that never openly signed the document did not become acknowledged for over one hundred years. With the assistance of two Stanford undergraduate students, Joshua Bushinsky and Tonyanna Borkovi, I went through the archival records of all eleven to select three as the main characters of my play *Calculus*, which by now

Chris Corner

Isabella Gregor

(a) (b)

(a) Leibniz and Newton confrontation played by Michael Fenner and David Gant at New End Theatre, London, 2004. (b) Carl Djerassi wearing Newton's wig, Dresden dramatic reading, 2004.

has been translated into five languages. One of the three men is Dr. John Arbuthnot (1667–1735), physician to Queen Anne, author of the political allegory, "History of John Bull," describing the prototypical Englishman, and friend of Alexander Pope, Jonathan Swift, John Gay and Thomas Parnell. Just as in *Oxygen* I again summarize the issue at hand through an excerpt of a conversation between Arbuthnot and his wife, because in each play the women are deliberately assigned key roles, in spite of the fact that historically they were almost totally ignored.

MRS. ARBUTHNOT *Well?*

ARBUTHNOT *It's done.*

MRS. ARBUTHNOT *Who was there?*

ARBUTHNOT *All eleven.*

MRS. ARBUTHNOT *No one else?*

ARBUTHNOT *Newton.*

MRS. ARBUTHNOT *That I expected... but who else?*

ARBUTHNOT *No one.*

MRS. ARBUTHNOT *That was clever.*

ARBUTHNOT *Newton is clever... but also cautious. Why invite unnecessary witnesses?*

MRS. ARBUTHNOT *What about other supporters? The kind Bernoulli called "Newton's servile sycophants."*

ARBUTHNOT *The Committee is already inundated with them. Besides, he called them "Newton's toadies."*

MRS. ARBUTHNOT *Hardly more complimentary. But would it not have been more politic to include in the Committee some Fellows less beholden?*

ARBUTHNOT *There were a few.*

MRS. ARBUTHNOT *Bonet?*

ARBUTHNOT *He's one.*

MRS. ARBUTHNOT *And you.*

ARBUTHNOT *(tired nod) And I.*

MRS. ARBUTHNOT *(impatient) Tell me what happened.*

ARBUTHNOT *I started out on the wrong foot.*

MRS. ARBUTHNOT *You mean with honesty?*

ARBUTHNOT *(nods) Does truth not bear the same relation to understanding as music does to the ear or beauty to the eye?*

MRS. ARBUTHNOT *Newton is concerned with understanding the universe. That truth concerns him… but no other music reaches his ear. I had warned you, John. (Reaches over to pat his hand or other gesture of affection). What did you say?*

ARBUTHNOT *I quoted Francis Bacon: "There is little friendship in the world… and least of all between equals."*

MRS. ARBUTHNOT *And?*

ARBUTHNOT *He stopped me.*

MRS. ARBUTHNOT *What did he say?*

ARBUTHNOT *Nothing.*

MRS. ARBUTHNOT *But you said he stopped you.*

ARBUTHNOT *He pointed to Hill.*

MRS. ARBUTHNOT *Why Abraham Hill?*

ARBUTHNOT *He's the oldest… almost eighty.*

MRS. ARBUTHNOT *(dismissive) And one of the toadies.*

ARBUTHNOT *(tired) Not any more than most of us.*

MRS. ARBUTHNOT *And what did the oldest toady say?*

ARBUTHNOT *That the Committee's concern was with superiority... not equality... of British science. Considerations of friendship were irrelevant. Newton and Leibniz were only surrogates in a larger struggle.*

MRS. ARBUTHNOT *He said that... in front of Newton?*

ARBUTHNOT *He didn't have to... most understood... though perhaps not Bonet.*

MRS. ARBUTHNOT *And that was it? (Pause). John! I've never had to push you like this. Do you not trust me?*

ARBUTHNOT *It's a matter of shame... not trust.*

MRS. ARBUTHNOT *Then confide in your wife.*

ARBUTHNOT *I could not help but think of John Flamsteed —*

MRS. ARBUTHNOT *Newton hates Flamsteed... in spite of his position as Astronomer Royal.*

ARBUTHNOT *(nods tiredly) In spite... and because of it. (Stronger). He once sent me a note, writing, "Those that have begun to do ill things, never blush to do worse to secure themselves." I hoped then that he meant Newton... and now I know he did.*

MRS. ARBUTHNOT *Because you were presented with the finished report before your Committee had even met?*

ARBUTHNOT *Even worse. Newton alone had written it —*

MRS. ARBUTHNOT *(shocked) What? That I cannot believe. Newton could not have been that brazen.*

ARBUTHNOT *He was... and termed the report* "Commercium Epistolicum Collinii & aliorum" *— an exchange of letters between Collins and others.*

MRS. ARBUTHNOT *Because much of the evidence came from the correspondence of John Collins?*

ARBUTHNOT *Letters written to Collins by Leibniz and Newton… and selected by Newton… (Pause)… to bolster his case in his own words.*

MRS. ARBUTHNOT *That is barefaced. And you were asked to sign… without further debate?*

ARBUTHNOT *All of us were.*

MRS. ARBUTHNOT *Thank God you yielded… even though I find it shameful. But I would not have wanted you to suffer Sir Isaac's wrath. We both know his unparalleled cunning —*

ARBUTHNOT *And cruelty.*

MRS. ARBUTHNOT *What happens now that you have signed Newton's Commercium?*

ARBUTHNOT *I didn't say that pen was set to paper!*

MRS. ARBUTHNOT *Oh! Does that mean you refused?*

ARBUTHNOT *No, the committee is reconvening tomorrow. By then we must decide.*

MRS ARBUTHNOT *So there is still time. (Pause.) John… I'm afraid.*

ARBUTHNOT *Of me… your husband?*

MRS. ARBUTHNOT *Not of you… but for you! I'm afraid of the consequences if you don't sign.*

ARBUTHNOT *Margaret, I'm a Fellow of the Royal Society —*

MRS. ARBUTHNOT *But he's the president.*

ARBUTHNOT *He'll understand when I explain —*

MRS. ARBUTHNOT *He may understand… but he will never forgive you.*

ARBUTHNOT *Nonsense!*

MRS. ARBUTHNOT *John… you're being foolhardy.*

ARBUTHNOT *I promise to be diplomatic… but honest. An untruth is best contradicted by truth… not another untruth.*

MRS. ARBUTHNOT *Have you not been listening to me, John? That will never work with him. Diplomacy? Perhaps. But honesty? Never!*

ARBUTHNOT *That is an unwarranted conclusion!*

MRS. ARBUTHNOT *However diplomatically delivered, Newton will never accept an honest explanation that criticizes him…*

ARBUTHNOT *I shall not criticize him.*

MRS. ARBUTHNOT *He will consider your refusal to sign public criticism.*

ARBUTHNOT *I shall explain in private.*

MRS. ARBUTHNOT *Your name's absence from that document alone will be sufficient insult.*

ARBUTHNOT *I shall prove you wrong*

MRS. ARBUTHNOT *John! Sign. You cannot afford the risk. He will spit on you… and then convince you it's raining.*

Perhaps the key question asked in my *Calculus* is whether a scientist can be ethically sullied or worse and yet be recognized as a great scientist. The sad answer is a resounding "yes!" Newton has not only been described as remote, lonely, secretive, introverted, melancholic, humorless, puritanical, cruel, vindictive and unforgiving, but also as spectacularly unethical. This was not demonstrated once, but three times in his great priority fights and manipulations with Leibniz, Flamsteed and Hooke. But as President of the Royal Society and at that time the most powerful scientific personality in England to appoint an anonymous committee of eleven fellows of the Royal Society as a jury to adjudicate his conflict with Leibniz and then to compose their report himself is clearly beyond the pale. Yet in a poll for the most important persons of the second millennium, the London *Sunday Times* in September 1999 ranked him first, even above Shakespeare, Leonardo da Vinci and Charles Darwin.

The world premiere of *Calculus*, directed by Andy Jordan, was held in London and proved to be outstanding, but the last staging, a Portuguese translation directed by Mario Montenegro in 2011 in Coimbra, the home of Portugal's ancient university, was equally memorable; aside from first-class acting, it featured some startlingly original music, composed and played by the Portuguese violist, José Valente, who subsequently spent some time as composer-in-residence at the Djerassi Resident Artists Program in California.

Calculus will always occupy a special place in my personal memory bank, because following a very original staging by Isabella Gregor of a German version in Vienna's *Museum Quartier*, the Austrian composer Werner Schulze approached us with the proposal to convert my play into a chamber opera adaptation with eleven instruments. Schulze, a bassoonist, is an unusual composer in terms of his wider interests which involve extensive work in classics, sophisticated mathematical knowledge, and most importantly, true scholarship with Leibniz's work. His timing in 2004 was truly fortuitous, because 2005 had been declared the International Year of Physics, which the Swiss Federal Institute of Technology (ETH) had also decided to combine with extensive centenary celebrations of Einstein's *annus mirabilis* of 1905. Members of the ETH faculty contacted me with the suggestion that my play *Calculus* be performed during those celebrations as a cultural interlude among the numerous scientific lectures and festivities. Naturally, I was delighted, but then countered with the rather shameless counterproposal to use instead the virtually completed chamber opera version by Werner Schulze for a world premiere in Zurich. The ETH authorities concurred and Isabella Gregor, who had earlier served as assistant director at the Zurich Opera for a celebrated production of *Wozzeck*, assumed the directorial role in addition to reworking the libretto. It was performed four times in succession in the Studiobühne of the Zurich Opera, while at the same time the main stage featured Mozart's *Zauberflöte*, Puccini's *Tosca* and Donizetti's *L'elisir d'amore*. I shall never forget that week in May 2005.

Isabella Gregor

Entire ensemble of chamber opera version of *Calculus* at Zurich Opera Studiobühne (2005) with composer Werner Schulze (standing, sixth from left) and director, Isabella Gregor (sitting on floor, center)

Ego

By 2003, the dialogic component of playwriting had really infected me. Instead of deliberately searching for further projects directly structured around science, I repeated what I had done in my fiction writing: departing temporarily from science and scientists to convert into a play the novel *Marx, Deceased* (*Ego*) with its concept of productive insecurity even among super-achievers. It was the territory my wife had advised her scientist–writer husband not to trespass upon, considering that I had chosen as my "hero" a highly successful writer rather than scientist. Yet the wonderfully cathartic feeling I had sensed when I had not followed her counsel led me — almost eight years after completing that novel — to embark on a play by the title *Ego* and to examine in a self-questioning manner the topics of suicide and therapy — with therapy as the subject and a therapist's function as two of the dominant themes.

The most economical and insightful way of presenting my take on these contentious issues is to reproduce herewith some selected excerpts from the play *Ego* (subsequently also renamed in certain productions as *Three on a Couch*) — until now, commercially speaking, the most successful of my nine plays. After its premiere in London, it was performed on tour in sixty-eight German theaters over the course of two years as well as broadcast by the West German Radio (WDR). Yet I am still not sure whether internally, I have resolved the issues I describe through the struggles of the main personage, Stephen Marx, clearly a complicated, but also solipsistic heartless character, who embarked on a staged suicide to test the limits of productive insecurity. When I was faced by the well-meaning advice that in a three-person play it might be wise to make the main character a bit more likeable, I simply replied, "not this time." Was this the creator of Stephen Marx or Carl Djerassi himself speaking?

Poster of German "Landgraf" tour of *Ego*, 2006–2007

STEPHEN *A legal question.*

SHRINK *I don't offer legal advice.*

STEPHEN *How confidential do you keep this?*

SHRINK *If you went to church for confession, would you ask a priest that?*

STEPHEN *I'm not here to confess. This is different.*

SHRINK *Therapy and confession aren't really that different. Call what usually happens here an unburdening.*

STEPHEN *In that case I could've saved a bundle by going to see a priest.*

SHRINK *Ah! But the difference is that we don't absolve… we help you understand yourself. That takes much longer….*

STEPHEN *And that's what you charge for?*

SHRINK *Well... if you're looking for bargains, perhaps you should go to church... but it's tougher on your knees. And while you'd certainly benefit from therapy... it's clear to me that you didn't come for that.*

STEPHEN *So why did I come?*

SHRINK *For justification... but packaged in the form of a private confrontation.*

STEPHEN *And why would I come to you for justification?*

SHRINK *Because for that you also needed assured confidentiality. You could have gotten that from a lawyer... but he would have charged more... and listened less.*

STEPHEN *(impatiently) Okay, okay! But you tell no one what we talk about? No exceptions?*

SHRINK *(getting annoyed) There are exceptions to everything. If you told me you had a gun in your pocket and were about to murder somebody, I'd call the police. I'd have to.*

STEPHEN *Even if I just killed myself?*

SHRINK *There is nothing I take more seriously than suicide. I'd do my utmost to persuade you not to do that.*

STEPHEN *But suppose you later learned that I'd actually done it?*

SHRINK *I'd feel terrible for not having prevented it. Personally... and professionally. But suicide doesn't go with your psyche.*

STEPHEN *Is that your diagnosis?*

SHRINK *We've only had five sessions... generally much too short for a diagnosis. But with you, I'm prepared to risk it: yours is a case of pure, unadulterated narcissism... and that may be untreatable.*

STEPHEN *Isn't that your job? To shrink big heads like mine down to normal size? Isn't that what you bill me for? You... of all people... must* *be used to that sort of talk: Suicide… justification… interpretation of the uninterpretable… unburdening. Pay your money, pick a neurosis. I might even paraphrase Descartes: "I'm analyzing myself, therefore I am."*

SHRINK *Exactly! Analysis is the key to self-knowledge.*

STEPHEN *Do you think I need to come here to find out who I am? Stephen Marx: great author who will be remembered for generations to come? Or a smart con man who peddles phrases for money? Or an original thinker? Do you think therapy can answer these questions, Doctor?*

SHRINK *An analyst is mostly a guide. It's the analysand who ultimately must deduce his present circumstances from his past history. But how did the idea of suicide come into your head?*

STEPHEN *Everybody thinks of suicide… sometimes. I even wrote about it.*

SHRINK *An article?*

STEPHEN *A novel...I don't do articles. (Long pause). The day after your death, all you'll read are the canned obits that were written a long time ago. What counts are the reflective critical commentaries that tell you what the people who count really thought of you.*

SHRINK *And that's what you're after?*

STEPHEN *Have you never dealt with people whose self-esteem depends on the opinion of others? Have you ever stopped to think how it feels to work in a field where success isn't something you can quantify? How much uncertainty that involves? How much insecurity? Even James Joyce was obsessed with such reviews. I call it productive insecurity*

SHRINK *Well put!*

STEPHEN *So now I'm getting complimented? Is that part of therapy?*

SHRINK *Call it encouragement rather than compliment.*

STEPHEN *But compliment or not, productive insecurity simultaneously nourishes and poisons us. (Pause). Have you ever heard of Fernando Pessoa?*

SHRINK *Should I have?*

STEPHEN *(spells it) P E S S O A.*

SHRINK *Now you're going to tell me who he is.*

STEPHEN *The greatest Portuguese poet of the last century… if not the last three centuries… but he didn't just write poetry… he wrote poets. He created alter ego authors… at least three of them… who wrote in totally different styles!*

SHRINK *Lots of authors write under pseudonyms.*

STEPHEN *Not pseudonyms. Heteronyms. One person… living simultaneously in different personalities… the heteronyms he developed.*

SHRINK *Psychiatrists have a term for that syndrome.*

STEPHEN *Don't they always? For me, he's a hero. And an integral part of my ongoing experiment. Can you imagine the literary freedom Pessoa enjoyed?*

SHRINK *He sounds like a candidate for life-long therapy.*

STEPHEN *Implying that he needs to be cured? How about emulated?*

SHRINK *To accomplish what?*

STEPHEN *Simple: to travel through space and time… forward to self-perpetuation… and simultaneously backward to self-immolation. I shall achieve what was always beyond Stephen Marx's reach. Imagine the glory of not just being a "great writer," but several? Imagine what people will say in the history books when they realize I was a literary genius — not just once but time and time again, but under a series of different names, styles… even personalities. Perhaps the public will never find out.*

SHRINK *You don't want to be part of the canon; you want to be the entire canon. I think you may be certifiable.*

STEPHEN *I'm not attempting to become Pessoa. What interests me is the Pessoa phenomenon. To start from scratch... each time with a blank* *canvas! To turn into your own creation and continue living as that person. I don't know of anyone that has truly managed it in fiction. Let alone anyone who has employed such a method in order to enter the canon repeatedly as two, three... four different authors!*

SHRINK *You're unstoppable, Stephen: a narcissist who sheds his identity.*

STEPHEN *Why not? What are we, Doctor, but the constructs we build around ourselves? What happens when we shed them? What are we at our core? That is what I'm discovering... that's where the real work... real literature... gets done. A new work by Stephen Marx would only be compared to what came before. To pull this off... to create a text unrecognizable as the work of Stephen Marx, but standing and maybe soaring in its own right... that's a real accomplishment. I'm testing the ultimate limits of productive insecurity. Raising the ante... surpassing the last success... but as another person, not just another name! But you are intrigued, aren't you? Some small part of you wants to know whether I can pull it off. Come on, admit it!*

SHRINK *You're delusional.*

STEPHEN *Which leads me to my reason for being here? I have a proposal.*

SHRINK *I can't wait.*

STEPHEN *This is the first proper conversation I have had in a month and already I feel more human. Theodore, I need someone to talk to... spontaneously, openly...*

SHRINK *Just "someone"? Or specifically a therapist?*

STEPHEN *The life I have chosen is to surround myself with heteronyms. They are real persons... in every sense of the word... but they are all creatures of my imagination. I need one living person... someone I can trust not to let the secret out... someone who has another voice than mine. I don't have anyone else but you. I propose that we continue our sessions...*

Phallacy

Within months of completing *Ego*, a Viennese acquaintance of mine, Professor Alfred Vendl of the University for Applied Arts, told me a story that instantaneously caught my fancy as the potential subject for a new play in which the borders of science and art overlap — two fields which in any event were and still are dear to my heart. Although the story he told me had happened two decades earlier, the passion and even indignation with which he presented it gave it a dramatic flavor that I immediately recognized as the stuff of theater. In addition, it illustrated another character fault — falling in love with a hypothesis that is simply too beautiful not to be true and hence ignoring all counterarguments — that all too often affects scientists as well as practitioners of inquiry in other disciplines. The background is outlined in the following foreword to the play which I named *Phallacy.*

> *Some time ago, Professors Alfred Vendl and Bernhard Pichler of the University for Applied Arts in Vienna drew my attention to a superb life-sized bronze of a naked young man that for several centuries had been described as a Roman original. Recent modern chemical analysis in their laboratory revealed that it was a Renaissance cast. The overnight loss of approximately 1400 irreplaceable years had many consequences for the museum that for over a century had displayed the sculpture as a jewel of its Antiquities collection. Aesthetically speaking, does that revisionist attribution make the sculpture less valuable? Does pricking the balloon of financial inflation automatically also diminish the art historical merit of the sculpture or the viewer's pleasure in its beauty? And what about the art historian's personal and professional response when an unblemished favorite suddenly becomes irretrievably tainted?*
>
> *For decades, I, a chemist-turned-playwright, have also been a serious art collector, who has been well aware of the disturbing fetishization of many art objects. But instead of addressing primarily the change in value as a well-known art object is reattributed — a situation quite different when a work, say a presumed Vermeer painting, is found to be forged by Van Meegeren — I decided to focus on what effect such reattribution might have on the behavior of the principals involved in the dispute.*
>
> *This dramatic lode has been mined before. Alan Bennett's play and subsequent BBC TV film,* A Question of Attribution, *uses the question of*

a Titian painting's authenticity to depict the relation between art historian (Sir Anthony Blunt) and owner (Queen Elizabeth II) as well as Blunt's behavior as a notorious Communist spy. And Simon Gray's more recent The Old Masters — though ostensibly covering the dispute whether a certain painting was created by Titian rather than Giorgione — really delves into the ethical and psychological conflict between art historian (Bernard Berenson) and art dealer (Lord Duveen). In other words, the principals and the art in those plays have a historical basis, which however has been altered to make a dramatic point.

And what is that point in my "Phallacy"? Here I concern myself with a conflict much closer to my professional competence: the quirks and idiosyncrasies of art historian and scientist, when they examine the age of an art object from their grossly different perspectives: aesthetic and art historical connoisseurship versus cold material analysis. In addition, I also wanted to explore the ramifications of a well-known character fault that transcends the gulf been art scholar and scientist: falling in love with a favorite hypothesis and defending it against all comers and new evidence.

Like other playwrights working with factual material, I have modified, manipulated, disguised or even deliberately misused many historical nuggets by claiming the authorial freedom that any playwright rightly exercises. Thus I request that any resemblance to the actual principals associated with the ongoing saga of the putative Roman sculpture in the Antiquities collection of a famous European museum be largely ascribed to coincidence and that in no respect have I attempted to damage the reputation of a living scholar. And if the explanation in my play of what has happened to that original sculpture should in the future prove to be correct, it is not a reflection of my art historical acumen but purely a playwright's dumb luck.

Aside from my interest as a scientist and art collector, there is a deeply personal reason why I chose this theme for my newest play. Born in Vienna, I immigrated after the Nazi Anschluss to the USA, where I became a research scientist. In 2004, the Austrian Government offered me Austrian citizenship. Since by that time, I had turned into a playwright, what better token of reconciliation than creation of a play that I situated in the city of my birth?

Carl Djerassi
London, Singapore, Eugene, and Hamburg, October 2004–March 2005

The sculpture mentioned in the foreword is a real one, called the *Jüngling vom Magdalensberg*, which is still displayed at the Kunsthistorische Museum in Vienna as one of the jewels in its antiquities section. Curiously, in spite of its utter relevance to the Austrian historical art scene, other than a single staged rehearsed dramatic reading directed by Isabella Gregor in the actual museum, no Austrian theater was prepared to undertake a full fledged staging, although first-class productions occurred in London, New York City and Porto (Portugal). As shown below in an excerpt from the first scene, as playwright I am not easy on either the chemist (Rex) or art historian (Regina) in terms of their foibles. But why should I be, considering that I would not have behaved all that differently if I had been in their respective shoes?

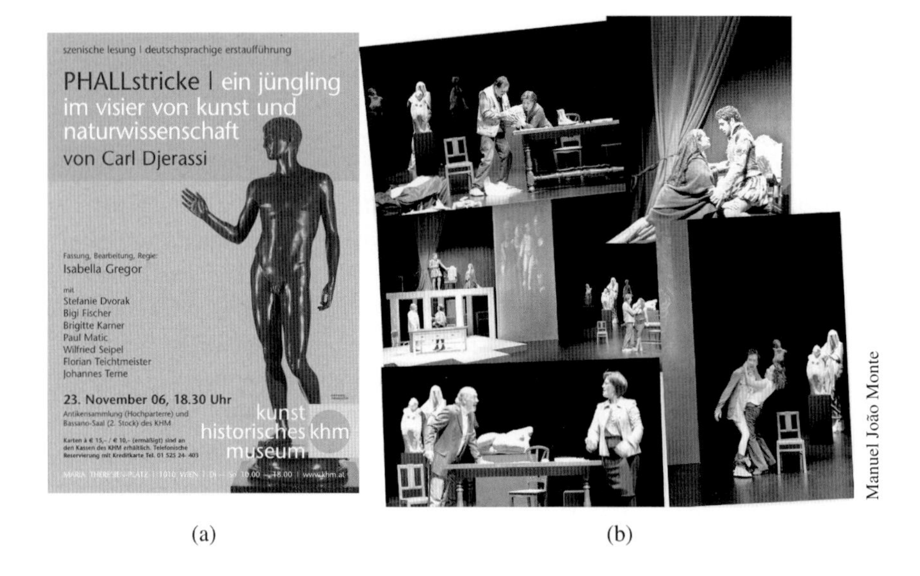

(a) (b)

Phallacy (a) Poster of German dramatic reading at Kunsthistorische Museum, Vienna 2006. (b) Views of theatrical production by Seiva Trupe, Teatro do Campo Alegre, Porto, 2011

REGINA *And you think that was appropriate? Telling your son about results that we're only discussing now?*

REX *It's not a state secret. Your museum director had asked me to take a look at your sculpture —*

REGINA *"Take a look?"*

REX *Yes. We developed some new chemical methods. We got some topnotch new equipment. State of the art. What's wrong with the museum commissioning a new approach to confirm the putative age of a sculpture?*

REGINA *(wounded) Putative?*

REX *It's not an insult. More often than not, age is considered putative until it's confirmed... even the age of a person. Take my son. In another couple of years, he'll have to produce a driver's license in order to buy a drink.*

REGINA *(sarcastic) And our museum director came to you for the driver's license of this bronze?*

REX *(enjoying the direction of the conversation) Just a parking permit. Doubts have been expressed whether it truly belongs in the Antiquities galleries.*

REGINA *Are you aware of the evidence I have amassed over years of research? Summarized in a scholarly book that has already been reprinted?*

REX *By your museum bookshop.*

REGINA *You've read the book?*

REX *I always read evidence before questioning it.*

REGINA *You read my book from beginning to end?*

REX *Eventually. But I started at the end.*

REGINA *You mean the last chapter?*

REX *The index.*

REGINA *The index?*

REX *Yes, the index. And looked for the words "trace analysis" and "nickel."*

REGINA *Why did you start with those words?*

REX *Because Roman bronze has a very low nickel content.*

REGINA *(sarcastically) I am delighted to hear that.*

REX *I wouldn't be if I were you.*

REGINA *Why not?*

REX *Your sculpture contains a lot of nickel. Rather typical of Renaissance bronze.*

REGINA *(interrupting) And you told this to our Museum Director?*

REX *Of course.*

REGINA *Instead of coming to me?*

REX *But he was the one who requested we examine your sculpture. Anyway, what matters here is the nickel content.*

REGINA *You're saying that our sculpture could not be of Roman origin? That all Roman bronzes, without exception, had low nickel content?*

REX *I didn't say without exception —*

REGINA *You see?*

REX *I'm saying it's extremely unlikely. And that's why I'm here. As a courtesy call. To tell you... before informing anyone else... what additional chemical tests we carried out to prove our assumption —*

REGINA *(pouncing on this) So, you're just making an assumption?*

REX *Well, no, because we carried out further tests —*

REGINA *Nonetheless, these tests were all based on your assumption. You assumed that the sculpture is a Renaissance work. That all the evidence in my book... all 345 pages... is hogwash.*

REX *Well... hogwash, no... I wouldn't say that, not exactly hogwash —*

REGINA *(infuriated) You see, this is what I find so I infuriating. You slavishly follow the rules of chemistry you learned as a student... lessons* *you now teach to your students… who will then teach it to* <u>*their*</u> *students, it's sterile crap —*

REX *(outraged) Crap?*

REGINA *(ignores interruption) I said "*<u>*sterile*</u>*" crap… consisting of rules promoted by art-hating boors, shielded from any sense of beauty by a dense fog spread from ear to ear. You disembowel every vestige of aesthetics… you ignore style, form, patina… in fact all connotative accompaniments. (Deep breath).*

REX *(outraged) Excuse me?*

REGINA *Someone really ought to prick that balloon of self-righteous… pompous… simplistic arrogance of yours.*

REX *You may live to regret those words.*

REGINA *(still steaming) No, not simplistic…*<u>*cocksure*</u>*. Transforming the wine of aesthetics into vinegar! How typical of you chemists. When chemists dabble with art, the best that can be said is the results are unpredictable.*

REX *Unpredictability is what science is all about…*

REGINA *Is it really? And even if it is, then why doesn't that teach you humility… rather than arrogance? And why not recognize the importance of visual beauty… a concept that barely exists in your chemical world.*

REX *For this discussion, the beauty of the sculpture is not important. Even the sculpture is not important —*

REGINA *So what is?*

REX *Truth.*

REGINA *That's all?*

REX *That's all.*

REGINA *How pathetic.*

REX *I'm trying to be collegial —*

REGINA *Collegial?*

REX *I wanted to explain how we arrived at our conclusion —*

REGINA *You think I need an explanation?*

REX *(sarcastic) Oh pardon me! I forgot. You have no use for trace metal analysis, but you're an expert in thermoluminescence... and scanning electron microscopy. In their scope and limitations —*

REGINA *Their limitations! Exactly.*

REX *I've had it! You're impossible! Here... (hands over the report he wrote). Read it.*

REGINA *I don't need to read this... I'll just file it in the only place I file such rubbish.*

REX *In that case... wait till it's published! And the shit hits the fan!*

REGINA *(taken aback) You plan to publish this?*

Without giving away the ending I invented, I cannot resist adding below a brief excerpt from the last scene of the play to show that even here, I was able to illustrate one behavioral quirk of scientists as well as of other academics, namely arguing about the order of authors in collaborative publications, an issue that may well seem trivial if not frivolous to non-academic outsiders.

REGINA *Which brings me to the reason for my peace overture. Why not publish our results jointly... our Spanish findings and your chemical studies.*

REX *Quite an offer! (Pause.) Why not? I can just see what people will say when they read a paper by Rex Stolzfuss and Regina Leitner-Opfermann. It sounds quite magisterial: "Rex and Regina."*

REGINA *No Opfermann. Just Leitner.*

REX *Even shorter. Stolzfuss and Leitner.*

REGINA *I was thinking of Leitner and Stolzfuss. Alphabetical order.*

REX *Since my name starts with an S, I've never believed in alphabetical order.*

REGINA *In that case, forget about alphabetical precedence. Call it chivalry.*

REX *That word does not exist in chemistry. Anyway, my name has always come first on every paper I ever published.*

REGINA *And mine has never appeared second. (Pause) Ah well, our rapprochement seemed promising while it lasted. (Pretends to look at her watch). All of four minutes... going on five. So I guess it's "goodbye." Anyway... it's time I got back to the museum,*

Taboos

Intellectually and pedagogically, the decade starting with the mid-1990s marked for me an increasing shift in attention from *contraception* to *conception*. The focus of my novel *Menachem's Seed* as well as that of my play *An Immaculate Misconception* was the *de facto* separation of sex and reproduction in the geriatric countries of the world, most notably in Europe and Japan. With the average-sized family in those countries hovering around 1.5 children, it is clear that sexual intercourse and reproduction are now almost totally divorced. Religions, such as the Catholic faith, still focus on the 1.5 sexual contacts leading to reproduction while neglecting the implications of the hundreds of acts of intercourse that occur prior and/or subsequent to the birth of a child. Since my life as a writer has always been interwoven with my extensive lecturing — at academic venues, to the broader public and increasingly also on radio and TV — it should not be surprising that even as a playwright I decided to revisit this contentious topic, but this time with a virtually exclusive emphasis on the societal rather than scientific implications of this immense shift in human reproduction. As I already recounted in some detail in the chapter on the Pill, my sixth play, *Taboos*, attempts to show how dramatically our views of "Family", "Baby,"

"Marriage" and similar apparently unambiguous terms have changed, largely through the monumental developments in reproductive medicine during the last three decades.

Perhaps the most contentious issue arising from these scientific discoveries is the politically charged question of why homosexual couples should not be permitted to have children or indeed marry. In my numerous lectures, no subject has caused more vigorous and even ugly debate than this one, which was one of the reasons why I decided to make a lesbian couple the lead characters in *Taboos*. The right of lesbian couples to have a child, in fact the motivation for their relationship, was one of the subjects that I chose to pursue. *Taboos* has been published in book form and performed in different versions in the UK, the USA, Bulgaria, Austria and Germany, which each time required substantial revisions to make it culturally acceptable to such diverse audiences. The ability to revise a text to fit a given cultural setting is an important aspect of playwriting, which is simply impossible when dealing with a film or a novel. I illustrate this below with an excerpt from an early draft of *Taboos* where I experimented with different approaches to introduce the motivation of my lesbian couple to the theater audience. The following introductory scene never made it into any of the staged productions and hence can be considered a tongue-in-cheek world premiere.

SALLY *I'm Sally Parker.*

ESTHER *(speaks mostly in friendly, motherly tone) I know, dear. (Points to chair.) First, make yourself comfortable. And then tell me how you heard of me.*

SALLY *If I told anyone that I'd asked for an appointment —*

ESTHER *They would laugh?*

SALLY *Half of them would. The others wouldn't believe it.*

ESTHER *Honey, that's what I always hear. But in the end, they all recommend me. Still, who told you about me?*

SALLY *Nobody.*

ESTHER *Now that I don't believe. I don't advertise… it's always word of mouth.*

SALLY *I overheard a conversation… but don't worry. They were practically whispering. Initially, I thought they were joking.*

ESTHER *But then it started to make sense, right?*

SALLY *Not at the outset. It seemed so absurd.*

ESTHER *If it were that absurd, why do you think I am in this business… and can charge what I charge? (Picks up a notebook and pen and looks at Sally expectantly.) Well?*

SALLY *Where do you want me to start, Ms Seligman?*

ESTHER *Call me Esther. Calling people by their first name is what my business is all about.*

SALLY *All right… Esther. Let's see… (Laughs self-consciously.) Of course, you can see what I look like.*

ESTHER *I've seen you on TV.*

SALLY *You watch my program?*

ESTHER *Occasionally.*

SALLY *I am 33 and unattached.*

ESTHER *(somewhat impatiently) Of course you are. Otherwise, why would you be here?*

SALLY *Yes… of course. I was raised in Florida —*

ESTHER *(interrupts and then impatiently speaks very fast) In Pensacola. Eventually a Master's degree in communication from Northwestern University… move to the West Coast… reporter for a local paper —*

SALLY *The San Francisco Chronicle.*

ESTHER *I know. I just didn't want to embarrass you, because I think it's a crummy paper.*

SALLY *But how do you know all that about me?*

ESTHER *Because you're somewhat different from the usual run of the mill.*

SALLY *In what sense?*

ESTHER *Lots of people have seen your face… and know your name. And if they're curious, they'll Google you.*

SALLY *Did you?*

ESTHER *As a matter of fact, I did.*

SALLY *You do that with every customer?*

ESTHER *I call them clients.*

SALLY *Clients then.*

ESTHER *I try… but not all are that Googleable. Not in the gay community. And especially among lesbians who haven't quite left the closet yet.*

SALLY *(somewhat archly) I do not hide my sexual preferences.*

ESTHER *Nor flaunt them, I presume. But I was just generalizing, because many of my clients do fall into that semi-invisible group. Otherwise… why would they come to me? A gay bar or disco is cheaper.*

SALLY *Before I heard about you, Esther, I didn't know there were lesbian matchmakers. I didn't mean a matchmaker who is lesbian. I meant a matchmaker for lesbians.*

ESTHER *Lesbians… at least some of them… need matchmakers. And since I've been around and seen it all, I'm effective… and expensive.*

SALLY *Is there enough of a market for such specialists?*

ESTHER *There are plenty of lesbians in the Bay Area.*

SALLY *Actually... I've never even met a matchmaker for straight people.*

ESTHER *Millions of straight couples use them.*

SALLY *Oh, come now!*

ESTHER *I mean worldwide. Here, we think people should just spontaneously fall in love. It's romantic all right... but it's not very efficient. Just think how many end up in divorce. Primarily because they haven't done their homework before falling in love... whatever that means. But even we are starting to change.*

SALLY *Who is "we"?*

ESTHER *A subset. Professionals... most of them successful... somewhat affluent... workaholics... already in their thirties or older... by which time they've become choosier... and have less opportunity to just (draws quotation marks in the air) "fall in love." They need someone to do the homework for them.*

SALLY *(laughs self-consciously) You're describing me.*

ESTHER *Of course I am. So let's get down to business. (Again picks up notebook, which she had put back on the side table). Well?*

SALLY *I'm looking for a serious long-term relationship.*

ESTHER *Aren't we all? Gay or straight.*

SALLY *Will you believe me that my problem is a social one?*

ESTHER *If you say it, I believe it.*

SALLY *I have trouble meeting women... I mean the type we're talking about.*

ESTHER *Why? You don't look shy... as a matter of fact, I'd call you self-assured. Most importantly, you maintain eye contact. I always tell my shy clients: Honey! Whatever you do, <u>don't</u> stare at the floor. If you can't look them in the eye, try their mouth or ear —*

SALLY *I'm okay with eye contact. At worst… I pretend it's the camera or the teleprompter.*

ESTHER *So what's the problem?*

SALLY *For the last couple of years, I've been anchorwoman for both the early news hour and the 6:00PM program. And I need 8 hours sleep… or I look and feel like a witch. So I'm in bed before 9:00. Such hours make any type of social life the pits.*

ESTHER *(kindly) That isn't all, is it?*

SALLY *Not quite. You talked about gays halfway out the closet. I'm not a liar… and when I'm asked directly, I answer directly.*

ESTHER *Nothing unusual there.*

SALLY *But I don't volunteer… anymore.*

ESTHER *So the problem is "anymore"?*

SALLY *Precisely. I come from a very, very Christian home… Christian in all caps. Where any word starting with "homo"… with the possible exception of "homo sapiens"… and that one only if you make it plain that it has nothing to do with evolution… is considered an abomination. It was only at Northwestern… a university I didn't select for its distance from home but because of its excellent Communications department… that I understood and then came to terms with my lesbian nature.*

ESTHER *While still very, very Christian?*

SALLY *That had gotten diluted earlier in College. Anyway, I made the mistake… the monumental mistake… during Christmas break of my senior year to tell my family about my sexual orientation. All I did was talk about feelings… not actions. All I wanted was understanding… not approval. Instead they threw me out. In the process, I lost both my family… and the rest of my religion. Anything that's that unforgiving and brutal was not for me.*

ESTHER *Even that has happened to all kinds of women… and men.*

SALLY *I suppose so. But since then, I stopped volunteering.*

ESTHER *So tell me… what sort of person should I find for you?*

SALLY *Well I want her to be kind —*

ESTHER *(interrupts impatiently) Of course you do. And intelligent! And with a wicked sense of humor! And romantic! And sexually compatible… physically attractive… non smoker… no drugs… bla, bla, bla! Come on Sally, everybody lists those desiderata.*

SALLY *What a precious word!*

ESTHER *(smiles) Every once in a while I like to show off. But you said you wished a long-term relationship. To offer you what?*

SALLY *The family structure I lost. To have a child and a partner who would enjoy parenting with me. Who wouldn't mind if I stayed home with my baby —*

ESTER *Not our baby? And stay home? Period? In other words, someone who'll support you? Like a traditional father… but in a skirt… who'll bring home the bacon while the wife cooks, cleans and raises the child?*

SALLY *Of course… our baby. But I want to give birth to it. And we'll share expenses. I've saved quite a bit and I'll work at home. I've had it with journalism or TV… I want to write. Books. Serious stuff.*

ESTHER *I'll tell you what. Go home and put it on paper: Your dream relationship… not only what you want from your partner, but what you're offering. How you'll convince her to say yes.*

SALLY *And then?*

ESTHER *My trade secret. Eventually, you'll hear from me and then you two meet… alone.*

SALLY *Where?*

ESTHER *Some neutral, public place. Café, park, restaurant you pick it. Just so either one of you can terminate the meeting without losing* *face in case something goes awfully wrong. (Grimaces). Every once in a while, even I make a mistake. By the way (reaches in desk drawer), here's my contract. Read it at home... fine print and everything... before signing. Half the fee is non-refundable. The rest is payable when the two of you move in together.*

SALLY *(rises) Why not?*

BRIEF BLACKOUT, long enough for scene shift.

HARRIET, in white doctor's coat, stethoscope around her neck, sits behind her desk. ESTHER sits in chair by side of desk.

HARRIET *Just a moment! You asked for the last appointment of the day... an urgent personal one... and now you tell me it's not for a patient? Not even a medical matter?*

ESTHER *I apologize... but I wasn't sure you would've been willing to see me.*

HARRIET *Before I ask you to leave... are you working for a drug company? I have specific hours — once a week, when I talk to their representatives.*

ESTHER *I'm not a salesman.*

HARRIET *Salesperson.*

ESTHER *I beg your pardon?*

HARRIET *Salesperson. Not salesman. You're not a man.*

ESTHER *(sarcastic) Oh... I see. Like fireperson... clergyperson... milk-person... henchperson —*

HARRIET *You think that's funny?*

ESTHER *Not particularly. It's just that I don't have much use for political correctness... at least not the verbal kind.*

HARRIET *(starting to laugh) You are trying to sell me something... but what is it? A new dictionary?*

ESTHER *I'm not trying to convince you to buy something… but I want to raise a possibility that may not have occurred to you.*

HARRIET *This better be good… and short.*

ESTHER *In your work, you're surrounded by men.*

HARRIET *All urologists are.*

ESTHER *And you work long hours.*

HARRIET *Please get to the point.*

ESTHER *You can't have much time to meet other women.*

HARRIET *What are you driving at?*

ESTHER *I'd like to introduce you to an interesting woman.*

HARRIET *(ironic) Not an interesting man… or person?*

ESTHER *An interesting woman of the right age would interest you more.*

HARRIET *And how do you know that?*

ESTHER *I have my sources.*

HARRIET *Who exactly are you? You don't look like a CIA or FBI type… or are they getting more sophisticated?*

ESTHER *I am a matchmaker.*

HARRIET *If I told anyone that a matchmaker asked for an appointment —*

ESTHER *Yes?*

HARRIET *And I told them that I fell for it…*

ESTHER *They would laugh?*

HARRIET *Half of them would. The others wouldn't believe it.*

ESTHER *That's what I always hear. But in the end, most of them listen.*

***HARRIET** I'm not interested in most. Why should I listen?*

***ESTHER** Because I am a discreet professional who has done her homework before coming here.*

***HARRIET** I see.*

***ESTHER** May I continue?*

***HARRIET** (looks at her wristwatch) I'll give you five minutes.*

***ESTHER** Too short.*

***HARRIET** Not to find out whether I want to hear more.*

Of the contentious topics that I presented in *Taboos*, few are more disturbing to a general public than the feasibility of sex predetermination. The following brief excerpt will serve to make the point.

***HARRIET** Why did you want to see me?*

***CAMERON** Because... (hesitates out of embarrassment)... you know why...*

***HARRIET** (sharply) No, I don't.*

***CAMERON** Jeez, Harriet.*

***HARRIET** (sharply) Tell me. Why?*

***CAMERON** (points at her stomach) I got you pregnant... so I felt responsible —*

***HARRIET** (quickly, almost angrily) Hold it, Hold it! You didn't get me pregnant... and you're certainly not responsible. (Brief pause.) Okay, listen. Listen very carefully: We used a few sperm of yours... seven to be precise... for injection into seven of my eggs. You wanted to know whether you were fertile... you didn't do it because you wanted a child with me! In your book that would've been adultery! And I agreed to use one of your sperm only because I wanted a baby of my own together with Sally. And since she's your sister, she's contributing to the baby's gene* *pool through your sperm. It was my decision… and it's my responsibility. You got out of the loop, once you masturbated.*

CAMERON *(dry with touch of irony) Much obliged Ma'am... for this clear explanation.*

HARRIET *I'm not finished. The moment the embryo implanted in my uterus, I offered to let you and Priscilla use some if you wished. That was the only bargain between us. That your wife became pregnant with one of those embryos is your responsibility… not mine. When this boy is born (She points to her stomach.) he will be my son. And when Priscilla gives birth that will be your son. Is that understood? Let's not confuse those two sons.*

CAMERON *How do you know they'll both be boys?*

HARRIET *Because I wanted mine to be a boy.*

CAMERON *But that's no guarantee. God decides what we get and we'll be grateful for whatever blessing He bestows.*

HARRIET *(gentler) Cam, I don't want to argue religion with you. This is biology. We used ICSI for the fertilization, right?*

CAMERON *Right.*

HARRIET *Injecting one sperm into each egg, right?*

CAMERON *Right.*

HARRIET *The sex of the child is always controlled by the sperm. A Y chromosome-bearing sperm leads to a boy, an X chromosome-bearing sperm to a girl. I'm sure you learned that in high school — even in Mississippi.*

CAMERON *So what are you telling me?*

HARRIET *That the technology… it's called flow cytometry… has now been developed to separate X- from Y-sperm —*

CAMERON *(taken aback) And you used that flow… thingamajig?*

HARRIET *Yes.*

CAMERON *And you didn't tell me?*

HARRIET *That wasn't part of the bargain. You wanted to know whether you're fertile. I wanted to have a son, and you wanted to have a child with your wife. There weren't any more eggs of mine left for new ICSI injections. I was generous enough to let you use some of the remaining embryos and all of those were potential males.*

CAMERON *Jeez!*

HARRIET *Cam, stop using that word. It's driving me crazy. And what's wrong with your having a boy?*

CAMERON *Nothing. But picking the sex of the child is so...*

HARRIET *Don't tell me... unnatural.*

CAMERON *Yes.*

HARRIET *And you think ICSI is natural? Most of modern medicine is full of interventions and materials that cannot be found in nature. You think "unnatural" is automatically "unethical?"*

The topics covered in *Taboos* range from science to religion, from politics to ethics, and, of course, from sex to reproduction without intercourse. On several occasions, following the theatrical performances, the audience was invited to remain for a discussion which at times became almost longer than the performance. For me the playwright, the length and vigor of these post-performance debates were the ultimate confirmation that "case histories" on the theater stage of contentious issues have merit.

Foreplay

In the chapter entitled *"Jew"*, I already quoted extensively from my autobiographical book *Four Jews on Parnassus — A Conversation: Benjamin, Adorno, Scholem, Schönberg*. I explained there why I had hit upon those characters and why I considered it my most important book. What I did not mention other than alluding to it through the word "Conversation" in the book's title but do so now, is that the totally dialogic style in which I composed that book would not have occurred to me, had I not embarked on it during the later stages of my playwriting career.

> *I chose to present this biographical material exclusively in dialogue, with the exception of the prefatory sections to each sketch. One reason lies in my own biography. In my former incarnation as a scientist over a period of half a century, I was never permitted, nor did I allow myself, to use direct speech in my written discourse. With very rare exceptions, scientists have completely departed from written dialogue since the Renaissance, when especially in Italy some of the most important literary texts were written in dialogue — ranging from expository or even didactic to conversational or satirical — that attracted both readers and authors. Galileo is an outstanding example. And not just in Italy. Take Erasmus of Rotterdam: his Colloquies are a marvelous example how one of the Renaissance's greatest minds managed to cover in purely dialogic form topics ranging from "Military Affairs (Militaria)" or "Sport (De lusu)" to "Courtship (Proci et puellae)" or "The young Man and the Harlot (Adolescentis et scorti)". This explosion of dialogic writing even stimulated literary theoretical studies. From the 16th century on, critics have attempted to exalt, defend, regulate, or — alas — abolish this genre of writing.*
>
> *One of these critics was the Earl of Shaftesbury, who in 1710 commented in his "Advice to an Author" that "dialogue is at an end [because] all pretty Amour and Intercourse of Caresses between the Author and Reader" had disappeared. Since my purpose is to present a humanizing view of my four subjects rather than theoretical insight into their work, I feel that dialogic "Intercourse of Caresses" rather than the more dispassionate third person voice may be the most effective way of accomplishing this. I can only hope that the intimacy of my caresses will convince the reader that at least in Four Jews on Parnassus: A Conversation I was justified in disregarding the Earl of Shaftesbury's counsel.*

I have never regretted this attempt to write an entire book in direct speech, yet I cannot help but point to one shadowy aspect: how many readers and even reviewers overlooked my introductory explanation why I had chosen a dialogic style for a carefully researched biographical account and have automatically assumed that the book represents a play, although even a cursory reading would immediately demonstrate that it did not follow the structure of traditional plays.

My irritation as well as disappointment triggered an internal "I'll show the jerks!" button that stimulated me to actually write a play dealing with some of the same characters. The product became *Foreplay,* with a focus on Theodor W. Adorno, his wife Gretel, Walter Benjamin and Hannah Arendt. Arendt, a famous political theorist, and Adorno disliked each other intensely, but both admired, even worshipped Benjamin. Adorno's life-long womanizing and the range of the deeply personal and extensive correspondence between Benjamin and Gretel Adorno are all well documented. It is also very likely that Benjamin carried a briefcase with him on his flight from France to Spain where he committed suicide in September 1940. The briefcase and its contents were never found. Those are the facts, as is the relationship between Hannah Arendt and the philosopher Martin Heidegger, which I wove into my play.

Using this and other background information, I started to speculate about aspects of their personal lives and actions, which I could only do if I discarded the shackles of a biographer and assumed the freedom of a playwright, focusing on the theme of jealousy — professional and personal — and of adultery. Both are topics I barely addressed openly in any of my past autobiographical writing other than fleetingly with reference to my three marriages. While adultery and its corollary, jealousy, have touched me as initiator, participant and even victim, I have never addressed them nor will I because it involves other people who obviously need to remain nameless. Therefore, it is precisely in the guise of fiction or plays that I return to them in my poetry, novels and plays. Here is an example from the first scene of *Foreplay* between Theodor Adorno and his wife.

GRETEL *First a question I've never asked you directly.*

TEDDIE *Yes?*

GRETEL *How jealous are you?*

TEDDIE *In general... or of you?*

GRETEL *Well... both.*

TEDDIE *Professionally, I'm very jealous.*

GRETEL *We both know that. I mean, otherwise.*

TEDDIE *Of you?*

GRETEL *Well, yes… for instance of me.*

TEDDIE *Never!*

GRETEL *Good. And of other women?*

TEDDIE *It depends.*

GRETEL *Could you elaborate?*

TEDDIE *I could, but I won't. First… what about this sudden SIE-DU ping pong with Walter?*

GRETEL *(smiling) I thought that would register.*

TEDDIE *(explosively) Register? Did I hear right: register? You mean this was deliberate?*

GRETEL *(coolly) I don't know what "this" refers to. But if you are fixated about my addressing Walter as DU —*

TEDDIE *(interrupts) And he doing the same!*

GRETEL *That you should discuss with him. Surely, you don't think that I would shift without serious deliberation from the formal SIE to the informal DU —*

TEDDIE *Calling someone DU within weeks —*

GRETEL *(interrupts) It wasn't weeks —*

TEDDIE *All right… months —*

GRETEL *(interrupts) It wasn't months —*

TEDDIE *Years?*

GRETEL *No plural.*

TEDDIE *One year?*

GRETEL *One week.*

TEDDIE *(outraged) One week? Seven days from SIE to DU?*

GRETEL *(mock innocence) Why not? Just because you always persisted calling our dear Walter "SIE"?*

TEDDIE *Because we are adults… and not children.*

GRETEL *In that case, ascribe the DU to a sudden childish informal gesture of mine.*

TEDDIE *One adult woman addressing an adult man within one week as DU instead of SIE is neither childish nor informal!*

GRETEL *So what would you call the motivation?*

TEDDIE *Postcoital!*

GRETEL *Teddie! So you are jealous!*

TEDDIE *We are discussing your behavior… not my jealousy.*

GRETEL *In that case, explain to me the operational feasibility of postcoital consummation between Walter and me while he was in Paris and I in Berlin?*

TEDDIE *You are limiting yourself to geographical coitus.*

GRETEL *Are we suddenly moving into coital dialectics?*

TEDDIE *No dialectics... just simple interrogation. Psychic coitus by definition must be more intimate than physical.*

GRETEL *I agree.*

TEDDIE *Is that all you have to say?*

GRETEL *What else do you wish me to add?*

TEDDIE *Well, if we are going to discuss the shift to DU at the coital level, how about telling me about the foreplay that Walter indulged in?*

GRETEL *This is becoming too personal.*

TEDDIE *Walter's foreplay is too personal for a personal discussion between husband and wife? A 7-day foreplay too personal for a marriage of years?*

GRETEL *Yes.*

TEDDIE *Yes?*

GRETEL *Yes.*

TEDDIE *I am astounded. No! Not astounded... shocked... wounded... and — to put it bluntly — royally pissed off.*

GRETEL *A sequence of adjectives I've never heard you use when you were dictating to me... be it in accounts of dreams or real events.*

TEDDIE *And you tell me this now?*

GRETEL *Even with your Gretel, on occasion the cup runneth over.*

Rather than citing some further excerpts, as I have done so frequently in this chapter, I shall end with an experiment I conducted with *Foreplay.* After completing a "book text" of it, I submitted the manuscript simultaneously to three literary and academic publishers in Austria, Argentina and USA, who rarely publish plays and in any event were not concerned about the theatrical merit of the manuscript, since it was offered to them as a book to be printed and distributed through conventional book channels. To my delight, all three accepted the manuscript within weeks of submission and coincidentally, all three published the play in March of 2011 under the titles *Foreplay*, *Vorspiel* and *Preludio*. For me, this was a triple vetting process of the literary merits of the play text by three independent judges in three different languages, who did not even know of each other's existence, and all that before any theater had even been approached for a potential production. Having received the confirmation that

Book covers of simultaneously published English (*Foreplay*), German (*Vorspiel*) and Spanish (*Preludio*) editions ((2011) of play prior to any staging

in terms of text and theme *Foreplay* now exists in three languages "in perpetuity" in the sense that any printed book does so, I proceeded to the next step of open dramatic readings before public audiences at some academic venues and literary festivals.

I am mentioning this last point since rehearsed dramatic readings (not to be confused with the American style of "theater workshopping") by professional actors, in contrast to full productions, are not that often performed even though they are cheaper, faster and, in terms of location, simpler to organize. Isabella Gregor, the by now often-mentioned theater director, has converted this medium into a specialty which she has applied to the German versions of all of my plays. The many dozens of such dramatic readings in numerous venues, which did not involve static readings in front of music stands, but extensive blocking and even inclusion of some scenery and costumes, have been performed in a range of public and academic venues before audiences ranging from less than one hundred to well over six hundred attendees.

Isabella Gregor

Dramatic reading at Semper Depot, Vienna (2007) directed by Isabella Gregor with (left to right) Wolfgang Pampel, Bernd Birkhahn, Peter Scholz and Johannes Terne

Insufficiency

The above described triple-submission book approach with *Foreplay* tempted me to repeat it on a much trickier front, namely with another true chemistry play. After all, if fear of scientific themes in the theater is widespread, chemophobia is truly rampant. Yet as I already reported, *Oxygen* managed to penetrate that chemophobic iron curtain — perhaps because of its connection with the Nobel Prize. This time, with *Insufficiency*, I decided to try as the smuggling device a "whodunit" to explore the topic of "fashion-in-science" on the stage. The very start, a prosecutor's monolog, outlines the "crime" whose nature, as is to be expected, is only disclosed toward the end of the play.

> ***DISTRICT ATTORNEY*** *Two men die within 21 minutes of each other's demise. Both are non smokers. Cholesterol levels below 180... according to their last physical exams. No particular health problems. One 47 years old, the other 54. What are the statistical chances of two apparently healthy, middle-aged men dying from embolisms in the same room within just a few minutes of each other? I don't know and I won't bother consulting a statistician, because we are not dealing with the entire country or even this state. We are dealing with two men, both professors in the same institution... and both having sipped champagne some two hours before their death. And not just any champagne. Not Dom Perignon or Veuve Cliquot... after all this is a University and not a banker's club. Not even non-vintage Piper Sonoma from California, because people are not known to drop dead drinking such champagnes for the simple reason that they would have been off the market long ago. No... they were drinking champagne from two unlabeled bottles. So what are the statistical chances of it <u>not</u> being an accident? I would say, one out of one... or at the most extreme, one out of two. Of course the toxicologist analyzed the remnants from those bottles and the pathologist performed detailed autopsies of the two bodies. The results? Nothing! Nothing... except that the cause of death was sudden embolism. So back to the toxicologist. Why didn't he find anything in the bottles? I mean, no poison or anything even resembling a poison? And why is Dr. Kroy here —*
>
> ***CROIX*** *(interrupts) Croix! I will not have my name murdered.*

DISTRICT ATTORNEY *(dismissive) Kroy, Croix... whatever. Why is the doctor suspected of double murder? Because Dr. Croix (Poor French pronunciation) is an expert on bubbles... first beer bubbles, but then focusing on champagne. Bubble formation, bubble shape, speed and expansion... I could go on with terms like turbothermodynamics, but I won't —*

CROIX *Don't forget turbokinetics!*

DISTRICT ATTORNEY *Your honor! May I request once more —*

CROIX *But this was no interruption! (Grins.) I was just trying to be helpful.*

DISTRICT ATTORNEY *I shall ask for help when I need it! As I already stated, the bottles of unlabeled champagne that killed Professor Aspinall and Professor Sehlig were not empty... they were both nearly half full. But unfortunately, the remaining liquid was not analyzed until three days after their death and since the bottles sat there uncorked, their contents were totally flat... so flat that it could not be called anymore champagne but only an innocuous liquid containing less than 12% of alcohol,* C_2H_5OH *to the chemist.*

CROIX *Bravo! (Raising hands in calming fashion.) Sorry, it was just meant in admiration of your chemical sophistication.*

DISTRICT ATTORNEY *In other words, nothing remaining in that bottle could have killed two apparently healthy men within two hours of having consumed some of its sparkling contents. But something did kill them... something that was in that bottle when it was freshly opened, but that had disappeared... poof into the air... two days later. I need not explain to you, ladies and gentlemen, what had disappeared in that interval, other than to remind you that Dr. Croix was known to have called some champagne bubbles "killer flowers." I suggest that they be renamed "murder flowers." And now to Dr. Croix's motive...*

As with *Foreplay*, before establishing any contacts with theaters, I first submitted the text to three literary and academic publishers, this time in Austria, UK and Mexico and again all three accepted it almost

Book cover images of *Chemistry in Theatre*: English and German editions (2012)

instantaneously for publication under the title *Chemistry in Theatre*. The book's long preface asked first "Why are there so few plays dealing with the world of chemistry?" before offering my explanation and then proceeding to the question of what the theater can do for science. Since then, *Insufficiency* has been presented in the form of dramatic readings in a number of major academic venues before its theatrical premiere in London in late 2012.

Coda

As I write these words, my ninetieth birthday has already passed. How much more time will I have for writing? If it is one or two years, *Insufficiency* is guaranteed to be my last play, since some sixteen years ago, I had set aside a partly finished "science-in-fiction" novel provisionally entitled *The Sleeping Beauty Syndrome* for what I expected to be a brief interregnum while writing *An Immaculate Misconception*. After a detour of ten plays (if I also count my classroom theatrical ventures), I decided that it was time to return to my long abandoned earlier literary love to complete that novel under a different title, *Sentenced to Life*. Is that new title related to my hope that I shall reach my father's age of ninety-six or that of my grandmother's of one-hundred-and-one? I wonder and keep dreaming.

In July of 1999, I was on my way to Spoleto to speak on "Science on Stage" at the Spoletoscienza Festival, in response to an invitation prompted by *An Immaculate Misconception*. My wife was with me because we planned to proceed from Spoleto to Sulmona, the birthplace of Ovid, on whose biography Diane was embarking. A driver was supposed to meet us at the Rome Airport, but once all waiting chauffeurs bearing placards with every kind of name but ours had departed, we were standing alone at the exit from the customs hall. Fifteen minutes must have passed, during which time my wife had gone to search for other transport while I manned the lonely gate in the hope that the driver from Spoleto might yet appear. At that moment, a woman pulling a heavy suitcase passed near me. When she saw me, she stopped. "You are Carl Djerassi, aren't you?" she said, still panting. "Yes," I admitted, nonplussed, "but how did you know that?" "Oh," she waved the question away, as if it were all too obvious. "I saw your play at the Eureka Theater in San Francisco." "And you remembered that?" I asked flattered. That's when she punctured my balloon of pride by pointing to the next exit about 50 meters away. "I did, when I noticed that man over there who's holding a sign with your name."

Somehow I dream of similar encounters to come, but ones where no prompting by a taxi driver is necessary. And why? Because I am not immune to approbation and especially not to the one so rampant in academic circles, namely honorary degrees. As a scientist, I have received my share — well over thirty — during the past half century. But in retrospect, the ones that have probably pleased me most are two which recognized solely my late-in-life literary contributions, whereas all others (including five recent ones from Austrian universities) represented appreciation for my scientific work. The first "literary" doctorate came in 2009 from the Technical University in Dortmund and was followed by a lovely two-day symposium organized by the vice rector, Walter Grünzweig, that has since come out in book form under the title *The SciArtist: Carl Djerassi's Science-in-Literature in Transatlantic and Interdisciplinary Contexts*; the second, some three years later, touched me in a different sense. It was bestowed by the University of Applied Arts in Vienna and to that extent constituted for me a very special recognition by my birth city from which I had been expelled over seventy years ago.

I admit that this wish for public praise is childish and not becoming someone of my advanced age, but in a way that is the reason why I chose to ignore two extraordinary warning signals. Indeed, if I were a superstitious person, I probably would never have continued writing fiction nor plays because of two powerful messages from the Gods on Olympus, not too far from Parnassus. The first was truly earthshaking in its dimension.

My first novel, *Cantor's Dilemma,* now in its twenty-ninth print run as a paperback, is my longest-selling novel. One would not have thought so considering its beginning starting with the publication in late 1989 of the hardback by Doubleday. Here is what long ago I have recorded elsewhere:

> *The inaugural public reading of the novel was scheduled for 8:00 P.M. at Cody's in Berkeley — at that time one of the most prestigious Bay Area bookstores, and seemingly the ideal launching pad for a local writer's career. My wife and I crossed the Bay Bridge shortly before 5:00 P.M. in order to have a leisurely dinner in Berkeley within walking distance of Cody's. I was about to feed a parking meter when the ground around us began shaking. Instantly, the air was filled with the hooting of car alarms, every Audi and BMW on the block hysterically heralding the apocalypse. We held on to the parking meters until the pavement stopped moving, and looked around us to find the city changed. We had just experienced the Loma Prieta earthquake of October 17, 1989. Across the Bay, back in the direction we had just come, a portion of the Bay Bridge had collapsed, and with it a stretch of elevated freeway. People lay dying under tons of concrete.*
>
> *There was no reading that night. Instead, it took us over four hours in horrendous traffic and virtual pitch darkness to get back across the Bay, where we found our home, to our relief, undamaged except for one broken pre-Columbian pottery figure and books strewn all over the floor.*

My next reading had been scheduled at the Palo Alto bookstore Printers Inc, but it was canceled: the building needed retrofitting to meet quake-safety standards. Meanwhile, on the other side of the continent, my publisher, Doubleday, was suffering seismic activity of its own, a corporate upheaval that included the sudden departure of both the editor and the publicist to whom my book had been assigned. When the unexpectedly wide review coverage of Cantor's Dilemma propelled it to the Bay Area bestseller list, Doubleday was caught unprepared. The modest first print run was quickly sold out and no one pursued a timely second printing until after Christmas, thus making Cantor's Dilemma a short-lived collector's item for the two busiest weeks prior to Christmas. The post-Christmas reprint by Doubleday, though personally gratifying, hardly made up for this marketing bungle.

A superstitious person would have concluded that some powerful god of fiction was advising an instantaneous return to chemistry. But that was nothing compared to what an even more powerful god of drama had in mind for me when in 2001 I was heading for the dress rehearsal for the New York premiere of my first play, *An Immaculate Misconception* at the Primary Stages Theater. For a budding playwright, expecting to see his first play staged at a respectable Manhattan theatrical venue was a dream come true. But that was when Osama bin Laden interfered.

On September 11, 2001, I was flying to New York City from London. Our plane was approaching northern Canada within a couple of hours distance from John F. Kennedy airport when I noticed that we seemed to be making wide circles rather than heading straight south. After about half an hour of circling, the pilot announced that we would be landing in Halifax, Nova Scotia "for technical reasons." It turned out that ours was the very first plane to be diverted when the twin towers of the World Trade Center were destroyed. When we arrived in Halifax, there was no other plane on the runway; seven hours later, as we were finally allowed to disembark, there were over forty jumbo jets parked behind us. It seems easiest to account for what happened by the following e-mail exchange with Casey Childs, the founder and executive producer of Primary Stages.

From: Prof. Carl Djerassi
Sent: Tuesday, September 11, 2001 3:48 PM
To: Casey Childs
Subject: News from Halifax
9/11/01 Halifax, Nova Scotia

Dear Casey,

For the past 7 hours I have been sitting in my plane on the tarmac of the Halifax airport with about 40 airplanes behind us. We have no idea when we shall be able to get off the plane and even less of an idea when we shall be able to leave and to where.

There probably is no purpose for me to even come to New York — I imagine there will only be bedlam and none of the appointments that were made so laboriously would have any chance of being realized. At this stage, I even wonder whether my play will start in New York at the indicated time if at all, but all of that is, of course, totally insignificant in light of the tragedies that must have fallen on hundreds or thousands of people. As you can imagine, we have only minimal news here on board where I have been occupying my time writing e-mail messages that may not even be sent for days.

If and when you get this message, try to answer by e-mail. At this stage, this looks like the only feasible means of contact.

All my best,
Carl

From: Casey Childs
To: Carl Djerassi
Subject: New Schedule for IMMACULATE MISCONCEPTION
Date: Wed, 12 Sep 2001 10:05:41 -0400

Dear Cast, Crew, and Staff for AN IMMACULATE MISCONCEPTION,

IF YOU RECEIVE THIS EMAIL, PLEASE CALL OTHER PEOPLE IN YOUR DEPARTMENTS AND LET THEM KNOW OF THESE CHANGES. I ONLY HAVE EMAIL ADDRESSES FOR CERTAIN PEOPLE.

Because of the tragedy that has occurred in New York City, I believe it is best to delay the first performance of our play until Wednesday, September 26 at 8 pm. This will give us time to regroup.

We will follow the same technical schedule except it will be one week later.

For example, instead of the first technical day being Friday, September 14,

it will be Friday, September 21 etc. Perhaps we can move the actors to the theater on Tuesday, September 18.

Tyler will do a completely new show schedule on Thursday. Press previews will be pushed back. Opening performance will be pushed back.

Thank you all for your help.

Sincerely,
Casey Childs

To: Carl Djerassi
Subject: An Immaculate Misconception in Tech
Date: Mon, 24 Sep 2001

Dear Carl,

Getting this show up in the wake of the recent tragedy has been EXTREMELY difficult and VERY expensive for us. What I worry about the most is filling our houses. Many of our mailings coincided with the disaster, and the box office phone has stopped ringing. Other off-Broadway theaters are having similar problems. Overrun on the set and the additional Equity and crew work week will be costly.

However, we persevere. Hope you are thriving.

Casey

In the larger perspective of what occurred on September 11, 2001, affecting and traumatically changing thousands of lives, bin Laden's sabotage of my play is, of course, of no significance whatsoever. Yet it is striking to note that I, a lover of anagrams and word plays, used the following one in Chapter 14 of my novel *NO*: *Pithecanthropus erectus*: "Pursue the person, catch it." My superb translator, Ully Mössner was even able to come up with an appropriate German equivalent: "*Ratetip: Sucht euch Person*." Somehow I like to pretend that President Obama subliminally got that message when he finally liquidated bin Laden. Let this be a warning to others who in the future would like to mess with my literary ambitions.

"Collector"

Collector of what? Matchbox covers are different from rare manuscripts, baseball cards very distinct from Impressionist paintings. And why collect? These are questions I have asked myself ever more frequently during the rather late introspective phase of my life. A succinct answer is provided in two stanzas from a long poem, "The clock runs backward," which was a fiercely honest poetic summary of my life that I had written on my sixtieth birthday:

> Forty-eight years, forty-five years,
> Then forty-one.
> Ah yes, the years of collecting:
> Paintings, sculptures, and women.
> Especially women.
>
> But wasn't that the time
> His loneliness had first begun?
> Or was it earlier?
> Why else would one collect,
> Except to fill a void?

In that last line, I was right on target. But voids at age sixty are very different from those faced thirty years later for the very simple reason that the means of plugging them are much more restricted late in life. Take people, for instance: lovers, friends, even social acquaintances, where a lonely middle-aged man has many more options than his aged counterpart. Here, as I have repeatedly pointed out, my current solution is producing rather than collecting; working rather than moping; moving rather than relaxing. My travel schedule, flitting about every week or two from city to city, which many with justification will categorize as mad, happens to be one of my idiosyncratic antidotes to loneliness.

But as stated in that poem, I have been an earlier collector of art for decades and the fact that I have now ceased that activity does not mean that a description of the role of art in my life is not worth recounting. The justification for my having stopped collecting is not just that I have run out of wall space — one reason why I continued collecting the work of my favorite painter, Paul Klee, longer than that of any other artist was because he worked on such a small scale — but because I concluded that direct support of the creation of art rather than acquisition of a completed work now makes more sense. Not only do I continue to give away art I have already collected — an activity that sounds simpler than it occasionally is as shown in the earlier described travails of my donating a valuable George Rickey sculpture to the city of Vienna — but my greatest focus is now the support of the Djerassi Resident Artists Program (DRAP). Though prompted by my daughter's suicide, it has also acquired a life of its own, because it has taught me that this ultimate form of art patronage largely separates evaluation of an artist's *product* from that of her or his *creativity*, thus reducing, if not totally eliminating, the effects of subjectivity and unfamiliarity. After all, aren't the most avant-gardist and daring artists likely to be the least patronized, primarily because they are exploring unfamiliar aesthetic or intellectual territory? Having been largely responsible for the recent construction of added facilities (named in memory of my late wife, Diane Middlebrook) at DRAP, I feel a degree of satisfaction which is based more on the anticipation of the work that will be produced there by the additional forty or more writers per annum whom we shall be able to house in these buildings rather than just contemplating the physical reality of the architecturally pleasing edifices.

My past art collecting activities can be divided into four phases. During the 1950s in Mexico, I became fascinated by pre-Columbian art, visiting all of the common and some of the less accessible archeological sites. In the process, I became a collector of pre-Columbian figures and continued to collect them until the exportation of such art works from Mexico became illegal in the 1960s. By that time, my personal affluence had started to grow as a result of my early investment in the shares of Syntex, the company to which I had contributed significantly through my scientific work, so that I could indulge in my next passion: focusing on the works of artists who were both painters and sculptors; individuals like Picasso, Giacometti, Degas, Marini, Moore and the like, who would have been famous even if they had stuck to only one of their skills. Professional bigamy or polygamy (not to be confused with promiscuity!) has always attracted me, which is probably the reason why Paul Klee so appealed to me and became the subject of my longest, largest and most sophisticated art collecting. The fourth phase involved contemporary, living artists: especially the American kinetic sculptor George Rickey, but also many artists that had been in residence at one time or another at DRAP. But in the 1980s, except for Klee, I sold my entire collection of high-powered dead artists from my middle collecting phase to fund living artists through DRAP, an act of liquidation that was wise, correct and logical in the sense of what the proceeds were used for, yet sufficiently traumatic that I shall not resurrect now its memory. That leaves Paul Klee, and writing about him is refreshing, instructive and on occasion also amusing. I have recorded my infatuation with Klee on several occasions and present below a collage from some catalog essays on Klee that I have authored in the past few decades.

Collecting Klee

Klee in German means *clover* and both — artist and plant — are pronounced as *clay*. In a snobbish way, this has always made me judge people as philistine when they require an explanation after following the botanical or ceramic route to Klee as shown in the following contretemps between the art historian Regina and the chemist Rex in my play *Phallacy*:

REGINA *What do you think of Klee?*

REX *I'm into bronze… not clay.*

REGINA *Paul Klee.*

REX *I see.*

REGINA *(dismissive) So you have heard of him?*

REX *I don't have to put up with this.*

REGINA *Well… how do you like Klee?*

REX *How is this relevant?*

REGINA *Because here's what Klee told a chemist…*

REX *(prickly) What kind of a chemist? Analytical? Organic? Physical? Or was it a cook that he mistook for a chemist?*

REGINA *A famous chemist.*

REX *What's his name?*

REGINA *A Nobel Prize winning chemist... who liked lecturing to artists about his scientific theory of color.*

Of the thousands of articles, chapters, catalogs and books that have been written about Paul Klee, few were authored by collectors. Aside from an unstoppable outpouring on the part of academics, it has been the art historians, art critics, artists, museum curators and even journalists, who have written so much about the influence that this towering European figure of art of the first half of the twentieth century has exerted on the cultural scene. I deliberately say "cultural," because although Klee has often been called a "painter's painter," even in the field of music, no visual artist seems to have influenced so many musicians as has Paul Klee. According to the latest (2012) count by the reigning expert in the field, Stephen Ellis, 602 composers have created 870 pieces of music that are based in one way or another on 1,113 items dealing with Klee or with a specific work of his oeuvre that itself encompasses in excess of 9,000 paintings, drawings or prints. His ever growing influence in music is demonstrated by the fact that while only a single composition appeared prior to his death in 1940 and only about two dozen in the following twenty years, over one thousand — encompassing every genre from modern classical to jazz, pop or rap — have since been created.

So why do we hear so little from the collectors? And who are they? In the case of Klee, there are two distinct groups. All of the major art dealers who for decades specialized in the works of Paul Klee, such as Beyeler, Kornfeld and Rosengart in Switzerland, Berggruen in Paris, and Saidenberg in New York, were also collectors of Klee, and each of them has left some written record of their passion for this artist. But what about the rest, the private collectors, for whom Klee has such an enormous advantage as the master of the *petit format*? He usually painted on such a small scale that even in limited quarters, one can always find some space to hang one more of his gems. Thus, the hurdles facing all but the most affluent Klee collectors are mostly financial limitations rather than restricted physical space. I suspect that one of the reasons for the paucity of public writings by private collectors is that in most instances, the very act of collecting a substantial number of works of a specific artist (unless done for financial reasons and eventual resale) represents a degree of intimacy, of personal connection, even of a silent conversation with the artist, for which open disclosure may appear inappropriate. Basically, it is a private love affair where more often than not, one of the partners is already deceased.

Given that I have been such a collector of a dead artist's work for nearly five decades, the question may well be asked why I have deigned to write several essays (and even two poems) about Klee. The answer is quite simple: around the late 1970s, I turned from private art collector, who could afford to purchase art for his private pleasure, to an actual as well as a potential museum donor who wished to share his collection with a wider public. The pivotal event was my divorce in 1976, when my wife first demanded sale of all of the art so that she could receive half the proceeds. Fortunately, my artist daughter, Pamela, agreed with me that we should instead gift it to two museums — the Stanford University Art Museum and the San Francisco Museum of Modern Art — and persuaded my wife to concur.

The impetus for my subsequent donations came from my long-held personal belief about private art collections, notably those of a dead artist, in that the serious collector of such an artist's work then also becomes that artist's interpreter. But if a significant portion of an artist's output is concentrated in one collection, it ought to be accessible to the general public — a museum then being the obvious choice. In my own case of collecting Klee's works on paper, I have gone beyond that point by actually wishing through frequent exhibitions to proselytize about my favorite artist even at the risk of displaying somewhat fragile works too often or too widely rather than focusing primarily on conservation. In view of the fact that my life expectancy is unlikely to extend beyond another decade, I feel that I can leave the question of long-term conservation to the two museums to whom I am donating my entire Klee collection in approximately equal parts: the San Francisco Museum of Modern Art in the city that had become my eventual home as a refugee from Nazi Austria; and the Albertina Museum in Vienna, the city of my birth and of gradual reconciliation late in life. In the process, my metamorphosis from collector to donor has actually influenced my criteria for acquisition.

As I just stated, my activities as a serious art collector started in the mid-1960s, when I first could afford to actually acquire rather than just dream of some drawings by Paul Klee, an artist who had already attracted me years earlier. It all started at a Klee exhibition at an important London gallery, where all of the works were for sale. I kept returning to two magnificent watercolors from his Bauhaus years in the 1920s — rather large ones for an artist who usually worked on a small scale. "Should I? Can I?" I asked myself, and realized for the first time that I actually could own one of them. Finally, I approached one of the gallery employees and asked about the price. "The 1925 *Pferd und Mann*?" he asked, looking me up and down. "Sixteen," he finally said.

"Sixteen what?" I wanted to ask, but didn't. I knew it could not be sixteen hundred, and was unlikely to be sixteen hundred thousand, so it had to be sixteen thousand. But sixteen thousand what? Dollars, pounds or even guineas? "And the other one, the 1927 *Heldenmutter*?" I asked hesitantly.

"Eighteen."

"Hm," I replied and went back to look at the pictures. A few minutes later, the man appeared by my side. "Which one do you prefer?" he asked in a slightly warmer tone.

"I can't make up my mind. Both are superb."

"Buy them both," he said matter-of-factly, "and maybe we can arrange a better price."

Bargaining, whether in a marketplace in Mexico or a bazaar in Cairo, always makes me uncomfortable; but this time I haggled by default. Every retreat of mine, every inspection and reinspection of first one and then the other Klee caused the price to drop. They were not big reductions, but given the overall sums — far above anything I had ever spent before on art — they were not insignificant. Finally, I said, "I'll have to think about it." A couple of days later, I was the owner of not one but two Klees. By now I own around one hundred and fifty of his works in various media, but these two are still among the *crème de la crème*.

My purchase of these watercolors was probably not very different from the experience of other novice collectors — in contrast to the acquisition of my third, *Was für ein Pferd!* (1929). I saw it on the walls of

New York's Guggenheim Museum on the occasion of its big Paul Klee exhibition with the notation "Collection Galerie Rosengart." It did not take me long to secure the address of that gallery in Lucerne, or to consummate by mail the purchase of that small gem — a description Klee would have agreed with, although I did not know it at the time. Galerie Rosengart was owned by a father–daughter pair, Siegfried and Angela Rosengart, who were among the most important Klee dealers and collectors. Eventually I got to know them well and made frequent pilgrimages to their gallery. A couple of years after I bought the watercolor off the Guggenheim walls, Rosengart *père* told me the significance of the small notation "*S Cl*" which Klee had marked in pencil in the lower left-hand corner of that watercolor. An abbreviation for *Sonderclasse*, "special class," it denoted his own favorites among over nine thousand works. Since that time, the *S Cl* notation has become a magnet in my collecting — essentially my first direct posthumous dialog with Paul Klee. "*What, Mr. Klee, made you anoint this particular work with your S Cl?*" It is really a subset of the more general question, I ask myself in front of every Klee drawing I inspect, be it in a museum, in a book, or on my own wall: *What was your initial impetus to create this particular image?*

Paul Klee, *Pferd und Mann* (1925), *Heldenmutter* (1927), and *Was für ein Pferd!* (1929)

But gradually I shifted my emphasis to works that would not be simply called "Klee-like" — in some respects a meaningless term since by definition all works by Klee should deserve this description — in fact on works that even experts might not recognize at first glance as "typically Klee." Where and how did my interest in "non-Klee-like" works first originate? In retrospect, I believe that this happened when I started to focus on Klee's graphic work — an important but very small subset of Klee's oeuvre, since only about 1% of his total output falls into that category. The early ones, done between 1901 and 1905, are Klee's first truly original creations and also among his rarest. He called them "inventions," and titled them ingenuously, such as the famous 1903 "sour" print, *Zwei Männer, einander in höherer Stellung vermutend, begegnen sich* [*Two men meet, each supposing the other to be of higher rank*], which I had acquired in the late 1960s. Their inherent sarcasm; their bizarre portrayals of the human figure; even their frequent perversity — these are some of the reasons why I continued to search for additional works from that period that eventually led me to my biggest single acquisition orgy.

In June of 1975, at the annual Kornfeld & Co. auction in Bern (Klee's home town), there was held possibly the largest-ever sale of early Klee prints of the 1903 to 1905 period. I had carefully examined the lots prior to the start of the auction. I had a mental art budget for the year, which promised — with luck — to suffice for the purchase of two of the early graphics. At an auction, however, one never can predict the outcome, since even one other eager bidder might cause the price to rise into the stratosphere. I informed two gallery owners specializing in Klee, whom I knew well and whom I saw in the audience that I was going for these two etchings; although I could not ask them openly not to bid in competition with me, I was reasonably sure that they would not drive the price up once they knew of my plans. But then my heart sank. In the distance, I saw Heinz Berggruen (then an important art dealer in Paris from whom I had purchased various paintings as well as a Picasso bronze over the years), who, I knew, not only had a magnificent personal Klee collection but bid actively at auctions. I approached him and, after exchanging the usual pleasantries, mentioned the two lot numbers I planned to bid on. "Good choices," he acknowledged, "I was thinking of buying these for the gallery." There was no question about Berggruen's financial ability to outbid me any time, but he must have read the disappointment in my face, for he offered to bid for me, saying he would just charge me a commission for any successful bids. I accepted immediately, convinced that such a commission was a bargain insurance premium in return for not having the art dealer as a competitor. I left the auction in shock. During the active bidding, Berggruen sat next to me and under his whispered persuasive urging I ended up buying not two, but seven Klees, blowing in minutes my art budget for five years. Yet I have never regretted following his advice. Two of the prints I bought at this Kornfeld auction have never again appeared at sales; and although I acquired the prints purely for delight, my pleasure is not tainted by the fact that they have increased greatly in value. With a few exceptions that I donated to the Albertina Museum in Vienna, the rest of my prints are now in the San Francisco Museum of Modern Art.

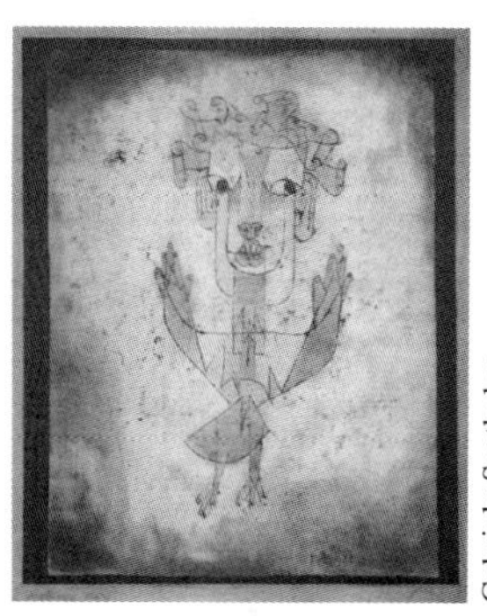

Gabriele Seethaler

Paul Klee, *Two men meet, each supposing the other to be of higher rank* (1903) and *Angelus Novus* (1920)

Possibly the most famous essay by a collector on a work of Paul Klee is Walter Benjamin's *Theses on the Philosophy of History*. His ninth thesis starts with the words:

> *There is a painting by Klee called* Angelus Novus. *It shows an angel who seems about to move away from something he stares at. His eyes are wide, his mouth is open, his wings are spread. This is how the angel of history must look. His face is turned toward the past. Where a chain of events appears before us, he sees only single catastrophe, which keeps piling wreckage upon wreckage and hurls it at his feet. The angel would like to stay, awaken the dead, and make whole what has been smashed. But*

> *a storm is blowing from Paradise and has got caught in his wings; it is so strong that the angel can no longer close them. This storm drives him irresistibly into the future to which his back is turned, while the pile of debris before him grows toward the sky. What we call progress is this storm.*

Starting already in the 1960s, these words of Benjamin's vision of Klee's *Angelus Novus* (1920) have made it one of Klee's most famous works. How is that possible, one might ask, when this 1920 drawing was acquired in 1921 in Munich by Benjamin, who had seen it earlier at a small exhibition in Berlin — a drawing that the general public had no opportunity to view for the following sixty-six years? Since its purchase in 1921, it had remained in Benjamin's personal possession until his suicide in Spain in 1940; it then resided with Theodor W. Adorno until his death in 1969, whereupon it passed to Gershom Scholem in Jerusalem until his death in 1982. Only in 1987 did it end up in its current public residence: the Israel Museum in Jerusalem. Yet long before then — not yet seen by the public except for an odd reproduction here or there during the numerous publications and re-publications of Benjamin's ninth Thesis on the Philosophy of History — it had already turned into a legendary work of art.

The explanation, of course, is quite clear: the three successive owners of that drawing, notably Walter Benjamin, whom many consider the most important European literary philosopher of the last century, were world renowned; Benjamin only posthumously so, starting in the late 1950s, largely through the efforts of his friends Adorno and Scholem, who published and promoted all of Benjamin's writings, including his *Theses on the Philosophy of History*. In other words, Klee's *Angelus Novus* was lionized by an intellectual public that had never seen the actual drawing, but was impressed by its ownership and by the words written by one of its acclaimed owners!

Benjamin and his followers were so obsessed by his metaphoric interpretation of the *Angelus Novus* that few addressed the question what Klee, its creator, had in mind; in other words, to ask the question I raised at the beginning of this chapter: *What, Mr. Klee, was your initial impetus to create this particular image*?

Even visually, Klee's drawing lacks many elements that Benjamin focuses upon, a typical example being "the pile of debris before him grows toward the sky", where Benjamin confused the brown color around the edges in Klee's oil transfer technique with debris. If correct, this would have meant that Klee was obsessed with debris in dozens upon dozens of works that he created around that time using his newly invented painterly technique. As a committed admirer of Klee, I was sufficiently irritated by this all-to-often assumed right of a critic to interpret a work of art without fair examination of the artist's purpose and motivation. Here is what Carl Djerassi in the assumed voice of Arnold Schönberg had to say on the subject in *Four Jews on Parnassus — A Conversation: Benjamin, Adorno, Scholem, Schönberg*:

> ***SCHÖNBERG*** *And you see all that in this image? I don't see any storm. I don't think his wings are caught in it... I don't see him driven anywhere. I didn't see any debris at his feet... and I certainly don't see any rising toward the sky. I see an angel, looking timidly sideways toward God, raising his wings to his praise —*
>
> ***BENJAMIN*** *That's all?*
>
> ***SCHÖNBERG*** *That's all. Now I do not question the text of your* Theses on the Philosophy of History... *not even this rather emotional ninth one. All of us, including Klee, have passed through this type of history, but putting these words into the image of* <u>*this*</u> *Angelus? You needed a metaphoric illustration and because this is the only Klee you owned —*
>
> ***ADORNO*** *So what is the argument?*
>
> ***SCHÖNBERG*** *That particular Klee... like this particular set of* Theses on the Philosophy of History... *rested almost unknown... except for a small group, the majority of which sits right in front of me... for years beyond the death of the author and the painter. I hope I am not insulting you, Herr Benjamin, but when I died I was already up here. It took some twenty years before you ascended to Parnassus — after they pushed you up here.*
>
> ***BENJAMIN*** *They?*
>
> ***SCHÖNBERG*** *Your friends right here as well as others such as Hannah Arendt. But I digress. I do not mean that you do not belong here. But it was* <u>*they*</u>*...* <u>*their*</u> *efforts, their publishing and translating your works... that made you famous... posthumously.*

***BENJAMIN** I know all that. But what is your point?*

***SCHÖNBERG** My point is the reputation of this particular Klee... the Angelus Novus. You became famous for the content of your writing... your own writing. Your friends just spread it around. But this Klee only became famous because your essay became famous... reproduced and re-reproduced and re-re-reproduced...often without even the accompanying figure of the Angel.*

***BENJAMIN** And you blame me?*

***SCHÖNBERG** Only indirectly, because your fame and your essay's fame was created by others. Yet what I find so intriguing is that it was the name, Walter Benjamin, that put this* Angelus Novus *on such an extraordinary pedestal. It practically became your logo. Not because you were the drawing's owner... although that helped... but because of what you wrote about it.*

I have already cited earlier other excerpts from my *Four Jews on Parnassus* because that book has served as my foil in presenting autobiographical issues dear to my heart. It should not surprise anyone, therefore, that Klee appears in it in a dominant role. In one section, I speculate in a long scene up on Parnassus — itself a metaphor for canonization — what Klee actually might have had in mind when he created the work twenty years earlier. In that connection, I address the point first raised by the art historian Konrad Eberlein that Paul Klee may well have thought of Adolf Hitler as the *Angelus Novus* — the ultimate irony for a work so loved by the three sequential Jewish owners, that now resides in the Israel Museum in Jerusalem.

***BENJAMIN** And what do you think Klee had in mind when he drew this angel and called it* Angelus Novus*?*

***SCHÖNBERG** A fair question... and one to which I gave some thought ever since I learned of your obsession with this work. Do you happen to know where Klee painted it?*

***BENJAMIN** Munich?*

***SCHÖNBERG** Exactly! And what other painters might he have encountered in Munich.*

BENJAMIN *(dismissive) All the Blaue Reiter, of course. Kandinsky … Franz Marc… Gabriele Münter… .*

SCHÖNBERG *(impatient) Of course.... and Alfred Kubin… and Jawlensky…. and so on and so forth. We all know their major exhibition in Munich. But what other painters were in Munich around that time? Minor ones, very minor ones. In 1920!*

BENJAMIN *How would I know? Ask Scholem. He lived there at that time.*

SCHÖNBERG *A very minor one… who then became notorious… actually infamous.*

BENJAMIN *No idea.*

SCHÖNBERG *Just think about it for a moment. Very minor… and then infamous throughout the world.*

SCHOLEM *(startled) You don't mean….?*

SCHÖNBERG *Precisely!*

BENJAMIN *Who are you talking about?*

ADORNO *I think he means Adolf Hitler.*

BENJAMIN *Hitler?*

SCHÖNBERG *(nodding) Adolf Hitler. And what did he wear then all the time?*

SCHOLEM *A trench coat.*

SCHÖNBERG *Exactly… like this miserable one when he still tried to support himself as a small time painter of street scenes while fantasizing about becoming an architect. And now compare Hitler, arms raised, in that trench coat with Klee's* Angelus Novus. *Even the way their feet resemble each other.*

BENJAMIN *Preposterous! Why would Klee have picked Hitler for this wonderful drawing?*

SCHÖNBERG *You are now switching to art... whereas I'm focusing on psychological verisimilitude... even metaphoric interpretation. If you permit yourself that luxury in your essay on history, why not extend that same privilege to me? You fell in love with your* Angelus Novus... *you wrote about him in the 1930s and again in 1940. You could not depart from your romantic infatuation with this drawing —*

SCHOLEM *Never mind! What about Klee's vision?*

SCHÖNBERG *He was the cynic realist... the visionary... who may well have encountered Hitler... still the disappointed, unsuccessful painter, but already ranting. Remember, it was late 1919 and 1920 when Hitler had just turned into a full-time political agitator and haranguer on the Munich beer hall fringes. Could Klee not have used "*Angelus*" in the Hebrew sense of "Messenger" and "*Novus*" in the bitter fearful way as the new messenger of the Germany to come? The equivalent of the Roman "*homo novus*" — the parvenu. Not so long ago, a scholar even called the image "a caricature of a priest pretending to be a hobo."*

SCHOLEM *Do you have any proof for this wild hypothesis?*

SCHÖNBERG *Not any more... or any less than Herr Benjamin for his wild view of the angel of history... and hardly a more plausible one than the Hitler likeness I just put forth. It's not as if Klee had never drawn any images looking like Hitler. Just look at these two.*

The little-known aspect of the earlier mentioned extraordinary effect Klee has had on composers prompted me to dramatize that topic in my *Four Jews on Parnassus* through an invented posthumous conversation between the musicologist Adorno and the composer Schönberg. Their somewhat contentious debate is lightened by including (through an attached CD) some actual musical compositions inspired by the *Angelus Novus*, including one especially composed by Erik Weiner for my book as a multi-disciplinary tribute to Paul Klee's genius which itself was so varied and multidimensional.

ADORNO *So let us finish with a quantitative musical analysis of the* Angelus Novus.

SCHÖNBERG *You mean count the composers inspired by* <u>*that*</u> *specific Klee? I suspect that the fingers of one amputated hand will suffice.*

ADORNO *That was nasty… and also dead wrong. So far, I have located 13 composers.*

SCHÖNBERG *More than a dozen composers inspired by the* Angelus Novus *to serenade him with their musical homage? Are you pulling my leg?*

ADORNO *(ironically) Maestro Schönberg! I wouldn't dare.*

SCHÖNBERG *In that case, prove it.*

ADORNO *Martin Bresnick… Vinko Globokar… Matteo d'Amico… Alex Nowitz… Erik Weiner —*

SCHÖNBERG *Stop! I've never heard of any of them.*

ADORNO *You asked for the names of composers… not of friends of yours. Besides, they came after your death.*

SCHÖNBERG *Then let me hear some of that angelic music.*

ADORNO *You mean all thirteen?*

SCHÖNBERG *Three will do.*

ADORNO *All right. Let's start with this.*

> *(Play about 1 minute from the beginning of Claus-Steffen Mahnkopf's* Angelus Novus*)*

SCHÖNBERG *(interrupts) This sounds almost Schönbergian! Who wrote that?*

ADORNO *Claus-Steffen Mahnkopf. He called his* Angelus Novus *a "music theater" based on Walter's interpretation of Klee's* Angelus Novus.

SCHÖNBERG *(laughs) This music certainly passed muster, but I wager he never even saw the image. At best, I'd call it music inspired by Walter Benjamin… not by Klee.*

BENJAMIN *(sarcastic) Thank you for the compliment.*

SCHÖNBERG *I repeat: I didn't criticize your writing... only its inapplicability to the drawing. (Turns to Adorno.) Anything else you can offer us?*

ADORNO *Let's move to Iceland to show you how far our Angelus's musical inspiration has reached.*

(Play about 1 minute from the beginning of the vocal portion of Egill Ólafsson's Angelus Novus*)*

SCHÖNBERG *(surprised) What on earth is this?*

ADORNO *An Icelandic pop song by Egill Ólafsson.*

SCHÖNBERG *What is a pop song?*

ADORNO *Call it post Schönbergian Lieder style. But since you are interested in new musical genres, let me offer you one more that will really surprise you. Written in what in the beginning twenty-first century has become the second most popular form of music in America.*

(*Play Erik Weiner's* Angelus Novus Rap of 2006)

There is a painting by Klee called Angelus Novus
And Benjamin interprets it and he supposes
That the angel moves away from what he sees ahead
Eyes wide, mouth open, and his wings are spread
This is how the angel of history must look
After war has been waged or the Earth has shook
When a series of events result in tragedy
The angel sees only one catastrophe
And feels responsible for all the bloodshed
And he'd like to stay, to awaken the dead
And make everything whole and make everything nice
But a cold storm is blowing from Paradise
And the wind in his wings is so strong he can't close them
Away from the horror, the angel's been chosen
To head to the future, more tasks to perform
And what we call progress — that is this storm

The wings of the angel and this chapter closes
"There is a painting by Klee called Angelus Novus."

BENJAMIN *Whose words are these?*

SCHÖNBERG *First... what kind of music is this?*

SCHOLEM *Whose voice is that?*

ADORNO *Erik Weiner's. He's the composer and artist. It's called a rap... or more precisely a hip hop rap and was commissioned by none other than yours truly!*

SCHÖNBERG *(ironic) And which Professor Adorno will now define for us.*

ADORNO *(laughs) Since you're urging me, why not? It's a lyrical form that doesn't just make use of rhyme but also of alliteration and assonance... and of course lots of percussion.*

Odds and Ends

Both auctioneers and collectors can tell all kinds of auction stories — amusing, bizarre, even dramatic — that are rarely raised, because they seem too envy-producing or even trivial, considering the sums of money that are often at stake. I have written an entire short story around one episode, which never actually happened to me, a story entitled "The Futurist" that eventually also became the title of my first published collection of short stories. But what about the story of a Klee purchase I made at a Sotheby's auction in London while stark naked? For reasons of convenience and anonymity (even at the risk of interrupted sleep) I prefer to bid at auctions over the telephone. Given the eight-hour time difference between San Francisco and London, I have on occasion been awakened around three o'clock in the morning to bid for an item in a sale conducted at a civilized morning hour in London. On this particular occasion, Sotheby's had notified me that they expected the Klee I was after to come up for bidding around three-thirty in the afternoon, London time. During those years, in San Francisco, I started each morning with thirty minutes of exercise on my cross-country skiing machine — one of the few forms of strenuous exercise I was able to do with my stiff left knee (fused from a skiing accident) — and I always did so naked before showering. It was barely seven in the morning when, still puffing and sweating, I was called away from my exercise machine by the insistent ringing of the telephone. The Oxbridge accent on the phone apologized for calling half an hour early, but the auction had progressed more rapidly than anticipated. I was panting heavily, which the man surely took for excitement. The moment bidding started on "my" Klee, I impatiently panted "yes… yes… yes." I soon found myself standing, sweating, shivering, stark naked, engaged in a *mano a mano* struggle with an invisible counterbidder across the Atlantic. Under the circumstances, it is perhaps not surprising that the bidding progressed at an exceptional speed to an unanticipated height: I had somehow lost my usual knack for waiting until the last moment before raising my bid. Still drenched in sweat, but now thoroughly chilled, I slammed down the phone the moment I had heard the final knock of the auctioneer's hammer to turn on the hot water in the shower.

How deeply Klee is woven into the fabric of my life is suggested by the truly unique set of coincidences that accompanied my acquisition of his *Schläfriger Arlecchino* [*Sleepy Harlequin*], a delightful gouache of 1933. The work did not sell at a 1995 autumn auction at Christie's in New York, whereupon I made a serious, though lower, offer to the owner, which was accepted after some minor haggling. I was pleased by my newest acquisition and promptly hung it by the entrance to my study; every time I headed to my computer I looked affectionately at my sleepy harlequin. A few weeks later, I flew to New York for a tête-à-tête with two editors from Penguin, whom I had never met before, about the forthcoming paperback publication of my second novel, *The Bourbaki Gambit.* The telephone message on my answering machine identified the restaurant for our lunch meeting as "Arlecchino" in lower Manhattan. A good omen, I thought, as I bade goodbye to my own Arlecchino on the way to the airport.

I was a few minutes early for our twelve-thirty appointment at what I was convinced would be my "lucky" restaurant. When I arrived, I found the place deserted. This could have been understandable in Madrid, where lunch starts at three in the afternoon, but surely not in New York. "I'm afraid we're closed today," the sole employee of the establishment announced, "the chef is sick." "Impossible," I countered, with all the arrogance of a proud though hardly best-selling author, "I have an appointment with my editors from Penguin." Somehow, I thought that the plural would impress him, but my conceit was promptly punctured by the announcement that the senior editor had just phoned, claiming to be sick, and that the junior member of the team would be late. When the latter arrived, the Arlecchino employee took pity and suggested a supposedly good Italian restaurant whose name I have long forgotten. My Penguin host and I started for the new location, but as soon as we turned the corner, I saw a sign for another establishment, "Rocco's." "We've got to go there," I exclaimed. As soon as we had sat down, I asked the maitre d' to look at what I was fishing out of my briefcase. It was the bound galleys of my next novel, *Marx, Deceased*, that I had brought along to show to the Penguin editor. Puzzled, the maitre d' looked at the cover of the elegant Magritte painting showing a double-faced man. "Why are you showing me that?" he asked. "It's a novel," I said. "I wrote it. Just read the first few sentences." One look at the opening sentence changed his expression:

> *"Rocco's." The voice on the telephone was brusque, the "r" rolling like a Ferrari in first gear.*
>
> *"I'm calling about a dinner reservation," he said. "For two. Next Thursday, 7 o'clock."*
>
> *"Name?"*
>
> *"Marx."*
>
> *"Spell it."*
>
> *God, not again, he thought. It's just a four-letter word.*

"Amazing," the maitre d' said. "Now, are you ready to order?" I waved away the proffered menus. "Just tell us what your special is for today's lunch." "Gnocchi," the man replied. It was my turn to look flabbergasted. "I don't believe it," I said as I turned to page 7 of my novel. "Just read this: '*The gnocchi on Marx's fork had nearly reached his mouth, but at the last moment he put it down.*'" By now, I was ready to believe in predestination, kismet or just plain karma. But alas, not even the charm of Klee–Arlecchino–Rocco–gnocchi was able to transcend the utterly poor judgment Penguin displayed in turning down the paperback rights to my most literate novel. The only balm to my hurt ego was provided by some reviews of the hardback edition, such as the *Washington Post* reviewer's impeccably perceptive "A classy, easy-reading page turner, light of heart and bright of mind… a literary novel to be reckoned with." Even my favorite European newspaper, *The Herald Tribune*, reprinted the review, without, however, any impact on the stony, penny-pinching Penguin operative in charge of paperback rejections.

There is one more story worth retelling, since it does not deal with the acquisition of art, but the potential risks collectors assume — one reason why so many wish to remain anonymous when lending works of art to a museum. My first and only encounter with such risks occurred on the Klee front with some rather slapstick dimensions.

On July 30, 1985, a *San Francisco Chronicle* front-page article bore the headline, "6 Stolen S. F. Museum Drawings Found." At that time, the San Francisco Museum of Modern Art, commonly known as SFMOMA, had not yet moved to its elegant new quarters south of Market Street. It still occupied the two top floors of the War Memorial Building, next door to the San Francisco Opera, its main entrance facing golden-domed City Hall across Van Ness Avenue. The War Memorial's ground floor entrance hall is a rather grand space that is occasionally rented out for anniversary celebrations, weddings, and other private functions. On the less fashionable side of the street, SFMOMA faces some apartment buildings, one of them the abode of a nineteen-year old psychology student attending San Francisco State University. When an art-student friend from out of town paid him a visit, his host suggested they crash a private party then in progress in the War Memorial's foyer.

"You know, being a student with no money, it's an easy way to get free food and drink," the newspaper quoted him. "So me and my friend went over." According to the thief, the reception was saving the champagne and food until later, so they started to get drunk on vodka and grapefruit juice. Some people get sleepy when drunk, others belligerent, but some art lovers evidently turn curious. "We wanted to see what the museum looked like in a different light," the instigator confessed, so they climbed the rear steps to the top floor and pulled on the door until "it just popped open." (Even now, a quarter of a century later, I get weak-kneed when I visualize that scene.) "I hadn't seen Paul Klee for a long time, so we went over to where his paintings are," said the discriminating burglar, "but it was dark, pitch dark, so we took four off the wall to a hallway so that we could look at them." On the way out, they passed some Picassos. "My God," he said, "Picassos! Let's take two of them."

Unbeknownst to the drunks, they had set off a silent alarm, but by the time the private security personnel had come to check, the thieves were gone with six drawings under their coats. The keystone cops working for the firm, who, I trust, fired them promptly, found nothing amiss. Around three in the morning, the two Klee kleptomaniacs had started to sober up and panicked. Deciding to return the stolen drawings, they made the return trip across the street with the four Klees and two Picassos under their coats, but found the front doors hermetically locked.

Upon reading in the morning's newspaper about the value of their loot and the magnitude of the sentence they were facing, the younger of the now cold-sober pair left the art works in a cardboard box on the third level of a downtown parking garage and called the police, who returned the paintings to SFMOMA the following afternoon.

Art thefts are one problem, but fakes of works by famous artists may be even more common. One could even look at them as some sort of back-handed compliment to which Paul Klee was not immune. In 1977, Felix Klee, the painter's only son, invited me to his flat in Bern, to show me his fabulous collection. As one of the greatest connoisseurs of Klee's work, he was also frequently called upon to authenticate works of his father. He showed me his collection of "fakes," which was both startlingly large as well as entertaining. At that time, I never thought that I would uncover such a sham in of all places Sotheby's, an auction house one would have suspected to be more diligent in confirming the authenticity of what they were selling.

Some decades ago, at one of their auctions, Sotheby's listed a hand-colored Klee graphic. From time to time, Klee had hand-colored some of his graphic works, which, of course, raised their value because of their relative rarity. This particular one, of which I owned a print, had never been listed as existing in a hand-colored variant, which piqued my interest. Fortunately, I was passing through New York prior to the auction date and thus could examine the work in person. Although attractive, something did not smell right. For one, such a hand-colored version was not listed in Kornfeld's book *Paul Klee's Graphic Work*, which did not preclude its existence but surely should raise a warning signal. I was also bothered by the provenance, which came from an unknown mid-western gallery, which I had never seen listed before in any Klee sales. The sheet even looked too clean. But what was most disturbing was the negative reply when I asked whether they had checked about this work with the Klee Foundation in Bern. Still, the idea of a newly discovered unique hand-colored graphic led me to pursue my research. On measuring the dimension of the putative Klee at Sotheby's, I noticed that its dimensions were not identical to those of the regular black and white print edition as listed in the Kornfeld reference book. How could Klee have painted one of slightly different dimensions? My suspicions turned out to be justified, since some days later Sotheby withdrew the work from the auction.

I shall conclude this account on the note on which I started this chapter: why did I collect art? Or even more pertinently, what is art? I answered this question in my play *Phallacy* from which I have already quoted once:

> ***REGINA*** *If the beauty of this sculpture is not important, what about art?*
>
> ***REX*** *Define Art.*
>
> ***REGINA*** *An image from the mirror of life.*
>
> ***REX*** *(derisive) Good God!*
>
> ***REGINA****: All right then. How about Art being everything* <u>*other*</u> *than what you see in the mirror?*
>
> ***REX*** *Better! But how necessary is that?*
>
> ***REGINA*** *Art is never necessary. It just happens to be indispensable.*

For me, not as a collector but as a human being, art happens to be indispensable. Isn't art what distinguishes us from all other species?

What If?

In ancient Greece, whenever lawyers met after some interesting case, they teased each other with "What if that had happened?" "What if he had done...?" "What if...?" Supposedly, that was the origin of fiction. "What would I now do differently if I could live my life all over again?" is, of course, also a form of fiction. It may also be the most common form of fiction indulged in by people as they approach the end of their life.

I was still a teenager in Vienna in the 1930s, when I recall first becoming interested in the "What if...?" query after reading Stefan Zweig's *Decisive Moments in History* and reflecting how a single step in a different direction by each of the principals in these fourteen historical episodes would have changed the course of history. Another incidence I still remember occurred a quarter of a century later when I had lunch with Fred Terman, the legendary provost of Stanford University and founder of Silicon Valley. He had invited me from Mexico City, where I was then serving as Syntex's vice-president in charge of research, to discuss an offer of a professorship at Stanford University. During our conversation, I mentioned to Terman a book on the Russian revolution by the Australian journalist Alan Moorehead that I had read on the plane trip to San Francisco. I remarked how well Moorehead had drawn the figure of Aleksandr Feodorovich Kerensky, and how dramatically the course of history would have changed if that moderate Russian revolutionary had just made a couple of different decisions, which might well have kept the Bolsheviks from assuming power. Terman's eyes twinkled as he leaned across the table. "Tell this to Kerensky. He's sitting right behind you." I had assumed that Kerensky was dead; now I found that the slightly built old man eating a sandwich by himself was a senior fellow at Stanford's Hoover Institution. I pondered the contrast between his fate and that of his fellow revolutionary Leon Trotsky (whose grandson had changed his name to Volkov and was then working with me at Syntex) whom Stalin's henchmen followed to Mexico and, in spite of high walls and bodyguards, assassinated.

By that time, I was thirty-seven years old, still too young to speculate what I might have done differently if the choice had been up to me, but old enough to reflect where I would have been now if two disasters had not struck earlier. First, what if the Nazi Anschluss had not occurred in 1938? What if the Jews had not been driven out of Austria? As the only child of two physicians, both of whom conducted their medical practices from home offices, I unquestionably would have studied medicine in Vienna to become a practicing physician rather than a research chemist. And what if around that time I had not had a tubercular infection as well as a skiing accident, which eventually resulted in a permanent knee fusion because the tubercle bacilli, though driven from my lungs, had settled in my knee joint? Like most members of that refugee generation, I would have served in the US army rather than having the luxury of breezing through college and then graduate school, getting my Ph.D. and thus ending my formal education in the same year in which the war ended and the surviving soldiers were returning home to start their education.

But these questions fade into insignificance compared with the monumental "What if?" that first arose on July 5, 1978 and has never left me since the suicide of my daughter, Pamela. Was there anything I could have done to prevent the greatest tragedy of my life? Toward the end of every earlier autobiographical account of mine I have reverted to that question and I must do so now since no darker shadow has ever covered my life — a shadow that has never receded. That, of course, explains why the first chapter in the present volume started with my personal attitude toward suicide. And since such acts of self-destruction invariably bear a message to the survivors, because the act itself is a message even if no other statement is left behind, I feel that I must reproduce below in large measure my account of Pami's suicide from my earlier autobiography, especially because it has long been out of print. I do not wish the totality of her message to be forgotten, especially since my reaction proved to be a dynamic one which is still evolving.

I

I always dread the question, "Do you have any children?" or, as phrased in Asia, "How many children do you have?" Should I answer, "A son," and let it go at that? Or should I say that once I also had a daughter? If I do, one out of two inquisitors will pursue the topic, usually after an embarrassed look of pity. "How did she die?" is almost bound to follow.

Should I tell them the whole story? How on July 5, 1978, I came home from the lab to the ranch, where I was living alone since my divorce from my second wife, Norma, to hear my son-in-law's panicky voice tell me my daughter, Pamela, had left a note that morning. It read:

> *After all my talking with myself, with you, and with others, I have come to the conclusion that today is the day to be my last one alive. I can't go on being useless — I've been paralyzed and inactive with my art for too long and cannot get started again… I've been chronically depressed for years and it's only been getting worse. I just don't want to feel it anymore, or my own guilt at being useless, or my loneliness and isolation… I'm leaving the premises to do my own dying in the woods somewhere because I don't want you to find me. Let someone else do that job…*

Ultimately, that job fell to me.

Pami's letter was headed "11 A.M." If she had decided on painkillers or sedatives — she had a large supply because of her chronic back pain — then it was crucial to locate her quickly: pumping out her stomach might still save her life. Steve (my son-in-law) reported that her green 1972 Opel station wagon was missing. I called the sheriff's office, where a sympathetic voice tried to calm me: "In the end, we almost always locate missing cars." "But when?" I wanted to shout. I tried to contact local radio stations to request that they broadcast a description of the missing car and announce a reward for finding it. But by then, it was past six in the evening, and every station I phoned had only recorded messages detailing programs and regular office hours. In my panic, it didn't occur to me to call Directory Assistance for alternative numbers. Instead, Steve and I used the remaining two to three daylight hours to search for the Opel — he on the west and I on the east side of Bear Gulch Road, which divided our ranch. I drove along that road and out to Skyline Boulevard, the main north-south access along the ridge of the Santa Cruz Mountains. No green Opel was parked there. Ever more frantic, I covered the interior ranch forest road, which — though unpaved — was accessible to automobiles except in the winter rainy season. When night fell, and we had found nothing, Steve and I realized that the situation was hopeless.

Since my former wife was then not on speaking terms with me, I phoned her attorney, who had impressed me during our bitter divorce as a warm human being. He discovered that Norma was in Hawaii and promised to locate her. My father was traveling in Europe, and I simply didn't have the heart to notify him. My son Dale was in Argentina, shooting footage for a documentary on South American soccer. Although I had the home telephone number of his film colleague in the provincial town of San Juan, I decided not to call him until we were more certain of Pami's fate. In the event, Dale's mother cabled him when she heard and he flew back immediately.

Years later, he told me about some startling coincidences. On 5 July, he was filming in the small town of Balcarce, the site of Argentina's first earth satellite station established to broadcast the soccer world cup events. That night, he read Georg Büchner's novella *Lenz*, based on the sad life and likely suicide of Goethe's friend, Jakob Lenz, and considered by some the first piece of "modern" prose in German. Büchner, who himself died at age twenty-three, summarizes Lenz's depression in the last sentences of his story. "He did everything that the others did; still there was a dreadful void inside him, he no longer felt any anxiety, nor any desire. His existence was an inevitable burden. And so his life went on..." Dale recalled the terrible nightmare from which he awoke the following morning, the sheets twisted, his body soaking wet. When he returned to Buenos Aires and found his mother's cable announcing a family emergency, he made plane reservations for the first flight out. A torrential rainstorm, the worst in years, nearly prevented him from reaching the airport. Two years later, when he met his future wife, he discovered that she had been at the Buenos Aires airport on the same stormy day, sitting at the same time in that departure lounge, before flying home from a visit to her brother in Argentina.

During the next four days, my terror and loneliness were so concentrated that I could not begin to face the deeper tragedy of Pami's decision. Early on the morning of July 6, I notified the few ranchers living in the neighborhood as well as Robin Toews, my son's former grade-school teacher who had been living with her daughter as a tenant in my former ranch manager's home since the end of the cattle operation in the year of my divorce. All agreed to search; by the end of the day, all had concluded that the green Opel was not on the ranch. In spite of the miles of trails and canyons and untracked areas where a body could be hidden, only a limited number of sites were accessible to a car. Pami must have driven elsewhere "to do my own dying in the woods." Her mother clung to the hope, reinforced by the advice of some of her friends, that she was just holed up somewhere, perhaps in a motel. Norma had even located a clairvoyant who had promised to use her powers to find Pami. But Steve and I were convinced that Pami had killed herself. But where? What if her car had been stolen after she had left it on the side of some forest road, and the infallible sheriff recovered it, only days or weeks later, at another site, hundreds of miles away? How would we even know where to start the search for her body?

My greatest fear was that I would never know what had happened to my daughter, who, during the past four years, had become my closest friend and only confidante. "In the woods somewhere" didn't have to mean our woods. The Santa Cruz Mountains encompassed miles and miles of uninhabited woods. And what about the Sierras? We used to backpack in Desolation Valley near Lake Tahoe and in the Tuolumne Meadows of Yosemite. Only two winters earlier, Pami and I had gone cross-country skiing for a long weekend near Donner Pass. We were blessed by one of the most magical winter settings of our experience: blue sky, newly fallen snow covering all but the most recent tracks, the temperature just cold enough to keep us from overheating even after hours of fast skiing. During hours of intimate conversation, in the wood-burning sauna or the hot tub out in the snow under the redwoods, my daughter had turned into a peer — indeed, my confessor and advisor. What if she had driven into those mountain woods, two hundred miles north of us?

On the evening of July 9, after four desperate days, I was standing at the sink in the galley kitchen of my house when I suddenly felt that I was not alone. The sun had not yet sunk, but dusk was sufficiently far along among the redwoods surrounding my house that inside it was dark except for the kitchen light. As I turned toward the glass door, I caught the outline of three persons. I couldn't make out the figures and their total stillness made my heart miss a beat. Walking toward them, I finally recognized my tenant, her young daughter and Bob Mann, the manager of the ranch bordering on Pami's side of our property. As I stood before them, the width of the glass separating us, I saw tears in Bob's eyes.

That evening, Bob related, as the falling sun shone horizontally across the Pacific on an area he had already passed twice during his searches, he noted in the grass the faint outline of car tracks that had escaped him earlier. Following the tracks down to the edge of the forest, he saw the green Opel partly hidden among some bushes. He then headed straight for my home on the other side of Bear Gulch Road. I immediately called the sheriff and then my son-in-law. Steve asked whether I would do what had to be done.

As a physician, Steve Bush had seen many a corpse, but I had led a remarkably sheltered life in that respect. Although I had escaped Nazi Austria and traveled all over the globe, I had never seen a person killed by a gun or any other form of violence; until then, in fact, I'd never seen a dead body. Except for the barely remembered death of my grandmother during my childhood, and my mother's passing that very year at age ninety-one in a nursing home three thousand miles away, I'd hardly ever thought of death.

I followed Bob's jeep to the spot where the faint tracks were barely visible in the dusk. He refused to accompany me further, so I walked alone down the golden-brown meadow. Through the Opel's windshield I saw my daughter's lovely face terribly misshapen and bloated; I fled, without opening the car door. Fighting my stiff left leg, I hobbled as quickly as possible to my car and drove off. Interminable minutes later, I saw in the distance the flashing lights of the police car and the ambulance behind.

Until then, I'd hardly ever cried as an adult. But that night, I cried for hours. Even now, over three decades after the event, tears well up when I think back to that night, to the horror of that first sight, to the rush of relief that the gnawing uncertainty was over, to the dawning despair that Pami's suicide was now irrevocable fact.

We agreed that, after the autopsy, Pami's body should be cremated and the ashes spread over the site she loved most — even though I had to lie to accomplish it. A peculiar California law prohibits the open scattering of human ashes anywhere but in the ocean, the progress of "ashes to ashes and dust to dust" being illegal in the Golden State. We picked a spot that years earlier Pami and I had declared the most beautiful one at SMIP: a small waterfall (see photograph on page 100), where Harrington Creek runs past many moss-covered rocks and onto a smooth glistening one, shaped like the orifice of a tilted amphora, whence the water shoots into a clear pool. Since Harrington

Creek eventually enters San Gregorio Creek, which terminates in the Pacific, our act was perhaps not wholly outside the law. The spot looks Hawaiian, with huge lush ferns on the sides of the pool and the trees forming a natural dome through which sunlight flickers. Years earlier, on one of our hikes, Pami and I had come upon it. I had decided that this was the place for the eventual disposition of my own ashes, never dreaming that I would pass through the bitterest experience of any parent: to be the survivor, to pick up at the crematorium the cardboard box of my child's ashes.

Pamela with parents in Mexico City (1950) and picking mushrooms at SMIP Ranch the year of her suicide in 1978.

Diane Middlebrook, who, though not yet my wife, had become one of Pami's few intimates, was present when, to the sound of some fifteenth-century Sephardic songs, Steve scattered the ashes onto the shaped rock at the mouth of the waterfall, while we dropped flower petals into the rapidly clearing pool. Dale and I held each other and I begged in a whisper, "Please don't leave me, Dalito." Later Diane wrote the elegiac lines "At the Scattering of Ashes," not realizing that they also forecast what I experienced when I scattered her ashes in the same spot some three decades later:

Pamela and Diane Middlebrook, 1977

> This cloud runs in my dreams,
> Part of my bloodstream now:
> Cloudy, then clear, the pool at the foot of the falls —
> A uniform gray: you . . .
> Let it be. But this pure image haunts me,
> Of water receiving your death;
> Being changed; flowing on.

II

Many a death — especially when caused by accident or sudden disease — is met by the bitter question "Why?" Invariably it is addressed to God or against God, meaning "Why did You let that happen?" But there is another "Why?" after a suicide that must be addressed to the person who is now dead. I was too depressed to ask at the time, nor was I ready to ask myself whether I could have done something to prevent it. On July 4, the day before she killed herself, Pami had hiked over to my home to spend a few hours with me in the sun, talking about her future. Nothing in her tone or conversation had given me any inkling that she was teetering on the edge of the precipice.

My immediate response to Pami's death was typical of how I coped at that period of my life with personal disaster: I drowned myself in work. Seventeen-hour workdays ensured that when I finally dropped into bed I fell immediately asleep. In addition, there was the legal and accounting work involved in being the executor of Pami's estate, which, like Dale's, had multiplied manifold with the rise of Syntex stock. But after eleven weeks of such work-induced anesthesia, I suddenly decided to travel, and invited Diane Middlebrook to join me. In all my trips to Italy, I'd deliberately avoided visiting Venice and Florence: I felt that these two jewels should not be visited as part of a tourist itinerary in which sites are strung together and no individual component means much. Especially in Florence, I wanted to focus on the art and do it with the right companion.

Three evenings in a row, Diane and I sat in an outdoor café on the Piazza della Signoria facing the Palazzo Vecchio, to relive the day's impressions — and to talk about Pami's decision. Was it an inevitable consequence of a person suffering from depression who had been unwilling to consider therapy? Was it the chronic physical pain that prevented her during the past two years of her life from doing any of the garden and animal work she loved? She could barely feed the horses that had meant so much to her, let alone ride them. Was it her disillusionment with the commercial art scene, with the humiliating compromises a young artist is called upon to make? Or was it the lack of professional peers resulting from her self-imposed isolation in the majestic but also overpowering setting of SMIP? Her husband, surrounded every day by a multitude of people in the hospital, had hardly a minute free for contemplation. Like me, he felt calmed upon returning in the evening to the solitude of the coastal mountains, often shrouded in the veil of the ocean fog pouring in through the canyons. The extraordinary silence, the absence of man-made sounds except for the occasional start of the refrigerator motor, was a soothing contrast to the cacophony of the workplace. But what about the person who remained behind all day? Does beauty in nature, when experienced in solitude, always calm and please, or does it also terrify? Some of Pami's bitterest poetry (published posthumously by her mother) was written during that period of her life.

Pami always read, and during those last few years she read a great deal. Much of it was women's writings and feminist literature, which provided fertile background to her ideas about the place of women in art. All her artistic role models, all her former teachers at the San Francisco Art Institute and at Stanford University, had been men. Indeed, until the time of her death, Stanford's art department had not a single woman faculty member among its studio art faculty. Diane, then head of Stanford's Center for Research on Women (CROW) in addition to her professorship in English, taught me a great deal about the daily affronts women sense in a male-oriented culture. These, and many other issues, were the subjects of our evening talks in the Piazza della Signoria, practically at the feet of the prancing bronze horse on which Cosimo de' Medici surveyed the splendor his family had sponsored. "It's hard to think what Florence would've been like without the Medicis," I mused, pointing to the Giambologna bronze. "But imagine what it would be today if their patronage had extended to women," said Diane.

In that hour, my own response to Pami's suicide finally took shape. Suicide is a message to the survivors, but the text must be read by each individual — parcnt, sibling, spousc, fricnd in the light of his or her past relationship with the deceased. An answer to the question "why?" can be provided only by the survivor; on that September evening in Florence, I decided that my answer, or at least my response, would be patronage of the type that would have benefitted Pami. Diane became my partner — intellectual and operational — in this endeavor. In the 1960s, when we acquired the land that became SMIP, my children agreed with me that it should be kept in an unspoiled state for the ultimate benefit of the public. Though we were not certain of the details, the decision contributed to the legal establishment of the Djerassi Foundation, a nonprofit entity that we envisaged as the eventual beneficiary of our respective testaments. Whatever philanthropic donations I made during the late 1960s and 1970s were funneled through that foundation, but any substantial activities were meant to await my death. In 1978, Pamela and Dale were the two trustees empowered to decide how my own estate — land, art and remaining assets — would be distributed by the foundation; but none of us was then thinking of death, and the full operation of the foundation still seemed many years away.

III

This changed on July 5, 1978, when a bottle of pills transformed the concept of the foundation into a real entity with a substantial financial equity and title to a significant portion of SMIP ranch land, as well as to Pamela's house and studio. Shortly after Diane and I returned from Italy, my son-in-law Steve joined a radiology research unit in Los Alamos, New Mexico, to take advantage of a new radiation source for cancer treatment. The resulting availability of Pamela's house and studio led us to decide to offer to Stanford University's art department the use of these facilities, as well as an annual stipend to underwrite one-year residencies at SMIP for women artists of accomplishment. The incumbent would have no formal teaching duties, but should be prepared for some open-studio events and for interaction with art students and faculty. Toward the end of her stay, an exhibition of the past year's work would be presented at the Art Museum of Stanford University. The foundation offered an initial commitment of four years to test the project's validity. We felt that this would accomplish at least two objectives dear to Pamela's heart: patronage through a public exhibition in a museum, free from commercial considerations; and exposure of mature women artists to the Stanford art community. But while the museum director agreed to the stipulation for an exhibition, and some art history professors, notably Albert Elsen, were supportive, the studio art faculty was not prepared for any obligations that would make such a visiting artist feel welcome. In the end, CROW became the Stanford sponsor, and members of CROW, the art department, and some outside art professionals and museum directors formed a selection committee that began to canvass nominators all over the world.

Helicopter transport of Barbara Greenberg's *Bird Nest* to the Stanford Museum (May 1981)

Out of some forty candidates, four were selected. In 1979 came Tamara Rikman, a graphic artist from Jerusalem, who was accompanied by her husband, the poet T. Carmi. She was followed by Barbara Greenberg, a New York fiber artist and sculptor, who had been nominated by her teacher, the renowned Polish fiber artist Magdalena Abakanowicz. Greenberg gave birth to her first child in Pami's house and also built a fifteen-foot-high "bird nest" out of twigs, branches and sisal. Since the nest was too big to be transported by truck along narrow Bear Gulch Road, a helicopter picked up the nest in a sling and deposited it in front of the Stanford Museum, where it gradually decomposed over the next six months. The third artist was a black writer and poet from Berkeley, Joyce Carol Thomas, who was in the final stages of completing her first novel, *Marked by Fire*. Pami's house provided the undisturbed concentration Thomas needed; and during her Djerassi Foundation residency, she completed also a substantial portion of her second novel. *Marked by Fire* won Thomas the National Book Award in the adolescent literature category on the same occasion that her friend Alice Walker won it in the adult category for *The Color Purple*. The fourth woman artist was Sue Gussow, a professor of painting at New York's Cooper Union, whose artistic output proved qualitatively and quantitatively outstanding, but who found herself too isolated from the Stanford art community.

We finally concluded that, although the driving distance from Pami's studio to Stanford University was only a few miles, the attitudinal gulf was simply too wide. The women artists got a lot done under close to ideal physical conditions, but their presence had only a minimal effect on the Stanford art scene. Barbara Greenberg had suggested during her stay that the buildings owned by me on my side of SMIP ranch — the ranch manager's residence and the twelve-sided barn — would lend themselves to conversion into a small artist's colony, which would overcome the sense of isolation and lack of peer interaction (also Pami's problem) encountered by the artists who had come alone for an entire year. I began to realize that free creative time without conditions may be the ultimate gift for an artist, but it is an incomplete gift if it denies the crucial human need: companionship.

Diane, in her capacity as a foundation trustee, visited the two oldest artist's colonies in the East, Yaddo and McDowell, as well as two smaller ones, the Edna St. Vincent Millay Colony and Hand Hollow — the latter founded by a friend of mine, the kinetic sculptor George Rickey, on his farm in East Chatham, New York. Diane's report and Rickey's counsel persuaded me to convert the four-bedroom former manager's residence into an eight-bedroom, five-bath house and to create two studios in the barn. Leigh Hyams, a San Francisco painter, was appointed interim director; and by late 1982, the Djerassi Foundation artist's colony was born — subsequently renamed "Djerassi Resident Artists Program" or DRAP. The gender requirement was dropped as the community expanded; and to make our program available to a larger number of artists, residencies were limited to periods of one to three months. Since all the bedrooms would be large and, with one exception, have balconies or direct access to the surrounding garden, they could easily double as workrooms for the writers. The two studios in the barn would be used by visual artists and composers. Rickey considered good food to be indispensable to the success of an artist's colony — advice we followed by hiring his chef, who had cooked at Hand Hollow with *cordon bleu* elegance.

Within a year, we had housed and fed fifty-two artists, twenty-eight of them women. The friendships and collaborations formed at DRAP extended beyond the professional: two sculptors, Patricia Leighton from Scotland and Del Geist from New York, and the Houston painter Josefa Vaughan and the San Francisco composer Charles Boone, got married after meeting at DRAP. (During the subsequent years, additional marriages as well as some divorces were spawned there.) Our first composer, John Adams, who later gained international recognition for his opera *Nixon in China*, spent three months at DRAP, where he composed the music to *Available Light*, a work commissioned for the opening of the Museum of Contemporary Art in Los Angeles. On several occasions, Adams remarked to me how difficult it was to collaborate with his choreographer, Lucinda

Childs, three thousand miles away. Adams's comments convinced me that we should encourage interdisciplinary collaboration by creating additional studio spaces within the twelve-sided barn.

By then, the cost of the program had started to exceed the endowment income of the foundation and my personal resources. Not only had my divorce reduced my assets by half, but much of these were tied up in real estate at SMIP and in art. Most of my SMIP property and the artists' residence and studio barn had already been gifted by me to DRAP, but more heroic steps had to be taken. Other than my Paul Klee collection — a story covered in the "*Collector*" chapter — I began selling much of my other art, mostly the work of dead artists in order to support the work of living ones. But even that was not sufficient: in addition to the chef, we had acquired a full-time director (who lived in my daughter's house), a resident manager and a grounds manager. Several local foundations — notably the Hewlett, the Irvine and the San Francisco — and later the MacArthur Foundation and the National Endowment for the Arts made generous grants, which, together with contributions from individuals and corporations, enabled us to complete additional studios dedicated to choreography and performing arts, to music, to photography, and to ceramics. The barn renovation included three sleeping lofts, which increased our housing capacity to nine artists. It has proved to be an ideal size: still small enough for everyone to sit around one table for dinner, yet large enough for real collegiality.

Thirty years later — under the successive directorships of Leigh Hyams, Susan Learned-Driscoll, Sally M. Stillman, Charles Boone (a former composer-in-residence), Charles Amirkhanian, Dennis O'Leary and now Margot Knight — the program has grown to such an extent that, with over two thousand artists from virtually every state and over thirty foreign countries, it has become the largest artist's colony west of the Mississippi. The artists have included a Nobel Prize laureate, several MacArthur Fellows, numerous Guggenheim fellowship holders and winners of literary and visual art awards, as well as many artists who had not yet received wide public recognition but were considered worthy of support by our selection panels, which change each year to assure diversity. During the first fifteen years while personally still spending significant time in my nearby ranch house, I met most of the artists at dinner — occasions that are usually followed by readings, slide presentations, music or dance performances, or just good conversation. As I listen to music just composed, or face a canvas, still moist on a studio wall; as I hear an author read lines that could only have been born here, such as those of the ninety-year-old poet Janet Lewis,

> One tone, one visible substance, a splendor
> Of multiplicity and self-abandon,
> Beneath that band of intense blue
> Lacking which, the hill is incomplete,

I catch myself wondering what my daughter would have thought of all this.

IV

Five years after Pami's death, I received a panicky phone call from an artist, saying she had found a note, deep in back of a drawer in her studio. She was shaking when she handed me the piece of paper with my daughter's handwriting on it, a ghost's message. That night I recorded some notes in my journal, which I published thirty-four years after Pamela's suicide in my poetry book under the title "I have nothing left to say."

> "I have nothing left to say,
> So I don't talk.
> I have nothing left to do,
> So I close up shop."
>
> Five years after your death,
> My only daughter,
> I find this note.
>
> No date.
> No address.
> No signature.
> Your handwriting.
>
> Written for whom?

Written when?
Hours,
Days,
Weeks,
Perhaps months
Before you entered the woods?

If only you had said these words to me.

Six years later, on Thanksgiving Day 1989, the sculptor David Nash and I went on a search for some felled redwood trunks at least five feet in diameter. It was the minimum size he required for the three-part sculpture he intended to site around some of the burned-out giant redwood stumps that could still be found here or there on our property from nineteenth-century logging. Nash is one of the most distinguished artists we have had in residence at the foundation. A British sculptor, now working in Wales, he first came to DRAP in 1987 at the time of his retrospective exhibition at the San Francisco Museum of Modern Art. While wood ("King of the Vegetables," he calls it) is his sole medium, and chain saw or ax his principal tools, he had never before handled redwood or madrone, the two most prevalent species in our forest. During his first stay, he had created a group of madrone sculptures for a highly successful show in Los Angeles; in addition, out of a huge redwood trunk that had lain for decades in Harrington Creek, Nash had fashioned *Sylvan Steps* — a Jacob's ladder rising at a steep angle out of the water into the sky. When he first selected that site — accessible only along the creek bed by clambering over rocks and fallen timber — he had had no inkling that a few hundred feet upstream we had, in 1978, scattered my daughter's ashes. Within a minute, some flecks of them must have floated past the spot where Nash's steps now rise into the air. *Sylvan Steps* was a magically simple sculpture which many subsequent artists have drawn, photographed or written about. Yet I must use the past tense, since roughly

David Nash sculpture *Sylvan Steps* in Harrington Creek, SMIP Ranch

a dozen years later, after a tremendous Pacific storm had passed over the Santa Cruz mountains, it was gone! I was the first to notice this during one my hikes and simply considered it impossible: a huge shaped log, weighing hundreds of kilos disappearing poof into the air? Days later, a group of us set out to search for it and discovered that it had been washed down the creek, which at that time had turned into a raging stream, and had finally come to rest against some boulders. Eventually, *Sylvan Steps* was recovered with some heavy equipment and now lies beached on dry land by the creek bed but its original unusual angle and Jacob's Ladder element could never be reconstituted.

But on Thanksgiving Day 1989, during Nash's second residency, we could not find the type of massive log he needed. On our morning hike, we did locate four sites in the forest where blackened trunks rise out of the bracken — just the right backdrop for the scorched pyramid, cube, and ball Nash planned to shape, but still missing was the right arboreal progenitor for these forms. Of course, we crossed the shadow of many a living redwood giant, but cutting one was out of the question. Then I recalled that some selective logging had just been completed on our neighbor's land across Bear Gulch Road; only a few days ago, I had followed impatiently a slow-moving truck stacked high with redwood logs. Perhaps "our piece" had not yet been removed.

I didn't expect anyone to be working there on Thanksgiving Day. But, after climbing over the locked gate and walking down the forest road, inches deep in dust (it had not rained for weeks), we heard in the distance the grinding of gears. Soon we came upon a mammoth tractor setting up erosion breaks to preserve the road bed during the winter rainy season.

"Have you moved out all the logs?" I shouted up to the bearded driver after he shut off the thundering engine. "We need…" I said, and then explained who David Nash was, and why we were searching for a special fallen redwood rather than a turkey for Thanksgiving.

"All gone," he said, but then remembered. "A big one fell across the fence near the property line. Probably years ago… in some storm." According to him, it was partly rotten — sufficiently so that it had not been worthwhile to haul it to the mill. Nash was dubious that it would do, but I said, "Let's look anyway."

We followed the man's directions to the fence a half-mile down the logging road. When we finally came upon it, I was dumbfounded. Eleven years earlier, I had hobbled here as fast as my stiff leg would carry me — but from the opposite direction, down the meadow from our side of the property, toward this fence across which the massive trunk was now lying, broken into three enormous pieces. It was the spot where my daughter had killed herself, where I had never dared to return. We found the rot to be only superficial; the wood was precisely what David Nash had been seeking all Thanksgiving long.

David Nash

David Nash sculpture *Charred Sphere* at SMIP Ranch

This brings me to the last "What if?". In retrospect, it seems almost certain that my daughter had been suffering from chronic depression and if properly diagnosed, might well have survived through appropriate treatment. But even though she even referred in her suicide note to having been "chronically depressed for years and it's only been getting worse," neither her parents nor her brother had recognized the symptoms. While I had been aware of her mood swings, my own interaction during the last few years of her life when she was living with her husband on the family ranch was of a nature where I mostly saw her during her emotional highs. I still remember the warmth and openness — just twenty-four hours before her suicide — when the two of us sat by the side of my swimming pool, our legs dangling in the water and laughing in the sunshine. Nor do I believe that her physician husband had been aware of her medical condition; if he had, he surely would have communicated that fact to the rest of the family. But that "What if?" is not productive, because it can only lead to recrimination and not solutions. Instead let me rephrase it in a forward-looking way by speculating what would never have been created by the two thousand odd artists, who otherwise would never have come to one of the most beautiful spots on the northern California coast and would never have done precisely the work that they ended up doing. I cannot count the number of artists who referred to the inspirational impact that the overpowering landscape has had on their work; who spoke about the empathy generated through their knowledge of the tragic reason that had made their residency possible; and the many works of art, notably poems and paintings, that paid tribute to the unique site and to the suicide that overhangs it to this day.

Perhaps the biggest and most moving tribute is the huge oil painting on which Jim Rosen had labored on and off for a year and which now hangs in solitary splendor on the biggest wall of my ranch house: a mysteriously veiled version of the famous Vermeer painting in London's National Gallery — *Young woman standing at the virginal* — where Pamela takes form if one focuses on the girl's face. It is so subtly gray that no photograph can do it justice. It requires eye-to-image intimacy.

I have finally come to the last "What if?" question: if Pamela would not have died, would I have founded the Djerassi Resident Artist's Program? Would I have sold most of my art collection and donated my portion of SMIP Ranch and its buildings for an artist's colony — in other words supporting the work of living artists rather than just collecting art? The honest answer is "very likely not."

My dearest Pamela. How I wish that your death had not been necessary before I took seriously the patronage of the living.

Caveat Emptor

Wikipedia/Wichsipedia — Google/Schmoogle

Considering that these autobiographical reflections started with the word *CAVEAT* (*let him beware*), do I now start the title of the final chapter with the same word for purposes of symmetry, to show off my infatuation for Latin phrases based on my early education in Vienna, or are the shadowy implications of *let him beware* and thus their relevance to the overall theme of this book the reason why I now end this way? While all three justifications apply, the key difference is that the book commenced with *caveat lector* — warning the reader of my book what (s)he is facing — but now ends with *caveat emptor* — a warning to the buyer, but not the buyer of this book. After all, anyone having stuck with my text until finally reaching this chapter cannot possibly benefit anymore from any warning and certainly won't get any refund from me. Instead, the buyer I am now addressing is all of us, who "buy" continually an unmeasurable amount of information in cyberspace without actually paying for what we download from Google, Wikipedia or other search engines.

The perceptive reader will have realized from my chapter on the Pill that when I stated that "in the time of Google and Wikipedia, where every news driblet, however preposterous, is soaked up in seconds and permanently preserved," I clearly had reservations about the ubiquitous use of these cyber-aids. I justified these concerns by stating that "by now, any error — inadvertent or deliberate... indeed any factoid — true or false — cannot be erased or corrected. It is simply fixed in cyberspace and usually picked up by slipshod journalists and the huge sloppy segment of the browsing public, who consider any cyberdetritus as truths set in stone or at least grist for journalistic mills." I would now like to indulge in some shadow treading by first expanding somewhat on my concerns and then end it with a hilarious example that specifically applies to me.

Is it fair to complain about the information we acquire from these search engines, since it is free? I believe it is fair, provided we do not address most of these complains to Google *et al.* but to their users of which I would foremost highlight students and the media, and, of course, also myself. I use these search engines all the time, but invariably as direct leads to the primary source — in academic jargon meaning the primary published references — or as hints for further searches that eventually point me to the originator of that particular informational nugget and the manner in which it was acquired. For persons like myself, whose (re)search skills were honed during the pre-computer age when hours were spent in the library for what can now often be found in seconds, the search engines are a spectacular blessing which, however, merit some healthy scepticism for the reason already stated above: *junk in implies junk out* and since Googling — at least the type practiced now by most students — does not guarantee veracity or confirmation, great caution needs to be exercised. Most journalists are simply hopeless in that regard — primarily because they operate under fierce time pressures and sound-bite constraints — and there is little the consumer of such news or the victim (meaning the interviewee) can do about it. I would not, however, give up on students, because there pressure can be exerted by the instructor through firm evaluation practices with insistence on primary sources. Since I still teach (part time) at my home institution, Stanford University, I can compare student papers over a period of half a century. My main complaint now is that in many courses, notably those outside the realm of the hard sciences, the bibliography is frequently reduced to a list of URLs whose accuracy or sometimes even relevance is assumed without further scrutiny. Hence playfully converting "Google" into "Schmoogle" does provide a phonetic criticism in both English or German through the words "smuggle" and "schmuggel" in that uncritical information is all too often smuggled into the searcher's mind and then sticks there.

In that regard, Wikipedia is somewhat different, since a mechanism for correction exists. Yet how often it is practiced is quite another question. I have encountered a sufficient number of Wikipedia errors or even major blunders, which have remained uncorrected that some caution must be exercised. However, Wikipedia does have the advantage that many entries list at least some primary or even useful secondary references, which can then be used for further refinement. Nevertheless, I wonder whether the initial inventor of the word *Wikipedia* knew any German and if so, whether that person purposely ignored the fact that *wichsen* is the colloquial German expression for masturbation. Phonetically, *Wichsipedia* does not sound all that different from *Wikipedia* and since masturbation is synonymous with self-gratification, searching the internet about oneself is done all too often for such a purpose. I myself have been guilty of that charge, although I have not been obsessive about it and with one exception — already related in the Pill chapter — I have not taken the time to attempt correcting mistakes.

But rather than continue on this slippery slope of pedantic wordplays or even whining, allow me to lead the reader through a Wichsipedia tour of Carl Djerassi starting with the Wikipedia URL http://en.wikipedia.org/wiki/Carl_Djerassi. The entire article with numerous references — some impressive, some incomplete, and some completely false — is too long for a detailed discussion; hence I shall restrict myself to the more entertaining aspects.

The Wikipedia entry starts with the following introduction, which, with the exceptions of the names of the persons who first carried out the animal studies on our norethindrone and a totally out-of-context quotation are factually correct:

> ***Carl Djerassi*** *(born October 29, 1923, Vienna) is an Austrian-American chemist, novelist, and playwright best known for his contribution to the development of the first oral contraceptive pill (OCP). Djerassi is emeritus professor of chemistry at Stanford University.*
>
> *He participated in the invention in 1951, together with Mexican Luis E. Miramontes and Hungarian George Rosenkranz, of the progestin norethindrone — which, unlike progesterone, remained effective when taken orally and was far stronger than the naturally occurring hormone. His preparation was first administered as an oral contraceptive to animals by Gregory Pincus and Min Chueh Chang and to women by John Rock. Djerassi remarked that he did not have birth control in mind when he began working with progesterone — "not in our wildest dreams… did we imagine (it)".*
>
> *Djerassi is also the author of several novels in the "science-in-fiction" genre, including* Cantor's Dilemma, *in which he explores the ethics of modern scientific research through his protagonist, Dr. Cantor.*

The start of the German version, http://de.wikipedia.org/wiki/Carl_Djerassi is not a translation of the above text, since the entire introduction is much shorter, although it makes the point that I am Jewish, which the much longer English entry does not.

These two informational items would seem to be completely irrelevant, were it not for the fact that when I was informed that there existed several web sites about a Carl Djerassi Glacier on the Bulgarian section of Antarctica — a factoid I already mentioned earlier — I started to search for related information and thus came across the following Wikipedia-associated URL http://www.zaped.info/C._Djerassi whose first two paragraphs I reproduce herewith:

> ***Carl Djerassi** (born October 29, 1923 in Vienna, Austria), is an American chemist, novelist, and playwright best accepted for his addition to the development of the first articulate contraceptive pill (OCP). Djerassi is emeritus assistant of allure at Stanford University.*
>
> *He alternate in the apparatus in 1951, calm with Mexican Luis E. Miramontes and Hungarian George Rosenkranz, of the progestin norethindrone — which, clashing progesterone, remained able if taken orally and was far stronger than the by itself occurring hormone. His alertness was aboriginal administered as an articulate contraceptive to animals by Gregory Pincus and Min Chueh Chang and to women by John Rock. Djerassi remarked that he did not accept bearing ascendancy in apperception if he began alive with progesterone — "not in our wildest dreams... did we brainstorm (it)".*

After recovering from my surprise and wiping away my tears of laughter, I decided to compare both articles, which, though thematically identical, seemed to point to preposterously parodistic transformations such as "first articulate contraceptive pill" instead of "first oral contraceptive pill;" "He alternate in the apparatus in 1951, calm with Mexican Luis E. Miramontes and Hungarian George Rosenkranz" instead of "He participated in the invention in 1951, together with Mexican Luis E. Miramontes and Hungarian George Rosenkranz;" or "His alertness was aboriginal administered as an articulate contraceptive to animals" instead of "His preparation was first administered as an oral contraceptive to animals;" and best of all, referring to me as "emeritus assistant of allure" rather than as "emeritus professor of chemistry" at Stanford University. Demoting me from *Professor* to *Assistant* seemed inconsequential when compared with elevating chemistry (or me???) to "allure."

This was simply too good to ignore and led me in Wichsipedia fashion to a careful scrutiny of the entire rather long http://www.zaped.info/C._Djerassi account of which partial excerpts of the more precious examples are herewith assembled in chronological order:

His parents met in medical academy at the University of Vienna, married, and confused to Sofia, Bulgaria. His mother alternate to Vienna for two months for the bearing of Carl, her alone child.... Until age fourteen, he abounding the aforementioned Realgymnasium that Sigmund Freud had abounding years earlier; he spent summers in Bulgaria with his father. Afterwards the Anschluss, his ancestor briefly remarried his mother in 1938 to acquiesce Carl to escape the Nazi administration and abscond to Bulgaria, area he lived with his ancestor for a year. Carl abounding the American Academy of Sofia and accomplished delivery in English, while his mother went to England to anticipate a acceptance to immigrate to the United States.... Djerassi wrote a letter to Eleanor Roosevelt, allurement area he should go to college. She beatific him an acknowledgment with "veiled advice", and he begin a academy and a scholarship. He abounding Tarkio College (now defunct) in Missouri, and afterwards Kenyon College.

Djerassi accelerating Phi Beta Kappa from Kenyon in 1942 with a B.A. in amoebic chemistry. He formed for Ciba the year afore and four years afterwards his alum studies. At Ciba, he got his aboriginal patent, for the antihistamine Pyribenzamine, which became a accepted decree drug. He affiliated his aboriginal wife, Virginia, an American, in 1943... and became an American aborigine in 1945.

In 1949, Djerassi was recruited to be the accessory administrator of analysis at Syntex in Mexico City... [where] he formed on a new amalgam of cortisone based on diosgenin, a steroid sapogenin acquired from a Mexican agrarian yam. His aggregation afterwards actinic norethindrone, a progestin-analogue that was able if taken by mouth. This became allotment of the aboriginal acknowledged articulate contraceptive, the combined articulate contraceptive pill (COCP). COCPs became accepted colloquially as the birth-control pill, or simply, the Pill....

Since 1959, Djerassi has been a assistant of allure at Stanford University and the admiral of Syntex Laboratories in Mexico City and Palo Alto, California. He afterwards started a aggregation alleged Zoecon, which acclimated acclimatized insect advance hormones to ascendancy fleas and added insect pests. Zoecon flourished for a few decades and again was bought by Occidental Petroleum; for a few years, Djerassi was on the lath of admiral of Occidental.

The Syntex affiliation fabricated Djerassi a affluent man. He bought a ample amplitude of acreage in Woodside, California, started a beasts ranch, and as well congenital up a ample art collection. His next-door acquaintance was artisan Neil Young, whose bandage could sometimes be heard call from several afar away.

With his additional wife, Norma Lundholm, he had a son, Dale, who is a documentary filmmaker; and a daughter, Pamela, who was an artist. His babe suffered from chronic pain as able-bodied as depression, and took a baleful balance of decree drugs in July 1978. Afterwards Pamela's suicide, Djerassi founded the Djerassi Resident Artists Program (DRAP) in her memory. Djerassi was affiliated to biographer and Stanford assistant emerita Diane Middlebrook until her afterlife in December 2007, alive in San Francisco and London.

Djerassi... founded DRAP [because] he would rather patronize alive artists than asleep ones (or art dealers and auctioneers). He afterwards bankrupt down his beasts ranch, acclimatized the barn and the houses to residential and plan amplitude for a amount of artists of abounding kinds, brought in a best chef, and confused to an appointment architecture he had acclimatized in San Francisco, converting one attic into a chic apartment, area he displayed allotment of his art accumulating and hosted a arcane salon).

Djerassi perceived the bolus as accepting a huge appulse on the amusing processes of women and men, which to a cogent admeasurement is afflicted through the sociobiology of sexual reproduction... Djerassi was awarded the National Medal of Science by President Nixon for his plan on the Bolus which was acrid to a degree, as he appear in his memoir, his name was on the abominable "Nixon's enemies list" at the time which was aggregate by Charles Colson and Nixon....

In 1991, he was awarded the National Medal of Technology for "his ample abstruse contributions to analytic ecology problems; and for his initiatives in developing novel, applied approaches to insect ascendancy articles that are biodegradable and harmless."

Austria has issued a postage stamp with Djerassi's account on it. The Austrian government as well beatific him a new Austrian passport. He was awarded the Austrian Cross of Honor for Art and Science, Aboriginal Class in 1999.

Djerassi Glacier on Brabant Island in Antarctica is alleged afterwards Carl Djerassi.

Selected references

1. *Djerassi, Carl (1990-05-01). Steroids Fabricated It Possible. An American Chemical Society Publication. pp. 205. ISBN 0-8412-1773-4.*
2. *The beasts ranch, alleged "The SMIP Ranch" (Syntex Fabricated It Possible), was acclimated for writers' retreats. For a photo of Richard Feynman at SMIP Ranch, taken by acclaimed atomic geneticist Esther Lederberg, see http://www.estherlederberg.com/ColleaguesIndex.html; bang "Feynman" in the larboard aeronautics pane. Composite Gazetteer of Antarctica: Djerassi Glacier.*

As I proceeded from one howler to another ("that he would rather patronize alive artists than asleep ones;" "Djerassi perceived the bolus as accepting a huge appulse on the amusing processes of women and men, which to a cogent admeasurement is afflicted through the sociobiology of sexual reproduction;" that I was an "amoebic" (rather than organic) chemist; that while I was demoted from professor to "assistant of allure at Stanford," I was simultaneously elevated to the position of "admiral of Syntex" and later even "of Occidental," it suddenly dawned on me that what had appeared on the web under www.zaped.info/C._Djerassi might well have been a Bulgarian translation of the original Wikipedia article which was then back-translated by a computer into English, not unlike the computer translation of the biblical quote, "the spirit is willing but the flesh is weak" which, via back translation from Russian, ended up as "The ghost is a volunteer, but the meat is feeble." Wikipedia, making fun of computer translations, actually cited the alternative "The vodka is good, but the meat is rotten" — an equally amusing warning about the dangers of such translations.

After relishing the comments received from members of my family and the many friends to whom I had sent that URL, I semi-seriously considered using it as my official "biography," but to my chagrin discovered a few months later that it had been totally sanitized by Wokwiki back into the original Wikipedia article. *Sic transit gloria mundi.*

Disappointed as well as miffed, I decided to throw a more critical eye at the references in the original (and still existing) English Wikipedia article and found — hardly a surprise — plenty of inaccuracies and outright errors. The

Carl Djerassi, Diane Middlebrook, and Eugene Garfield at SMIP Ranch (May 1979)

most startling one was the reference[17] to a photograph purportedly documenting a visit to me by Richard Feynman — one of the greatest physicists of the twentieth century — at SMIP Ranch in 1979 ("Esther M. Zimmer Lederberg, Djerassi, Carl; ?; Feynman, Richard May 6, 1979"). As yet I show no discernible symptoms of Alzheimer's disease and hence was sure that I would have remembered such a visit, especially since I was equally certain never to have met Feynman in person. Checking the original reference, I found a photograph showing me, my future wife Diane Middlebrook and Eugene Garfield (alias Richard Feynman), the founder of the Institute for Scientific Information and inventor of the Citation Index and other important bibliometric and scientometric tools. When I turned to Google under "Richard Feynman and Carl Djerassi at SMIP Ranch" I found over a dozen references all leading to the same misleading picture.

As I reread these concluding musings, I find that my seemingly entertaining Wichsipedia self-indulgence runs the danger of turning into simple wishy-washiness, which in German leads to a more meaningful wordplay. The first half of the German equivalent, *Wischiwaschi*, refers to *wischen,* in other words *to wipe*. Is not one facet of autobiographical writing a form of wiping dirt or dust to produce a cleaner slate? Although a good part of this very last autobiography of mine displays a fair amount of self-criticism — how could it fail to do so when I have emphasized throughout the autopsychoanalytical nature of my last two decades? — I would be fooling myself if I did not realize that at my age I am anxious to leave behind a reasonably clean slate. Perhaps the entire book is simply a form of pre-emptive strategy against Wichsipedia and Schmoogle, since Wikipedia and Google are likely to keep alive in cyberspace at least some references to Djerassi's own views about himself. "The past cannot move into the present uncorrupted" is the epigraph to this book as well as its epilogue. Why not let the corruptions be my own and not those of some anonymous source in cyberspace?

[17] http://www.estherlederberg.com/EML%20LME%20AG%20GA%20RFeynman%20(right)%2079.html

By The Same Author

Fiction

How I Beat Coca-Cola and Other Tales of One-upmanship
Cantor's Dilemma
The Bourbaki Gambit
Marx, Deceased
Menachem's Seed
NO

Poetry

The Clock Runs Backward
A Diary of Pique

Plays

An Immaculate Misconception
Oxygen (with Roald Hoffmann)
Calculus
NO — A Pedagogic Wordplay for 3 Voices (with Pierre Laszlo)
Ego (Three on a Couch)
Phallacy
Sex in an Age of Technological Reproduction: ICSI and Taboos
Foreplay
Insufficiency

Nonfiction

The Politics of Contraception
Steroids Made It Possible
The Pill, Pygmy Chimps, and Degas' Horse
From the Lab into the World: A Pill for People, Pets, and Bugs
This Man's Pill: Reflections on the 50th Birthday of the Pill
Four Jews on Parnassus — A Conversation (Benjamin, Adorno, Scholem, Schönberg)
Chemistry in Theatre

Scientific Monographs

Optical Rotatory Dispersion: Applications to Organic Chemistry
Steroid Reactions: An Outline for Organic Chemists (editor)
Interpretation of Mass Spectra of Organic Compounds (with H. Budzikiewicz and D. H. Williams)
Structure Elucidation of Natural Products by Mass Spectrometry (with H. Budzikiewicz and D. H. Williams)
Mass Spectrometry of Organic Compounds (with H. Budzikiewicz and D. H. Williams)

Index